Design of Multi-Frequency CW Radars

Design of Multi-Frequency CW Radars

M. Jankiraman

SciTECH
PUBLISHING, INC.
Raleigh, NC
www.scitechpub.com

Printed in the U.S.A.
10 9 8 7 6 5 4 3 2 1

ISBN: 1891121561
ISBN13: 9781891121562

SciTech President: Dudley R. Kay
Production Director: Susan Manning
Production Coordinator: Robert Lawless
Cover Design: Kathy Gagne

This book is available at special quantity discounts to use as premiums and
sales promotions, or for use in corporate training programs. For more information
and quotes, please contact:

Director of Special Sales
SciTech Publishing, Inc.
911 Paverstone Dr. – Ste. B
Raleigh, NC 27615

Phone: (919)847-2434
Fax: (919)847-2568
E-mail: sales@scitechpub.com
http://www.scitechpub.com

Library of Congress Cataloging-in-Publication Data.
Jankiraman, Mohinder.
  Design of multi-frequency CW radars/M. Jankiraman.
    p. cm.
  ISBN-10: 1-891121-56-1 (hardcover : alk. paper)
  ISBN-13: 978-1-891121-56-2 (hardcover : alk. paper)
  1. Radar. I. Title.

TK6575.J35 2007
621.3848–dc22
2007015687

Dedicated to:
The friendly people of Holland

Many thanks for your hospitality and the opportunity

About the Author

Mohinder Jankiraman received his B.Tech. degree in electronics and telecommunications from the Naval Electrical School in Jamnagar, India, in 1971. Subsequently, he served as an electrical officer in the Indian Navy for many years. In 1982, he was seconded to research work in military electronics. He took part in a number of military research projects and won a number of awards for technology development in India. In 1985 he was awarded "Scientist of the Year" by the Prime Minister of India for his work in developing a mine for the Indian Navy. His research has spanned several disciplines, emphasizing signal processing, development of naval mines, torpedoes, sonars, radar, and communication systems. He retired from the Indian Navy in 1995 with the rank of a Commodore and joined the International Research Centre for Telecommunications and Radar (IRCTR) in the Delft University of Technology, The Netherlands, in 1997. During this period, he took part in the research and development program for high-resolution radars at IRCTR.

Dr. Jankiraman completed his Master of Technology in Design (MTD) degree in 1999 from Delft University of Technology, The Netherlands, graduating *cum laude*, and went on to complete his Ph.D. from Aalborg University, Denmark, in September 2000. He is currently living in the United States, based in Dallas, Texas. He is a senior member of IEEE.

Contents

Preface

One of the great properties of continuous wave (CW) radars is its ability to operate stealthily. This property lends itself to designing the so-called low probability of intercept (LPI) radars. In such radars, stealth is realized by bringing together many properties into play, such as low power, wide bandwidth, and frequency hopping which make it extremely difficult for such radars to be detected by means of a passive intercept receiver. Moreover CW radars, yield extremely high resolutions realized easily in terms of hardware and signal processing. However, an unmodulated CW signal cannot measure a target's range. Therefore, the transmitted signal needs to be imparted a bandwidth. This is achieved through various means such as frequency modulation, phase coding and many variations of these. A detailed analysis of these waveforms and their properties and use in designing radars has received limited attention in the literature as well as the methodology of designing such radars and the hardware considerations which go into it. Thus, the endeavor of this book is to impart a working knowledge of the subject not just for postgraduate students and researchers but for the entire radar community.

It is assumed that the reader is familiar with basic digital signal processing, RF system engineering, and probability theory. The style of writing has been kept as simple as possible and technical jargon has been kept to a minimum. All effort has been made to explain the basics in a cogent and conversational manner.

The book is organized into three parts. This first part covers the basic CW radar theory. The second part introduces the reader into how this theory is utilized in the design of an actual radar called Calypso. The third and final part discusses the theory and design of a more complex multifrequency CW radar called the PANDORA. The PANDORA radar was born out of the author's research work done at the International Research Centre for Telecommunications and Radar (IRCTR) at the Delft University of Technology, in the Netherlands. This radar in fact formed part of the author's Master thesis [1]. It is pointed out that to the best of the author's knowledge and belief, *this is the first time such a radar has been developed anywhere*. The PANDORA radar yields extremely high resolutions even for CW radars and has found many applications, the first being in its use as a ground penetrating radar (GPR) for landmine detection, an area where extremely high resolutions are vitally required. Moreover, the architecture of the PANDORA radar is such that it can also be applied to pulsed radars and is not just confined to CW radars. Hence, it is expected that this type of radar will find wide application.

The teaching effort in this book is aided by a set of accompanying software. This software is distributed on the basis of chapters and directly pertains to topics discussed in those chapters. The entire coding has been deliberately kept simple and sometimes very low-tech to enable readers to clearly understand the various steps involved in the implementation of the program. The software is a very basic one and the reader is encouraged to experiment with it and modify it in any manner to suit one's convenience. This is an excellent method to learn the subject. The software presupposes a sound understanding of MATLAB® and has been tested on MATLAB® Version 6.0 (with Signal Processing Toolbox) and above. Some software has been written in C++ and can be run on any C compiler like Visual C++ as a console application. It is pointed out that in a technology of this nature the best way to assimilate a subject is by programming. Coding an operation forces the user to look at all aspects of the subject. This is similar to learning mathematics through solving problems.

A consistent set of notations has been used throughout the book and excessive mathematics has been avoided. Emphasis is placed on imparting a physical understanding of the subject so that the reader has a clear grasp of the processes involved.

There will be errors even though every effort has been made to detect and eliminate them. Any inconvenience to the readers as a result is deeply regretted. The author has a web site: http://www. jankiraman.com/errata_radar. This web site will be kept up to date based on reader inputs as well as software corrections. The readers are advised to visit this site from time to time.

Reference

1. Jankiraman, M., *Pandora Multifrequency Radar—Project Report*, IRCTR-S-014-99, Delft, The Netherlands, April 1999.

ACKNOWLEDGMENTS

This work would not have been generated without the help of many other participants. While it is not possible to list all of them, I gratefully acknowledge those who helped in the preparation of this book. I gratefully acknowledge the help rendered by Artech House in the person of Eric Willner for granting permission to use some of the material in Chapter 1. I am also grateful to Bertil Jonson of Saab Microwave Systems, Sweden, for granting permission to use the material on Pilot radar in Chapter 1. I am also grateful to Dr Bassem Mahafza of deciBel Research Inc., Huntsville, Alabama and Zoe Arey of Taylor & Francis for the permission to use materials in Chapters 2 and 4. I was helped immensely by IEEE in the person of Jacqueline Hansson who granted permission to use figures and materials used throughout this book. Her help and cooperation is gratefully acknowledged. I also acknowledge the help rendered by Georgina Baglin of IET, Stevenage, UK, for granting permission to use material from IEE papers. Thanks are also due to Sheik Safdar of John Wiley & Sons Inc., Hoboken, New Jersey, for granting permission to use material from their publications. I am also grateful to Dr Graham Brooker of the University of Sydney, Australia, for granting permission to use his material in Chapter 2. I am also grateful to Ad de Ridder of the International Research Centre for Telecommunications and Radar (IRCTR), Delft University of Technology, The Netherlands, for granting permission to use their material in Chapters 8 through 10. Thanks are also due to Sally Barone and Louis Lupin of Qualcomm, San Diego, for granting permission to use their materials in Appendix G. I am very grateful to all of you.

I also thank Dr David A. Lynch Jr. for kindly reviewing the manuscript and offering many suggestions toward its improvement. I am truly grateful for his very patient and very painstaking work. I also thank Andy Stove of Thales Aerospace Division, UK, for reviewing my manuscript and giving me his many valuable suggestions and help. Andy also authored Chapter 7 of this book on the Calypso radar. I am deeply indebted to Andy for his help and permission to use material from his papers. I am also grateful to Prof Piet van Genderen of IRCTR, Delft University of Technology, The Netherlands, for his help in authoring Chapter 10 on the PANDORA ground penetrating radar. Prof. van Genderen was involved in the PANDORA program from its very inception. I also thank Prof. L.P. Ligthart, Director, IRCTR, Delft University of Technology, The Netherlands, for his encouragement in the PANDORA radar program. Both Prof. van Genderen and Ligthart were instrumental in making the PANDORA radar a reality. I am truly grateful to you both and to the staff at IRCTR for building the PANDORA radar. It has been a memorable experience. I am also grateful to Prof. Nadav Levanon of Tel Aviv University, Israel, for his patience in answering my many questions during the course of writing this book. I also acknowledge with gratitude his permission to use his software on ambiguity functions from his website. Many thanks for your cooperation.

A work of this nature requires a lot of grit and determination. I thank my children, Pavan and Pallavi, for their patience and encouragement.

Finally, I wish to acknowledge the help rendered by my publishers. Thanks are due to Dudley Kay, President and Editorial Director of SciTech Publishing Inc., for encouraging me to complete this book. I also thank Susan Manning of SciTech Publishing Inc., for being patient with me and with my repeated "final versions." I acknowledge their superb support and efficient handling of this publication project.

M. Jankiraman.
Dallas, Texas
mjankiraman@ieee.org

List of Acronyms

ADC/A/D	Analog-to-digital converter
AGC	Automatic gain control
AF	Ambiguity function
AM	Amplitude modulation
AP	Anti-personnel (landmines)
BT	Bandwidth time (product)
CFAR	Constant false alarm rate
COTS	Commercial off-the-shelf
CPI	Coherent pulse interval
CPT	Coherent pulse train
CRT	Cathode ray tube
CW	Continuous waveform
DAC	Digital-to-analog converter
DAS	Data acquisition system
DDS	Direct digital synthesizer
DFT	Discrete Fourier transform
ECM	Electronic counter measure
EMI/EMC	Electronic mutual interference/Electronic mutual compatibility
ESM	Electronic support measure
FCP	Fast convolution processing
FFT	Fast Fourier transform
FH	Frequency hopping
FM	Frequency modulation
FMCW	Frequency modulated CW
FMICW	Frequency modulated interrupted CW
FSK	Frequency shift keying
FT	Fourier transform
GPR	Ground penetrating radar
HRR	High range resolution
IDFT	Inverse discrete Fourier transform
IEE	Institution of Electrical Engineers
IEEE	Institute of Electrical and Electronic Engineers
IF	Intermediate frequency
IFFT	Inverse fast Fourier transform
LFM	Linear frequency modulation
LNA	Low noise amplifier
LO	Local oscillator

LPI	Low probability of intercept
MF	Matched filter
MMI	Man machine interface
MMIC	Monolithic microwave integrated circuits
MMW	Millimeter wave
MTI	Moving target indication
ODN	Own Doppler nulling
OFDM	Orthogonal frequency division multiplexing
OLPI	Omnidirectional LPI
PACF	Periodic autocorrelation function
PAF	Periodic ambiguity function
PPI	Plan position indicator
PRF	Pulse repetition frequency
PSD	Power spectral density
PSK	Phase shift keying
PSL	Peak side lobe level
QFM	Quadratic FM
RCS	Radar cross section
RF	Radio frequency
RMS	Root mean squared
RPC	Reflected power canceller
SAR	Synthetic aperture radar
SAW	Surface acoustic waveform (generator)
SFW	Stepped frequency waveform
SFWF	Stepped frequency waveform
SFCW	Stepped frequency CW waveform
SNR	Signal-to-noise ratio
SRI	Set repetition interval
SRF	Set repetition frequency (a term used in GPR technology)
SRF	Sweep repetition frequency
SSB	Single sideband
UAV	Unmanned aerial vehicle
UWB	Ultra wideband
VFM	Vertical FM
WBLNA	Wideband LNA
WRF	Waveform repetition frequency
YIG	Yttrium, Iron, and Garnet (oscillators)

Description of Software in CD

The following software is supplied in the enclosed CD with the book: **"Design of Multifrequency CW Radars"**

Chapter 2: lfm.m

Chapter 3: lfm_amb.m, sfw.m, cohopulsetrain.m, singlepulse.m

Chapter 4: lfm_resolve.m, stretch_processing.m

Chapter 5: barker.m

Chapter 6: sfw_resolve.m

Chapter 8: This directory contains all the programs which went into the development of the Pandora radar and the SFCW GPR. These are:

1. fmcw2.m : This file simulates the Pandora radar. It works with fmcw.m file. This is the main script file.
2. sfcw.m : This file simulates a sfcw radar.
3. sfcw2.m : This file simulates the sfcw GPR. It works with up_mix.m and down_mix.m files and sfcw.m. This the main script file.
4. ambf.m : This file calculates the AF of the FMCW Pandora radar.
5. delay.m : This file calculates the delay and Doppler cuts of the AF of the Pandora FMCW radar.
6. mixer.m : This file calculates as to which signal harmonics are in the pass bad of a bandpass filter.
7. The CW.cpp and pandora.cpp along with their EXE files are also included.

Appendix K: zip file containing GUI based software.

Appendix L: Pandora Excel spreadsheet.

Design of Multi-Frequency CW Radars

PART I

Fundamentals of CW Radar

1

Frequency Modulated Continuous (FMCW) Wave Radar

1.1 FMCW RADAR CHARACTERISTICS

The need to see without being seen has been the cardinal principle of military commanders since the inception of warfare. Until the advent of the World War II, the only means available to commanders from this point of view was espionage and intelligence gathering missions behind enemy lines. Just prior to World War II, the allies came up with a groundbreaking invention, the pulsed radar. This invention radically altered the equation and for the first time in the true sense of the term one could see without being seen. The word "RADAR" is an acronym for **Ra**dio **D**etection **A**nd **R**anging. As it was originally conceived, radio waves were used to detect the presence of a target and to determine its distance or range. The pulsed radar could sight the German fighter formations well before they reached the English coast and could, therefore, concentrate allied fighter groups where they were most needed. The German fighters were not even aware that they were detected. In effect, the pulsed radar acted as a force multiplier and helped the allies defeat the vastly superior Luftwaffe in the Battle of Britain. The allies pressed home their advantage of having the radar, by going on to win the Battle of the Atlantic against the German U-Boats by catching them unawares on the surface at nighttime when they were charging their batteries. This was truly stealth warfare in the purest sense of the term. The German reaction to these events was slow and by the time they came up with their own radars and radar emission detectors (now called intercept receivers) it was too little, too late to influence the outcome of the war in their favor.

After the war, understandably, radar engineers around the world concentrated on developing radar emission detectors (intercept receivers), during the Cold War that followed. This effort in turn led to what are today called *Low Probability of Intercept* or LPI radars, which as the term suggests, are radars that can be intercepted, but with a low probability. This is achieved by resorting to *continuous wave* (CW) transmissions instead of the usual pulsed transmissions. CW transmissions employ low continuous power compared to the high peak power of pulsed radars for the same detection performance. To achieve this, this class of radars employ a 100% duty cycle, because it is the average power that determines the detection characteristics. However, CW radars using unmodulated waveforms cannot measure a target's range. The transmitted signal needs to be imparted a bandwidth. This is achieved through various means to be discussed in this book. The most popular among these techniques is frequency modulation of the transmitted signal. In this method the transmit frequency is varied in time and the frequency of the return signal from the target is measured. Correlation of the return signal with the transmit signal yields information on both range and Doppler information of the target. The most common method of frequency modulation is linear modulation. Consequently, the waveform can have a sawtooth shape or a triangular one. Each has its advantages/disadvantages, which are discussed elsewhere in this book. Such radars are called frequency modulated continuous wave (FMCW) radars.

FMCW is an effective LPI technique because of the following principal reasons [1]:

- The frequency modulation spreads the transmitted energy over a large modulation bandwidth ΔF providing good range resolution, critical for target discrimination in the presence of clutter.
- The power spectrum of the FMCW signal is nearly rectangular over the modulation bandwidth, making noncooperative interception difficult.
- Since the transmitted waveform is deterministic, the form of the return signals can be predicted. This makes it resistant to jamming, since any signal not matching this form can be suppressed.
- The signal processing required to obtain range information from the digitized intermediate frequency (IF) signals can be done very quickly with fast Fourier transforms (FFTs).

An additional advantage of FMCW technique is that it is well matched to simple solid-state transmitters, which lead to systems with low initial cost, high reliability, and low maintenance costs. The technique allows a wide transmitted spectrum to be used giving good range resolution without the need to process very short pulses. The frequency modulation required is easier to produce than the short pulse modulation for magnetrons. Furthermore, like for any CW radar, FMCW technique is a mean power system, wherein the same performance can be achieved from FMCW radar transmitting in the order of a few watts of CW power as that from a conventional magnetron-based radar transmitting tens of kilowatts of peak power. This makes FMCW radar much more difficult to detect by electronic support measures (ESM) systems.

This exciting area of radar technology is scarcely discussed in radar literature and there are very few books on this topic. Furthermore, to the best of the author's knowledge, there is no book in radar literature that discusses the analysis and design of FMCW radars, in detail on a stage-by-stage basis, both for the transmitter channel as well as the receiver channel. This book has been written expressly to fulfill this need.

1.2 RANGE EQUATION FOR FMCW RADAR

In this section, we examine the performance of a typical CW radar in terms of maximum achievable range. The CW radar has low continuous power with a 100% duty cycle. We shall initially consider the case of a classical pulse radar and then later extend it to FMCW radars.

This section is based on the work done by Pace (From [2] © 2004, Reproduced by permission of Artech House). Consider a pulse radar which emits high power (P_T) short electromagnetic pulses using a directional antenna of gain G_T. The power density at distance R from the transmitter is equal to (ignoring multipath effects) [2]

$$p(R) = \frac{P_T G_T}{4\pi R^2} \tag{1.1}$$

The total power illuminating the target of effective cross-section σ_T is equal to

$$P_A = \frac{P_T G_T}{4\pi R^2} \sigma_T \tag{1.2}$$

The target reflects all illumination power omni-directionally (which is usually the case), the power received by the receiving radar antenna with effective surface A_R is equal to

$$P_R = \frac{P_T G_T}{16\pi^2 R^4 L}\sigma_T A_R \tag{1.3}$$

where L is the overall loss in the radar system, including transmission, propagation, and receiving losses. If we substitute the antenna gain for the effective surface in (1.3), we obtain the classical radar equation of the form

$$P_R = \frac{P_T G_T G_R \lambda^2}{(4\pi)^3 R^4 L}\sigma_T \tag{1.4}$$

where G_R is the gain of the receiving antenna and λ is the wavelength of the transmitted frequency.

The receiver's equivalent noise can be expressed as

$$P_N = kTF_R B \tag{1.5}$$

where T is the effective system noise temperature (dependent on the receiver's temperature), F_R is the receiver noise figure (see Appendix "L"), B is the receiver's bandwidth (assuming that at the receiver's end match filtering is used), and k is the Boltzmann's constant ($1.3806505 \times 10^{-23}$ [JK^{-1}]). If we consider using a Neyman–Pearson detector, it is reasonable to assume that there is a target echo in the signal when the echo power is greater than the noise power multiplied by the detectability factor D_0 which usually has a value somewhere between 12–16 dB, depending upon the probability of false alarm. The radar range equation can then be written as

$$\frac{P_T G_T G_R \lambda^2}{(4\pi)^3 R^4 L}\sigma_T > kTF_R B D_0 \tag{1.6}$$

and the maximum detection range is equal to

$$R_{\max} = \left(\frac{P_T G_T G_R \lambda^2}{(4\pi)^3 LkTF_R B D_0}\sigma_T\right)^{1/4} \tag{1.7}$$

In the case of pulse radars, the receiver's bandwidth is approximately inversely proportional to the pulse width τ. Substituting in equation (1.7), we obtain

$$R_{\max} = \left(\frac{E_T G_T G_R \lambda^2}{(4\pi)^3 LkTF_R D_0}\sigma_T\right)^{1/4} \tag{1.8}$$

where $E_T = \tau P_T$ being the transmitted pulse energy. This implies that the per pulse detection range depends upon the transmitted pulse energy E_T, transmitter and receiver antenna gains, and wavelength and *does not* depend upon the pulse length or the receiver's bandwidth. Equation (1.8) is a general radar range equation applicable to all kinds of radars. Specifically in the case of CW radars, E_T equals the product of the transmitted power and the time of target illumination or coherent signal integration. This result of equation (1.8) can also be explained by the matched filter theorem (see Section 2.4), which states that the best sensitivity against noise is obtained by using a filter whose frequency response is the complex conjugate of the spectrum of the signal.

This has two consequences:

(a) All signals with the same mean power and time duration, that is, the same energy, have the same sensitivity, providing it is practical to implement the receiver.
(b) Since the filtering removes the phase of the Fourier components of the signal (by multiplying the signal by its complex conjugate) all signals having the same spectrum have the same range resolution, whatever the time-domain waveform.

We rewrite equation (1.7) in a different form for CW radars of average power P_{CW} in watts. The detectability factor D_0 can be expressed in terms of the output signal-to-noise ratio as (SNR_{Ro}), while the bandwidth can be expressed as the output bandwidth B_{Ro}. Therefore, in terms of these output parameters we obtain

$$R_{max} = \left[\frac{P_{CW} G_T G_R \lambda^2}{(4\pi)^3 LkTF_R B_{Ro} (SNR_{Ro})} \sigma_T \right]^{1/4} \tag{1.9}$$

The processing gain of the radar is defined as

$$PG_R = \frac{SNR_{Ro}}{SNR_{Ri}} \tag{1.10}$$

where (SNR_{Ri}) is the SNR at the input of the receiver.

We now derive another form to equation (1.9) that radar designers generally find useful. Using equation (1.10) in equation (1.9) and using B_{Ri} for the receiver input bandwidth, we obtain

$$R_{max} = \left[\frac{P_{CW} G_T G_R \lambda^2 \sigma_T}{(4\pi)^3 LkTF_R B_{Ri} (SNR_{Ri})} \right]^{1/4} = \left[\frac{P_{CW} G_T G_R \lambda^2 \sigma_T}{(4\pi)^3 LkTF_R B_{Ri} (SNR_{Ro}/PG_R)} \right]^{1/4} \tag{1.11}$$

We also know that the processing gain of FMCW radar is given approximately by the time–bandwidth product

$$PG_R = B_{Ri} T_s \tag{1.12}$$

where T_s is the sweep time.

Substituting in equation (1.11) for PG_R, we obtain

$$R_{max} = \left[\frac{P_{CW} G_T G_R \lambda^2 \sigma_T B_{Ri} T_s}{(4\pi)^3 LkTF_R B_{Ri} (SNR_{Ro})} \right]^{1/4}$$

or

$$R_{max} = \left[\frac{P_{CW} G_T G_R \lambda^2 \sigma_T T_s}{(4\pi)^3 LkTF_R (SNR_{Ro})} \right]^{1/4}$$

or assuming that the sweep repetition frequency $SRF = 1/T_s$, we finally obtain

$$R_{max} = \left[\frac{P_{CW} G_T G_R \lambda^2 \sigma_T}{(4\pi)^3 LkTF_R (SNR_{Ro}) SRF} \right]^{1/4} \tag{1.13}$$

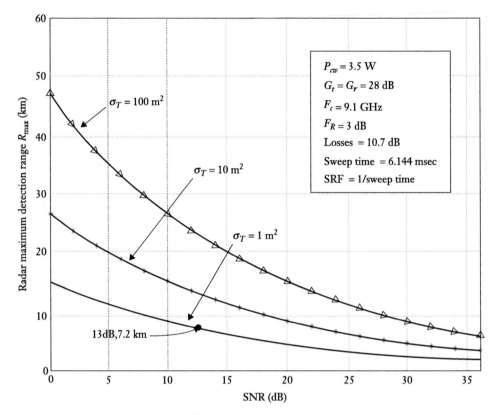

Figure 1–1 Calypso radar performance. Maximum detection range for $\sigma_T = 1, 10, 100$ m^2.

We have expressed the radar range equation in terms of output SNR and the sweep repetition frequency. Equations (1.13) and (1.9) are the most popular forms of this equation.

We now introduce the reader to a radar called Calypso. The name is a pseudonym for an actual X-band LPI navigation radar based on FMCW transmission [3]. It is necessary at this stage to introduce such a radar, because it is useful to illustrate the design approach of such radars based on numerical examples. A detailed study of this radar and its salient parameters have been taken up in Part II of this book, wherein we have discussed the design considerations which go into firming the parameters of this radar and the additional problems that arise in the development of such radars including problems like noise control and calibration. Once the reader understands and appreciates such issues, he/she will be in a position to take up design work of similar X-band radars. Furthermore, the reader will then be in a position to understand the additional complications that arise in the design of multifrequency radars, which is the final goal of this book. However, for now, Figure 1–1 shows the performance of this radar with the parameters shown. The losses are 10.7 dB. It can be seen that at an *output* $SNR_{R_0} = 13$ dB and an *output* bandwidth $B_{R_0} = 160$ Hz, a 1 m^2 target can be detected at a range of 7.2 km. Performance of this kind is pretty good for such a radar. However, the LPI advantage is not apparent in this example. To understand this, we need to compare this range with that of an intercept receiver.

1.3 INTERCEPT RANGE OF FMCW RADAR

Figure 1–2 shows the block diagram of an intercept receiver. We note principal stages, viz. the predetector stage and the postdetector stage. The three major components include the RF amplifier with bandwidth B_{RF}, the detector (e.g., square law), and the postdetection video amplifier with bandwidth B_{video}.

During the design of the intercept receiver, the front-end RF bandwidth B_{RF} is matched to the largest coherent radar bandwidth expected and the video bandwidth B_{video} is matched to the inverse of the smallest radar coherent integration time expected. This usually requires *a priori* information about these parameters of the signal. To achieve 100% of intercept, many systems use wide beam width antennas of the order of 0 dB gain and receiver bandwidths of the order of several gigahertz. Such a system will typically have a minimum detectable signal in X-band of around −60 dBm and an effective receiver aperture of around −40 dBm². The minimum detectable power density will then be about −50 dBWm⁻². This means that

$$T_{threshold} = -60 \text{ dBm} - 40 \text{ dBm}^2 = -20 \text{ dBm/m}^2 = -50 \text{ dBW/m}^2$$

Suppose we use the Calypso radar parameters of antenna gain of 30 dB and power output of 3 W, then we obtain our corresponding detection range as

$$R = \sqrt{\frac{P_{CW}G_t}{T_{threshold}4\pi}} = \sqrt{\frac{3 \times 1000}{10^{(-50/10)} \times 4\pi}} = 4886 \text{ m} \tag{1.14}$$

where $T_{threshold}$ is the minimum detectable power density.

If instead of the Calypso, we had used a pulsed radar with a peak power of about 50 kW [3] (this FMCW radar, Calypso, was intended as a replacement for a 50 kW magnetron-based

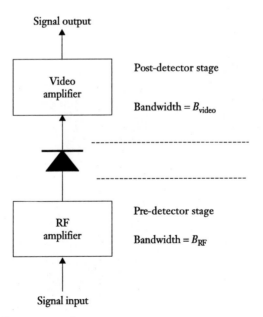

Figure 1–2 Block diagram of intercept receiver.

pulsed navigational radar on a similar performance basis; see Chapter 7), the same system would have detected it at 630 km range, if propagation effects are neglected (as we have done in these calculations). If we view these facts against the design detection range of 7.2 km against a 1 m² target, our Calypso radar is truly LPI.

Hence, an FMCW radar like the Calypso can detect a 1 m² aircraft at a range of 7 km, but the target aircraft's ESM system cannot detect the radar emission till it is 5 km from the Calypso. Hence, if the aircraft carrying the ESM system cannot come within the radar's detection range, for its own safety, then the radar becomes in practice undetectable to the ESM system.

1.4 COMMERCIAL FMCW RADARS

1.4.1 The PILOT Radar

The PILOT radar operates on FMCW principle and was developed by Philips Research Laboratory in 1988. Hollandse Signaal Apparaten B.V. of Holland, since taken over by Thomson-CSF (now Thales) upgraded this radar to Scout.

The PILOT is a well-established example of an FMCW tactical navigation LPI radar [4, 5] (see Figure 1–3). It can be added to existing navigation radar, while retaining its I-band antenna, transceiver, and display system. If extra sensitivity or accuracy against high-speed targets is needed, and LPI is not important then, the PILOT can be switched out and pulsed radar switched in when higher SNRs are required. The PILOT uses an FMCW 1 kHz sweep repetition frequency (SRF) with a low noise figure ($F_R = 5$ dB) and very low output power to ensure that it is undetectable by hostile intercept receivers. It uses a 1,024-point FFT (512 range cells) high range resolution (2.7

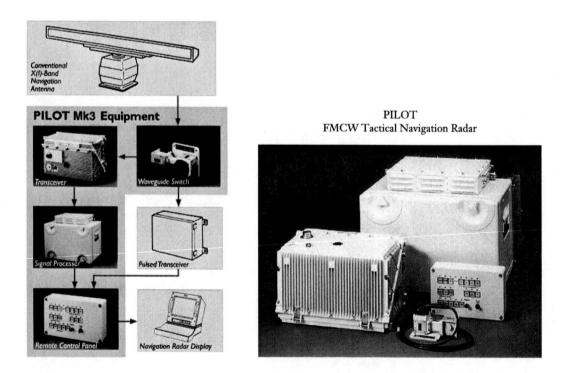

Figure 1–3 Equipment that makes up the PILOT Mk3. (From [5], © Reprinted with permission)

m to 86 m) capability and is constructed for ease of installation. The technical parameters of this radar are given in Table 1–1. The equipment details are shown in Figure 1–5. The radar employs FMCW waveform with frequency agility.

The PILOT as an LPI radar employs only one low side lobe antenna, transmits a maximum CW power of 1 W and uses an FMCW waveform with a variable modulation bandwidth ΔF to vary the range resolution. Table 1–2 gives the intercept capability of this radar. Note that in Tables 1–1 and 1–2, 1 nautical mile (nmi) = 1.852 km. Also, in Table 1–2, dBmi represents dB in mW with reference to a system containing an isotropic antenna $G_I = 1$.

FMCW principle is outstanding for this purpose since it makes possible a very low output power level (1 W continuous wave, selectable down to 1 mW). This gives an ESM system a very short detection range while the PILOT has the same navigation radar detection range as conventional pulsed radar with peak power levels of several kW.

1.5 EXPERIMENTAL AIR SEARCH CW RADAR

This radar was primarily developed to achieve a search mode with low vulnerability against the anti-radar missile (ARM) threat. To achieve this, the radar would need to

- avoid using a scanning transmit beam.
- transmit low power.
- use a phase-coded LPI waveform for the determination of target range.

These requirements were fulfilled by this radar called omnidirectional LPI (OLPI) radar [6, 7]. The transmitter and receiver are shown in Figure 1–4. It achieves these objectives by employing a

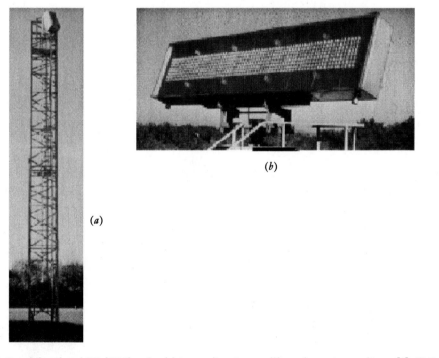

(b)

(a)

Figure 1–4 Omnidirectional LPI (OLPI) radar (a) transmit antenna; (b) receive antenna. (From [6], © Reprinted with permission)

Table 1–1 Technical Characteristics of the PILOT Mk3 Radar

Antenna

Type	Single or dual slotted waveguide
Gain	30 dB
Side lobes (w.r.t. peak gain)	< −25 dB
	< −30 dB
Beamwidth (3 dB)	
horizontal	1.2°
vertical	20°
Rotational speed	24/28 rpm
Polarization	Horizontal

Transmitter

Output power	1.0, 0.1, 0.01, or 0.001 W (CW)
Frequency	9.375 GHz (I-band/X-band)
Range selection	24, 12, 6, 3, 1.5, 0.75 nmi
Frequency sweep	1.7, 3.4, 6.8, 13.75, 27.5, 55 MHz
Sweep repetition frequency	1 kHz

Receiver

IF bandwidth	512 kHz
Noise figure	5 dB

Processor unit

No. of range cells	512 (1024 point FFT)
Range resolution	< 75 m at 6 nmi scale
Range accuracy	< ± 25 m at 6 nmi
Azimuth accuracy	± 2°
Azimuth resolution	1.4°

Display system

Type	Color
Minimum effective PPI diameter	250 mm
Resolution	768 × 1024 V
Tracking capacity	40 targets
Range ring accuracy	1.5% of selected scale or 50 M, whichever is greater

From [5], © Reprinted with permission.

wide transmit beam and preformed receive beams as shown in Figure 1–5(*a*). The radar employs eight dipoles in a column, which are combined by a microstrip feeding network resulting in a horizontal fan beam pattern with a beam width of about 20° in elevation and 120° in azimuth. The antenna is fed by a transistor amplifier with an output power of 5 W. The wavelength is 11 cm (S-band). The

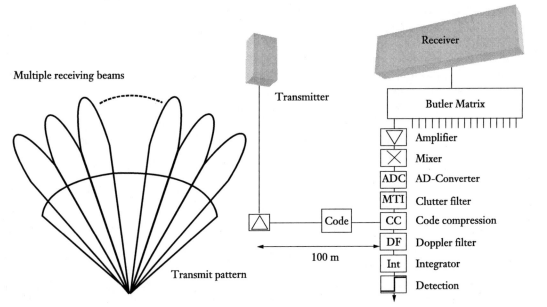

Figure 1–5 (*a*) OLPI antenna pattern; (*b*) OLPI block diagram. (From [6], © IEEE 1996)

Table 1–2 Summary of PILOT Detection and Intercept Range Calculations

Radar Output Power	Radar Detection Range (km)		Intercept Range (km)		
	100 m² Target	1 m² Target	Intercept δ_I (−40 dBmi)	Intercept δ_I (−60 dBmi)	Intercept δ_I (−80 dBmi)
PILOT Mk2					
1 W	28	8.8	0.25	2.5	25
0.1 W	16	5	0	0.8	8
10 mW	9	2.8	0	0.25	2.5
1 mW	5	1.5	0	0	0.8
LPRF Radar					
10 kW	49.6	15.7	25	254	2.546

From [5], © Reprinted with permission.

receiving antenna consists of an array of 64 columns, each column containing 8 dipoles combined by a network as for the transmit antenna. The 64 beams are generated using a microstrip 64-Butler matrix [8]. Each output of the Butler matrix represents one beam and is connected to a receiver channel. Each receiving channel delivers digital I and Q signals with 12 bits each at a sample rate corresponding to the signal bandwidth. At present, 16 receiving channels in the experimental system are realized out of a total of 64 possible channels. This also makes sense, because (in accordance with theory) the beams at the edges will necessarily be distorted (broadened). The received beams have a nominal beam width of 2° in azimuth and cover the illuminated sector of 120° in azimuth.

The received target echoes are integrated over a time, for example, 2 second. The side effect of this radar is that (since it processes in the Doppler domain) it is handy for detecting flashing of helicopter blades that typically have a tip velocity of 300 m/sec. To ensure that the target stays within one resolution cell for the 2 second integration time, it becomes necessary to keep the range resolution cell to a width of 600 m. The cross-range resolution for a 2° azimuth beam width is 1,000 m at a range of 30 km. This is nearly matched to the range resolution cell of 600 m and is therefore, considered adequate. The length of the subpulse is therefore chosen as 4 μsec. The transmitter array is located 100 m from the console by cable and is also separated from the receiving array by a similar distance. The signal processor is shown in Figure 1–5(b). It is advantageous to process in the Doppler plane so as to notch out the nearby ground clutter located on zero Doppler. The radar employs the polyphase Frank code (discussed in Chapter 5) which is more Doppler tolerant than most phase-coded signals. Using a subpulse of 4 μsec and a code length of $N = 64$, the resulting code period of $T = 256\,\mu$sec gives an unambiguous range of 38.4 km.

1.5.1 Miscellaneous Uses of LPI Radars

Altimeters: LPI radars also find extensive use in altimeters for measuring the flying height of aircraft. Before the advent of CW technology, altimeters used to employ pulsed radars. These altimeters worked well at high altitudes. However, this became a problem for low-flying platforms like cruise missiles. This is because pulse radars have a blind zone area surrounding their installation, where no targets can be detected. This blind zone area is a function of the transmitted pulse width. For a pulse width of 0.1 μsec, no target within 50 feet of the radar can be detected.

Hence, for low flying vehicles we would like to measure altitudes down to zero feet. FMCW radars provide the answer. In a typical FMCW altimeter [1], the transmitter's carrier frequency changes linearly over a 120 MHz modulation bandwidth that ranges from 4.24 to 4.36 GHz. The transmitter works continuously to generate a CW output and changes frequency at a constant rate based on a sawtooth pattern or a triangular pattern. A fixed broad beam antenna system is used to illuminate a large area of underlying terrain. The broad beam allows for correct operation over the normal range of missile pitch and roll. The signal reflected from the surface is correlated to a sample of the transmitted signal. The difference produced after mixing, is a low frequency beat signal proportional to the range being measured. A limiter then selects the strongest signal from the surface directly below the vehicle. This yields the height information, that is, range. The system also provides LPI capability by carefully controlling the transmitted power, so as not to alert enemy intercept receivers. Different types of altimeters are covered in [1].

Landing Systems: Landing systems for automatic and precision landing of UAVs (unmanned aerial vehicle), transmit a beacon and aid landing operations. These systems must remain necessarily LPI as they remain active on a battlefield. Similarly, battlefield surveillance equipment and fire control radars are also designed with LPI capability. There is not sufficient space in this work to cover these aspects, but the interested reader is invited to check the references [1].

1.6 A SURVEY OF THIS BOOK

Part I: Fundamentals of CW Radar Design

This chapter introduced the reader to the concept of CW radar systems and their advantages/ disadvantages. The factors that went into calling a CW radar an LPI radar were then examined. The reader was then introduced to the CW radar range equation. The concept of intercept receiver was then examined. The LPI capability against such interceptors was demonstrated through worked examples. The reader was then introduced to some well-known LPI radars like the PILOT radar and the OLPI experimental radar. The salient features of these systems were then discussed. Further applications of LPI techniques like altimeters and landing systems, and so forth, were then briefly examined.

Chapter 2 of this book examines the various types of radar waveforms, including CW, pulsed, and LFM. High range resolution (HRR) waveforms and stepped frequency waveforms are also analyzed. The concept of matched filter (MF) is introduced and analyzed.

Chapter 3 presents in detail the principles of radar ambiguity functions. This includes ambiguity function for single pulse, LFM pulses, coherent pulse train ambiguity function, and stepped frequency waveform ambiguity function.

Chapter 4 deals with the theory of FMCW radars. There have been a lot of books written on pulse radars, but practically none on this topic. The reader is introduced to the FMCW radar concept and its advantages/disadvantages as compared to pulsed radars. The CW radar range equation is revisited and various antenna configurations and waveform designs are examined. The receiver–transmitter isolation problems (for single antenna systems) and dual antenna systems are then examined. The reader is then introduced to problems pertaining to nonlinearities in LFM waveforms and waveform generation. We then examine LFM pulse compression techniques and *stretch processing*.

Chapter 5 discusses the design of phase-coded radars employing phase coding techniques. Details on Barker sequences and Frank codes are presented and their spectral and ambiguity properties investigated. The reader is then introduced to the concept of Periodic Ambiguity and Periodic Autocorrelation Functions peculiar to CW waveforms.

Chapter 6 discusses frequency hopped (FH) radar waveform design. We then examine a unique waveform design which is a combination of LFM and FH. This has found an interesting application in automobile engineering [9]. The technique enables the measurement of range and Doppler of multiple targets very quickly and unambiguously, a requirement which is very essential in automobile collision avoidance systems. Otherwise FMCW systems take too long a time in a multiple target environment (like road traffic), requiring the processing of different chirp gradients due to range-Doppler coupling.

Part II: Theory and Design of Calypso FMCW Radar

In Chapter 7, the reader is introduced to an FMCW X-band navigational radar called Calypso. The parameters of this radar are based on an actual FMCW navigational radar. It is necessary at this stage to introduce such a radar to the reader, because it is useful to illustrate the design approach of such radars based on numerical examples. Calypso is a dual antenna system. Based on this radar, the reader is introduced to such issues as calculation of the basic parameters based on user requirement, noise figure calculation for an FMCW radar, AM noise cancellation problems, FM noise cancellation problems. We also examine IF amplifier design criteria, LNA selection criteria, AGC design requirements, ADC selection and control circuitry requirements. Finally, radar performance measurement, radar calibration and verification, range resolution issues, ESM and ECCM problems, moving target indication (MTI), and single antenna operation issues as well as trials and testing are also examined.

Part III: Theory and Design of Pandora Multifrequency Radar

This part brings us to the design of multifrequency radars. The specific radar is called the Pandora and it is a *range profiling radar*. This implies that it is a cut above high range resolution radars that emphasize on target discrimination. The Pandora, on the other hand, was expressly developed to take this requirement to an order beyond high range resolution, to range profiling. The radar was intended to profile ground mines, so that (based on the radar image of the mine) one can determine in advance as to the type of ground mine and the approach one needs to adopt to disarming it. The design procedure, however, is based on the radar's ability to profile a missile/ fighter aircraft over a range of approximately 7 km. Hence, the calculations involved center around this requirement and should, therefore, interest a wider audience. This radar employs dual antennas, one for transmission and one for reception. The signal processing comprises eight parallel channels required to enhance signal processing speed. In view of this, this part of the book has been divided into three chapters, one for single channel design, one for single channel hardware implementation, and one for the overall radar with the eight parallel channels. This radar was developed at the International Research Centre for Telecommunications and Radar (IRCTR), Delft University of Technology, The Netherlands, and formed part of the author's Master thesis [10]. It is pointed out that to the best of the author's knowledge, *this is the first time such a radar has been developed anywhere.*

Chapter 8 deals with the nuts and bolts of radar design at the very basic level. There is a very considerable amount of RF engineering discussed here. Hence, this chapter is common to all CW radar development requirements. It starts with the FMCW radar equation and waveform selection. It then examines the entire design on a stage-by-stage basis, starting with the selection of a low noise amplifier (LNA), to passive filter design, to the choice of mixers (including power levels of signals at the mixer inputs) for low intermodulation distortion, and the choice of ADCs all the way to baseband signal processors. The basic signal processing in this radar revolves around stretch processors for pulse compression. Stretch processing yields very high signal processing gains. This is very necessary for high radar resolutions. Since this is a range profiling radar we need to go a step beyond conventional CW radars to control generation of time side lobes. Toward this end, issues like filter group delays and nonlinearities in LFM generation are also examined. Unless properly addressed, these factors will also contribute to time side lobe generation. Finally, we develop level diagrams to determine signal levels at each stage. The reader is also introduced to *spreadsheet calculations* typical to any radar development. Calculations of noise figures and controlling of AM and FM noise are also examined.

Chapter 9 deals with the hardware implementation details of this radar for a single channel. Results from verification measurements are also included.

Chapter 10 addresses the system integration problems of the overall radar with the eight parallel channels as well as field results.

Admittedly, it is advisable for an LPI radar designer to study radar emitter interception aspects also. However, this is a vast field and it is not possible to include it in a work of this nature. The interested reader is referred to [1] and the references listed therein. This work is intended to solely concentrate on the design and development and testing of LPI radar emitters.

Finally, a word on the nomenclature used in this book. The term "pulse radars" imply unmodulated (gated CW) radars that are magnetron based. Radars using chirp pulses or phase-coded pulses will be referred to as pulse modulated radars or chirp pulse radars or phase-coded pulse radars as appropriate. In discussing CW radars, the term "pulse" is out of place. In such cases, it will be more appropriate to use the term "step" as in frequency step or signal step.

References

1. Pace, Phillip E., *Detecting and Classifying Low Probability of Intercept Radar*, Artech House, Norwood, MA, 2004.
2. Skolnik, M. I., *Introduction to Radar Systems*, 3rd edn., McGraw-Hill, Boston, MA, 2001.
3. Barrett, M., Reits, B. J., and Stove, A. G., "An X-band FMCW navigation radar," in *Proceedings of the IEE International Radar Conference on Radar '87*, IEE Conf. Publ. No. 281, 1987, pp. 448–452.
4. Beasley, P. D. and Stove, A. G., "PILOT—an example of advanced FMCW techniques," in *IEE Colloquium on High Time-Bandwidth Product Waveforms in Radar and Sonar*, May 1, 1991, pp. 10/1–10/5.
5. http://www.naval-technology.com/conchannelors/weapon_control/celsiustech/

6. Wirth, W. D., "Long term coherent integration for a floodlight radar," in *Record of the IEEE International Radar Conference*, 1995, pp. 698–703.

7. Wirth, W. D., "Polyphase coded CW radar," in *Proceedings of the IEEE Fourth International Symposium on Spread Spectrum Techniques and Applications*, Mainz, Germany, September 22–25, 1996, Vol. 1, pp. 186–190.

8. Balanis, C. A., *Antenna Theory Analysis and Design*, Harper and Row, New York, 1982.

9. http://www.smartmicro.de/Automotive_Radar_-_Waveform_Design_Principles.pdf.

10. Jankiraman, M., *Pandora Multifrequency Radar—Project Report*, IRCTR-S-014-99, Delft, The Netherlands, April 1999.

2

Radar Waveforms and Processing

2.1 RADAR SIGNALS

There is a need to choose the correct type of waveform based on a radar's mission. A correct decision in this respect is always a cost-effective decision. Broadly, radars can use pulsed waveforms or continuous waveforms (CW) with or without modulation. The quality of range or Doppler resolutions required, influence our choice of a suitable waveform or signal. We define low pass signals as signals that contain a significant frequency composition at a low-frequency band including DC. Those signals, which have a significant frequency composition around some frequency away from the origin are called band pass signals [1, 2]. We express this band pass signal as

$$x(t) = r(t)\cos(2\pi f_0 t + \phi_x(t)) \tag{2.1}$$

where $r(t)$ is the amplitude modulation or envelope, $\phi_x(t)$ is the phase modulation, and f_0 is the carrier frequency. The frequency-modulated signal is defined by the rate of change of phase of the signal as

$$f_m(t) = \frac{1}{2\pi}\frac{d}{dt}\phi_x(t) \tag{2.2}$$

The instantaneous frequency is, therefore, given by

$$f_i(t) = f_0(t) + f_m(t) \tag{2.3}$$

If f_0 is very large as compared to its bandwidth B, then we call it a narrowband signal.

In the case of narrowband modulation, we can represent the band pass signal of (2.1) as comprising two low pass signals known as quadrature components. We can then write (2.1) as

$$x(t) = x_I(t)\cos 2\pi f_0 t - x_Q(t)\sin 2\pi f_0 t \tag{2.4}$$

where

$$\begin{aligned} x_I(t) &= r(t)\cos\phi_x(t) \\ x_Q(t) &= r(t)\sin\phi_x(t) \end{aligned} \tag{2.5}$$

Equations (2.4) and (2.5) are in fact approximations to (2.1) for narrowband modulations. Figure 2–1 shows the extraction of these quadrature components from the band pass signal.

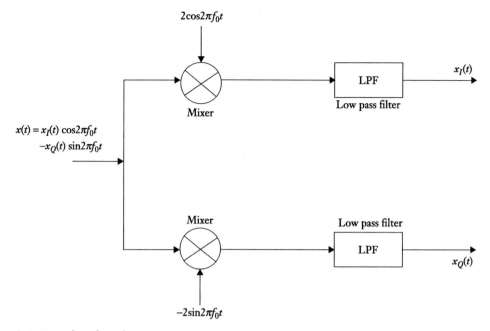

Figure 2–1 Extraction of quadrature components.

2.2 CW WAVEFORM

The spectrum of a signal describes the spread of its energy in the frequency domain. We have two types of spectrum classification. They are:

- Energy spectral density (ESD): J/Hz
- Power spectral density (PSD): W/Hz

The former pertains to energy signals and the latter to power signals.

Signal bandwidth is the range of frequency over which the signal has a nonzero spectrum. A time-limited signal (of finite duration) has an infinite bandwidth, whereas a band-limited signal has an infinite duration, for example (as an extreme case), a continuous sine wave, whose bandwidth is infinitesimal. A time domain signal $f(t)$ has a Fourier transform (FT) $F(\omega)$. A signal can be freely converted to its FT and vice versa (inverse Fourier transform (IFT)). Similarly, a signal's autocorrelation function $R_f(\tau)$ is related to its PSD $\bar{S}_f(\omega)$ through its FT. These important relationships are shown graphically in Figure 2–2.

A CW waveform is defined by

$$f_1(t) = A\cos\omega_0 t \tag{2.6}$$

The FT of $f_1(t)$ is

$$F_1(\omega) = A\pi[\delta(\omega - \omega_0) + \delta(\omega + \omega_0)] \tag{2.7}$$

where $\delta(\bullet)$ is the Dirac delta function and $\omega_0 = 2\pi f_0$.

The signal $f_1(t)$ is shown in Figure 2–3 to have infinitesimal bandwidth, located at $\pm f_0$.

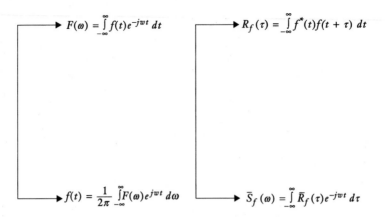

Figure 2–2 Relationship between signal $f(t)$ and its FT $F(\omega)$, and between signal autocorrelation function $R_f(\tau)$ and its PSD $\overline{S}_f(\omega)$.

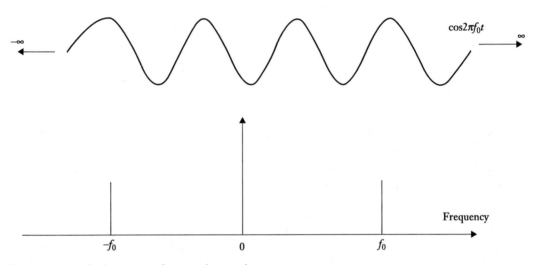

Figure 2–3 Amplitude spectrum for a continuous sine wave.

2.3 RANGE RESOLUTION

There are many ways one can define range resolution. For the purposes of this section we define range resolution of a sensor as the minimum separation (in range) of two targets or equal cross section that can be resolved as separate targets. It is determined by the bandwidth of the transmitted signal.

The bandwidth Δf is generated by widening the transmitter bandwidth using some form of modulation

- Amplitude modulation
- Frequency modulation
- Phase modulation

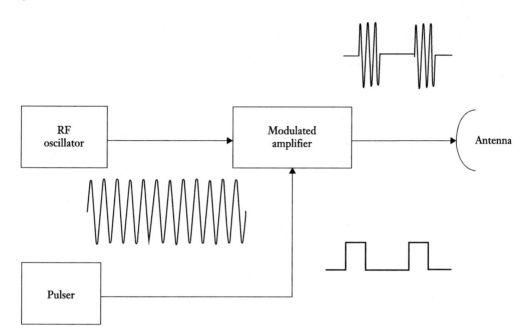

Figure 2–4 Generation of transmit pulses from a sine wave. (From [3], © Reprinted with permission.)

This section is based on the work done by Graham Brooker [3] and is reprinted with permission.

2.3.1 Amplitude Modulation

A special case of the amplitude modulation technique is the classical pulsed radar where the amplitude is 100% for a very short period and 0% for the remaining time (see Figure 2–4).

It can be shown using Fourier analysis that for a pulsed system, the relationship between pulse width τ sec and the effective bandwidth Δf Hz is given by

$$\Delta f \approx \frac{1}{\tau} \tag{2.8}$$

In addition to this, Figure 2–5 shows the fast Fourier transform (FFT) of a pulse of 10 second duration from which it can be seen that the minima on either side of the peak are displaced from the peak by $1/\tau$ Hz.

The range resolution is determined as

$$\Delta R = \frac{c\tau}{2} \tag{2.9}$$

where c is the velocity of light (3×10^8 m/sec).

It can be rewritten in terms of effective bandwidth as

$$\Delta R = \frac{c}{2\Delta f} \tag{2.10}$$

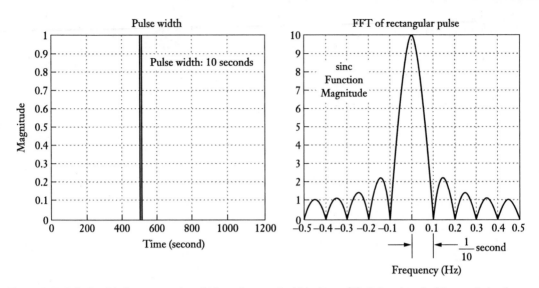

Figure 2–5 Relationship between pulse width and spectral width. (From [3], © Reprinted with permission.)

Narrow pulse systems require a large peak power (>5 MW) for long-range operation and so special precautions must be taken to minimize the problems of ionization and arcing within the waveguide for radar systems.

2.3.2 Frequency and Phase Modulation

Solutions involve lengthening the pulse width to achieve large radiated energy, while still maintaining the wide bandwidth for good range resolution.

The received signal is processed using a matched filter (see Section 2.4) that compresses the long pulse to duration $1/\Delta f$. The time–bandwidth product $\Delta f \times \tau$ of the uncompressed pulse is used as a figure of merit for the system.

The following techniques are used in CW radars to obtain large time–bandwidth products:

- Frequency modulated continuous wave (FMCW)
- Stepped frequency
- Phase-coded signal compression
- Stretch
- Interrupted FMCW (FMICW)

We now briefly examine each of these techniques [3]. Some of these will be reexamined elsewhere in this book in more detail.

2.3.2.1 Frequency Modulated Continuous Wave   The inability of a CW radar to resolve range is related to the narrow spectrum of its transmitted waveform. An unmodulated CW radar has in fact a line spectrum. Frequency modulation of the carrier is one of the most common techniques used to broaden the spectrum. Such radars are called FMCW radars.

FMCW is an effective low probability of intercept (LPI) technique for various reasons. The frequency modulation spreads the transmitted energy over a large modulation bandwidth Δf. This large bandwidth helps us detect targets with high resolution. The power spectrum is nearly

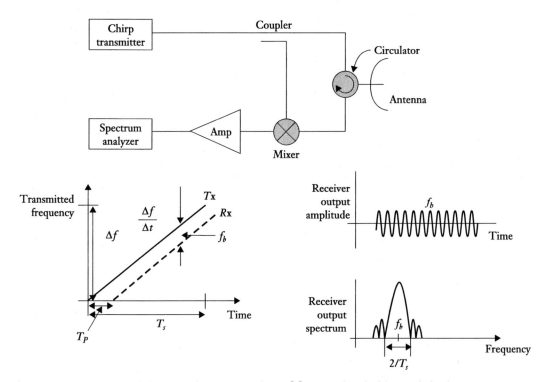

Figure 2–6 Frequency modulated continuous wave. (From [3], © Reprinted with permission.)

rectangular over the modulation bandwidth. This makes interception difficult. It is resistant to jamming because it is a deterministic signal and its form is only known to the user. It is readily compatible with solid-state transmitters unlike pulsed/chirp pulsed radars that require vacuum tubes, high voltages, and so forth. This type of signal is called linear frequency modulated (LFM) signal.

Figure 2–6 shows the principle. A ramp signal is transmitted at a ramp rate of $(\Delta f / \Delta t)$, where Δf is the maximum sweep bandwidth and Δt is the sweep time. The return signal after a time $T_p = 2R/c$, where R is the target range and c is the velocity of light, is then fed to a homodyne receiver as shown in Figure 2–6 and correlated with the reference signal. Because of the delay T_p, there will be a difference frequency called the beat frequency f_b, which is proportional to range. This beat signal is then given to an FFT processor, which identifies the frequency with the range. We will look at the FMCW waveform as a background to the other techniques. This waveform will be discussed in detail in Chapter 4.

2.3.2.1.1 Derivation of the Swept Bandwidth   It can be proved that, for an FMCW waveform,

$$\Delta f = \frac{c}{2\Delta R} \tag{2.11}$$

Equation (2.11) implies that the higher the required range resolution, the more the required signal bandwidth. This inference is the same as for pulse radars.

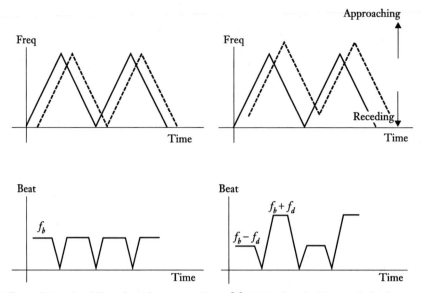

Figure 2–7 Effects of Doppler shift on beat frequency. (From [3], © Reprinted with permission.)

2.3.2.1.2 Calculating the Range In general, the range in FMCW radars is calculated from the measured beat frequency using the following relationship (see Chapter 4 for more details)

$$R = \frac{f_b T_s c}{2\Delta f} = \frac{Nc}{4\Delta f} \qquad (2.12)$$

where f_b is the beat frequency,

Δf is the sweep bandwidth,

T_s is the sweep time, and

N is the number of samples/sweep ($N > 2 \times f_{b\,\text{max}}$).

2.3.2.1.3 Effect of Target Motion A moving target will impose a Doppler frequency shift on the beat frequency as shown in Figure 2–7.

One portion of the beat frequency will be increased and the other portion will be decreased. For a target approaching the radar, the received signal frequency is increased (shifted up in the diagram) decreasing the up-sweep beat frequency and increasing the down-sweep beat frequency

$$f_{b(\text{up})} = f_b - f_d$$

$$f_{b(\text{dn})} = f_b + f_d$$

The beat frequency corresponding to range can be obtained by averaging the up and down sections $f_r = [f_{b(\text{up})} - f_{b(\text{dn})}]/2$.

The Doppler frequency (and hence target velocity) can be obtained by measuring one half of the difference frequency $f_d = [f_{b(\text{up})} - f_{b(\text{dn})}]/2$.

The roles are reversed if $f_d > f_b$.

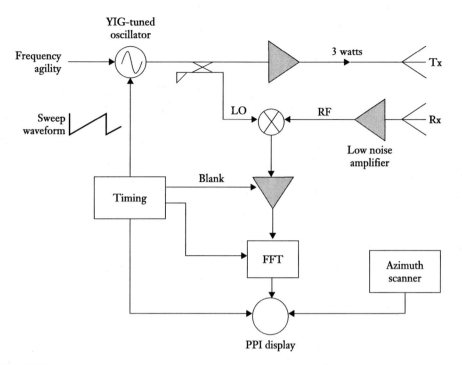

Figure 2–8 FMCW radar schematic.

2.3.2.1.4 FMCW Generation and Reception   There are many ways to generate LFM signals. One way is to use a Yitrium Iron Garnet (YIG) oscillator to produce the ramp signal shown in Figure 2–8. The schematic for this approach is shown in Figure 2–8.

The system timing for the start of the sweep comes from a timing circuit which is usually processor based. The rising ramp signal (also called up-chirp; a falling ramp is called down-chirp) is generated by the timer which then modulates the YIG oscillator. As the ramp value rises, so does the frequency. This generates an LFM signal. This signal is then amplified and transmitted. The power level shown in this schematic is 3 W average. The system employs two antennas, one for transmission and one for reception. As the signal is transmitted, a portion of the power is coupled to the receiver mixer input as the local oscillator (LO) reference signal. The RF return from the target is routed via the receiver antenna and a low noise amplifier (LNA) to the receiver mixer as RF input. This mixer is what is called a "stretch" processor. This will be discussed below. Essentially, it correlates the target return with the LO reference, and obtains the difference frequency, called the beat frequency, proportional to target range. This is then amplified and given to an FFT, called the range FFT, which identifies the beat frequency with the corresponding range. The FFT output is sent via a threshold detector (not shown) to a plan position indicator (PPI) display. This display is basically a cathode ray tube (CRT) that rotates syn-phase with the antenna assembly. This is accomplished through an azimuth scanner. When the target return crosses the threshold it is painted on the display at the corresponding bearing, which is the direction at which the antenna is pointing *at that moment*. The timing circuit blanks the IF amplifier at the ends of the sweep. This controls nonlinearities, which are at the maximum at the ends of the sweep, that is, at the start and end of a sweep. The waveform should essentially be a pure-rising ramp signal, that is, a rising frequency for an up-chirp case.

In practice, it becomes extremely difficult to ensure this, especially for large sweep bandwidth signals. This impure ramp gives rise to spurious frequencies or nonlinearities, which affect the quality of the target return by creating large side lobes. These side lobes are called time side lobes and can be seen in Figure 2–6. They are not desirable and should be extremely low, or else they will occlude nearby target returns and introduce clutter into the main lobe via the side lobes. Blanking pulses control the nonlinearity at the ends of the sweep, but to control the nonlinearity during the sweep is extremely difficult and requires excellent oscillators with very low-frequency drift. The timing circuit starts the FFT to process the target returns at the end of the blanking pulse. It also controls the automatic gain control (AGC), to operate at the end of the blanking pulse. The AGC circuit is not shown in the schematic. It is intended to control the return signal strength so that it does not saturate the IF amplifier. The frequency agility input is basically to control the sweep frequencies used by the radar from a set menu of frequencies. This is to ensure that the radar is not jammed. There are also other signal sources, other than the YIG oscillator. A popular one these days is a digital source or direct digital synthesizer (DDS). This basically mathematically generates an LFM signal. The advantage with DDS is that it generates LFM signals with low nonlinearities and is also very frequency agile for resistance against jamming. This will be examined in detail in Part III of this book.

2.3.2.1.5 Problems with FMCW   The primary problems with FMCW all relate to transmitting and receiving simultaneously as the transmitted power can be more than 100 dB higher than the received echo, so if even a small fraction of the transmitted power leaks into the receiver it can saturate or even damage the sensitive circuitry.

The performance of even well-designed systems used to be degraded by 10–20 dB compared to that which is achievable with pulsed systems. This limitation can be minimized by ensuring that there is good isolation between the receive and transmit antennas by separating them and by using low antenna side lobe levels and also using short-range clutter cancellation techniques. Modern signal processing techniques and hardware can also be used to cancel the leakage power in real time, and good performance can be obtained.

A lingering problem with FMCW signals is that the FMCW signal suffers from what is called range–Doppler coupling. This means that a target having velocity (a moving target) will have a frequency shift due to target motion. This is called Doppler shift. This causes an error in the measurement of its range.

2.3.3 Stepped Frequency Waveform

This is a popular waveform with both pulsed and CW radars. When using a real or compressed waveform to achieve high-range resolution, there is a challenge to receive the acquired data at the required rate (or bandwidth). A technique that avoids the data acquisition problems associated with wide bandwidth signals is the stepped frequency mode that shifts the transmitted frequency from step-to-step.

A frequency step of τ is selected to span the range of interest, for example, a 100 ns width will span a range of 15 m. The frequency of each step is shifted by a small amount Δf from that of the previous pulse, where Δf is selected to be about $1/2\tau = 5$ MHz in the example to ensure that there is no phase ambiguity in the returned signals.

After each step is transmitted, the received echo at a particular range is coherently detected (to maintain the phase information) and the amplitude and phase information stored. For transmit frequency F_1 the phase of the received echo will be

$$\phi_1 = \frac{4\pi F_1 R}{c} \tag{2.13}$$

For a static target, the phase of the next step echo transmitted at a frequency of F_2 will be

$$\phi_2 = \frac{4\pi F_2 R}{c} \tag{2.14}$$

For a sequence of steps equally spaced in frequency, there is a predictable step-to-step phase shift of $\Delta\phi$ that is a function of the frequency difference $\Delta F = F_2 - F_1$

$$\Delta\phi = \frac{4\pi R\Delta F}{c} \tag{2.15}$$

This step-to-step shift appears as an apparent Doppler frequency which is a function of the range to the target. If multiple targets appear in the same range bin, then each will produce a unique frequency that can be extracted from the time domain signal using the FFT process.

The total unambiguous range after processing is $c/2\Delta F$ and the range resolution is $c/2F_{tot}$, where F_{tot} is the total frequency excursion of the transmitted signal. For a sequence of N pulses $F_{tot} = N\Delta F$.

The primary difficulty with using stepped frequency is to maintain the stability of the transmitter and local oscillators for the whole period that a measurement is being made.

If all the targets spanning the full unambiguous range R_{max} must be sampled into bins 15 m wide, a total of $R_{max}/15$ gates will be required, each of them will have to be processed by the FFT to produce range bins.

2.3.3.1 SFW Generation and Reception   Stepped frequency waveform (SFW) generation and reception is the same as for FMCW and is as shown in Figure 2–8. It is normal to provide a switchable option between FMCW and SFW signals.

2.3.3.2 Problems With SFW Signals   A further difficulty that must be considered is that if the target is a moving vehicle, then Doppler effects will shift the apparent range of the targets as with the FMCW techniques due to range–Doppler coupling. This means that a target having a velocity (Doppler) will cause an error in measured range. Furthermore, for large bandwidth signals, SFW exhibits grating lobes in addition to the main lobe. This occurs only if we use unmodulated pulses. This will be examined in greater detail in Chapter 6.

SFW signal processing is necessarily slow. This causes *Doppler smearing* [4] wherein the target changes range bin by the time the SFW transmitter has finished transmitting. To avoid Doppler smearing, it would be advisable to devise a method to quickly generate the waveform. Just such a method viz. the Pandora signal processor is discussed in Part III of this book.

2.3.4 Phase-Coded Signal Compression

2.3.4.1 Phase-Coded Signal Generation and Reception   The Binary Phase Shift Keying (BPSK) technique discussed here is one of a number of phase-coded waveforms that can be generated. It is the most widely used for radar work because it is relatively easy to biphase modulate the carrier.

The carrier is switched between ±180° according to a stored digital code (see Figure 2–9). It can be implemented quite easily using a balanced mixer, or with a dedicated BPSK modulator.

Demodulation is achieved by multiplying the incoming RF signal by a coherent carrier (a carrier that is identical in frequency and phase to the carrier that originally modulated the BPSK signal). This produces the original BPSK signal plus a signal at twice the carrier which can be filtered out. However, a more common technique that is used widely by radar designers is shown in Figure 2–10.

The received signals are bandpassed by a filter matched to the chip rate, the outputs are then demodulated by I and Q detectors. These detectors compare the phase of the received signal to the phase of the LO which is also used in the RF modulator.

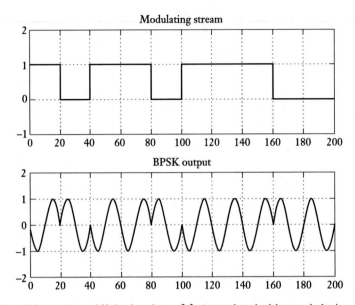

Figure 2–9 Example of binary phase shift keying. (From [3], © Reprinted with permission)

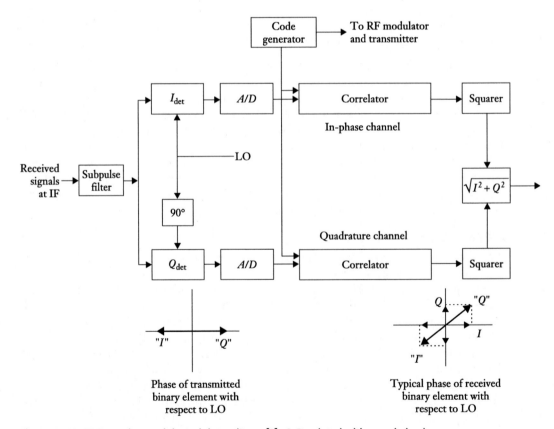

Figure 2–10 BPSK receiver and demodulator. (From [3], © Reprinted with permission.)

Though the phase of each of the transmitted signals is 0° or 180° with respect to the LO, on receive the phase will be shifted by an amount dependent on the round trip time and the Doppler velocity. For this reason, two processing channels are generally used, one which recovers the in-phase signal and one which recovers the quadrature signal. These signals are converted to digital by the analog to digital (A/D) converters, correlated with the stored binary sequence and then combined.

The primary advantage of this configuration is that it utilizes the coherence of the system to produce two quadrature receive channels. If only one channel is implemented, then there is a loss in the effective signal to noise ratio (SNR) of, on average, 3 dB.

The echo is compressed by correlation with a stored reference. This is the discrete equivalent of the convolution process.

Special cases of these binary codes are the Barker codes, where the peak of the autocorrelation function is N (for a code of length N) and the magnitude of the maximum peak side lobe is 1. The problem with the Barker codes is that none with lengths greater than 13 have been found (see Table 2–1).

Barker code sequences are called optimum, because, for zero Doppler shift, the peak to side lobe ratio is $\pm n$ after matched filtering (where n is the number of bits).

Table 2–1 Barker Codes

Code Length	Code Element	Side Lobe Level (dB)
2	$+-$ or $++$	-6
3	$++-$	-9.5
4	$++-+$ or $+++-$	-12
5	$+++-+$	-14
7	$+++--+-$	-16.9
11	$+++---+--+-$	-20.8
13	$+++++--++-+-+$	-22.3

In Figure 2–11, we illustrate the concept of phase-coded pulse compression for a 5-bit Barker code. We generate a 5-bit Barker code as a sequence $+++-+$ and use a filter matched to the chip length τ_0 with a bandwidth $B = 1/\tau_0$ having a *sinc* transfer function. This is followed by a tapped delay line having four delays τ_0, the outputs of which are weighted by the time reversed code $+-+++$ and summed prior to envelope detection. The output consists of $m-1$ time side lobes of unit amplitude V and a main lobe with amplitude mV each of width τ_0. The ratio of the transmitted pulse width to the output pulse width is $\tau/\tau_0 = B\tau$ which is called the pulse compression ratio. The relative side lobe power level is $1/m^2 = -13.9 \approx 14$ dB.

2.3.4.2 Correlation Detection of Phase-Coded Signals For binary sequences where the values are restricted to ± 1 the following approach often is taken (see Figure 2–12).

The transmitted sequence is loaded into the reference register, and the input sequence is continuously clocked through the signal shift register. A comparison counter forms a sum of the matches and subtracts the mismatches between corresponding stages of the shift registers on every clock cycle to produce the correlation function. This method is also called fast convolution processing (FCP).

Circular Correlation: For correlation of two long sequences, their Fourier transforms may be taken, followed by the product of the one series with the complex conjugate of the other, and finally, the inverse Fourier transform completes the procedure as shown in Figure 2–13.

Phase-coded signals will be examined in detail in Chapter 5.

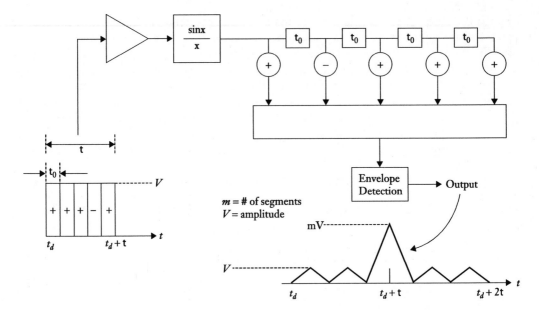

Figure 2–11 Generating a 5-bit Barker code sequence. (From [3], © Reprinted with permission.)

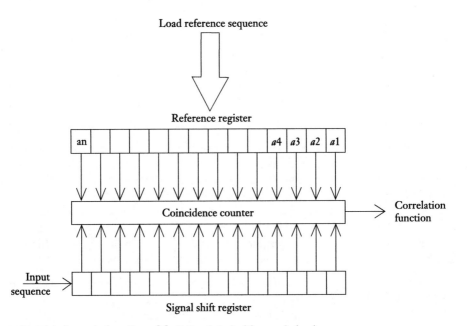

Figure 2–12 Digital correlation. (From [3], © Reprinted with permission.)

2.3.5 Stretch

In Stretch [1, 4, 5], a linear FM signal is transmitted and then the return echo is demodulated by down-converting using a frequency modulated LO signal of identical or slightly different FM slope. If the identical slope is used then the echo spectrum corresponds to the range profile. This is a form of a signal compression intermediate between standard signal compression and FMICW.

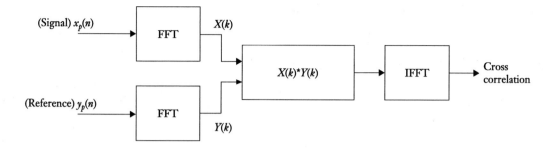

Figure 2–13 Cross correlation using the Fourier transform method. (From [3], © Reprinted with permission.)

If the slope of the LO is different to that of the transmitted chirp, then the output of the stretch processor comprises signals with a reduced chirp. These can then be processed using a standard SAW signal compression system to produce target echoes as described in the previous section (see Figure 2–14).

Stretch processing will be discussed in detail in Chapter 4.

2.3.6 Interrupted FMCW

Known as IFMCW or FMICW, this involves interrupting the FMCW signal to eliminate the requirement for good isolation between the transmitter and the receiver. It is generally implemented with a transmission time matched to the round trip propagation time. This is followed by a quiet reception time equal to the transmission time.

A duty factor of 0.5 reduces the average transmitted power by 3 dB but the improved performance due to reduced system noise improves the SNR by more than the 3 dB lost (see Figure 2–15).

2.3.6.1 Disadvantages   The major problems are the limited minimum range due to the finite switching time of the transmitter modulator and the need to know the target range to optimize the transmit time.

For imaging applications where a whole range of frequencies are received, maintaining a fixed 50% duty cycle is suboptimum due to eclipsing except at one range. The optimum duty ratio is approximately 1/3.

FFT processing of the interrupted signal results in large numbers of spurious components that can interfere with the identification of the target return as shown in Figure 2–16.

To overcome this problem, we need to so design the radar that the ICW rate must be maintained greater than the largest expected beat frequency.

2.3.6.2 Optimizing for a Long-Range Imaging Application   The Tx time is optimized for the longest range of interest (where the SNR will be lowest) (see Figure 2–17).

The shorter ranges will suffer from the following problems:

- Reduced illumination time → lower SNR
- Reduced chirp bandwidth → poorer range resolution
- Suboptimal windowing → higher range side lobes

The degradation in range resolution at short range is compensated for by the improved cross-range resolution (constant beamwidth) so the actual resolution (pixel area) remains constant (see Figure 2–18).

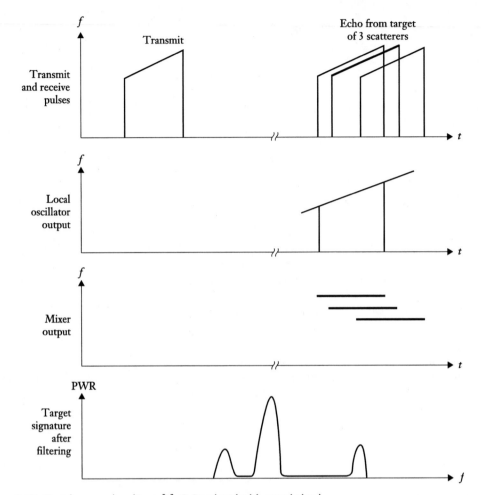

Figure 2–14 Stretch processing. (From [3], © Reprinted with permission.)

2.3.7 Side Lobes and Weighting for Linear FM Systems

The spectrum of a truncated sine wave output by an FMCW radar for a single target has the characteristic $|\sin(x)/x|$ shape as predicted by Fourier theory.

The range side lobes in this case are only 13.2 dB lower than the main lobe which is not satisfactory as it can result in the occlusion of small nearby targets as well as introducing clutter from the adjacent lobes into the main lobe. To counter this unacceptable characteristic of the matched filter, the time domain signal is mismatched on purpose. This mismatch generally takes the form of amplitude weighting of the received signal.

One method to do this is to increase the FM slope of the chirp signal near the ends of the transmitted signal to weight the energy spectrum which will result in the desired low side lobe levels after the application of the matched filter. A more conventional method that is often used in digital systems is to apply the function to the signal amplitude prior to processing to achieve the same ends as shown in Figure 2–19.

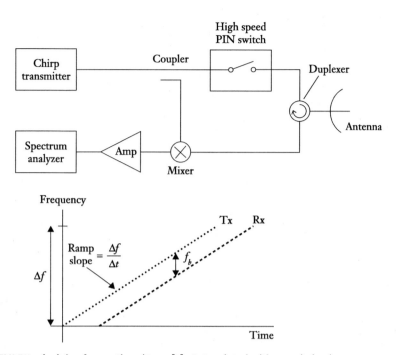

Figure 2–15 FMICW principle of operation. (From [3], © Reprinted with permission.)

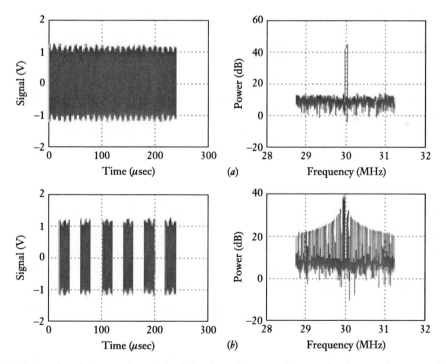

Figure 2–16 Comparison between the received signals and spectra for two closely-spaced targets of different amplitudes for (a) an FMCW radar and (b) an FMICW radar with a deterministic interrupt sequence. (From [3], © Reprinted with permission.)

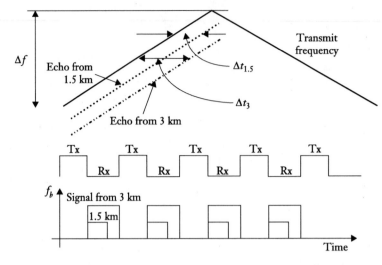

Figure 2–17 FMICW waveform optimized for 3 km. $\Delta t_{1.5}$ and Δt_3 are timing offsets in the echoes (From [3]: © Reprinted with permission).

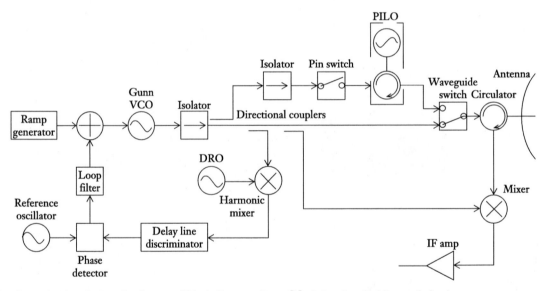

Figure 2–18 FMICW seeker front-end block diagram. (From [3], © Reprinted with permission.)

The reduction in side lobe levels does come at a price though: the main lobe amplitude is also marginally reduced in amplitude, and it is also widened quite substantially as summarized in Table 2–2.

The rectangular, or uniform, weighting function provides a matched filter operation with no loss in SNR, while the weighting in the other cases introduces a tailored mismatch in the receiver amplitude characteristics with an associated loss in SNR which can be quite substantial.

In addition to providing the best SNR, uniform weighting also provides the best range resolution (narrowest beamwidth), but this characteristic comes with unacceptably high side lobe

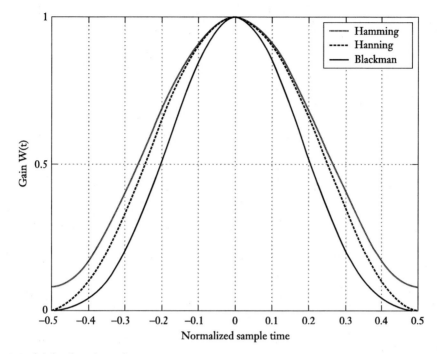

Figure 2–19 Weighting function gains.

Table 2–2 Properties of Some Weighting Functions

Window	Rectangle	Hamming	Hanning	Blackman
Worst side lobe (dB)	−13.3	−42.2	−31.5	−58
3 dB Bandwidth (bins)	0.88	1.32	1.48	1.68
Scalloping loss (dB)	3.92	1.78	1.36	1.1
SNR loss (dB)	0	1.34	1.76	2.37
Main lobe width (bins)	2	4	4	6
$a0$	1	0.54	0.50	0.42
$a1$		0.46	0.50	0.50
$a2$				0.08

Weighting function is given by:

$$W(n) = a0 - a1\cos[2\pi(n-1)/(N-1)] + a2\cos[4\pi(n-1)/N-1]$$

levels. The other weighting functions offer poorer resolution but improved side lobe levels with falloff characteristics that can accommodate almost any requirement as seen in Figure 2–20.

Of particular interest are the Hamming and Hanning weighting functions that offer similar loss in SNR and resolutions, but with completely different side lobe characteristics. As can be seen in Figure 2–20, the former has the form of a cosine-squared-plus-pedestal, while the latter is just a standard cosine-squared function. In the Hamming case, the close-in side lobe is suppressed to

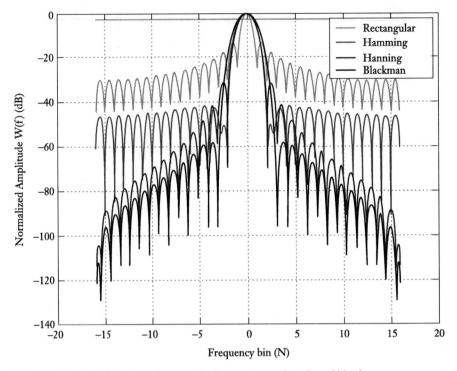

Figure 2–20 Normalized weighting function amplitude spectra as a function of bin size.

produce a maximum level of −42.2 dB, but that energy is spread into the remaining side lobes resulting in a falloff of only 6 dB/octave, while in the Hanning case the first side lobe is higher at −31.5 dB, but with a falloff of 18 dB/octave.

For most FMCW applications, the Hamming window is used as it provides a good balance between side lobe levels (−42.2 dB), beamwidth (1.32 bins), and loss in SNR compared to a matched filter (1.34 dB). For imaging applications where a large dynamic range of target reflectivities is expected, then the Hanning window with its superior far-out side lobe performance is the function of choice.

2.3.8 Linear Frequency Modulation Waveforms

This section is based on the work done by Mahafza (From [2] © 2004. Reproduced by permission of Taylor & Francis, a division of Informa plc.). In our quest for high resolution we are driven toward higher bandwidth signals. Linear frequency modulation (LFM) is one such signal [1, 2, 4]. We had briefly examined it in the preceding sections. It is extremely popular and for a good reason as we shall see. We have basically two broad classes of LFM signals defined by their linear sweep characteristic, up-chirp or down-chirp. The matched filter bandwidth is proportional to the sweep bandwidth and is independent of the pulse width. Figure 2–21 shows the two types of LFM signals.

The LFM up-chirp instantaneous phase is expressed by [2]

$$\psi(t) = 2\pi\left(f_0 t + \frac{\mu}{2}t^2\right) \quad -\frac{\tau}{2} \le t \le \frac{\tau}{2} \tag{2.16}$$

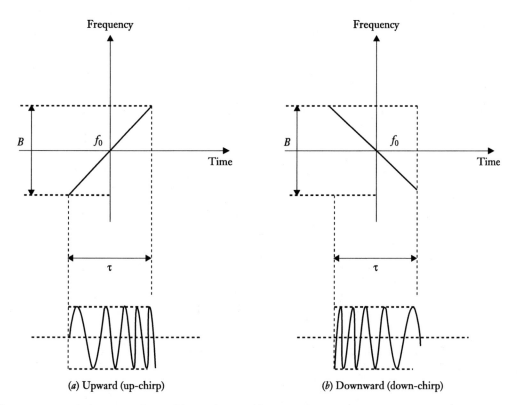

(a) Upward (up-chirp)   (b) Downward (down-chirp)

Figure 2–21 Typical LFM waveforms: (a) up-chirp and (b) down-chirp.

where f_0 is the radar center frequency and $\mu = (2\pi B)/\tau$ is the LFM coefficient. Thus the instantaneous frequency is

$$f(t) = \frac{1}{2\pi}\frac{d}{dt}\psi(t) = f_0 + \mu t \qquad -\frac{\tau}{2} \le t \le \frac{\tau}{2} \tag{2.17}$$

Similarly, for down-chirp,

$$\psi(t) = 2\pi\left(f_0 t - \frac{\mu}{2}t^2\right) \qquad -\frac{\tau}{2} \le t \le \frac{\tau}{2} \tag{2.18}$$

$$f(t) = \frac{1}{2\pi}\frac{d}{dt}\psi(t) = f_0 - \mu t \qquad -\frac{\tau}{2} \le t \le \frac{\tau}{2} \tag{2.19}$$

A typical LFM waveform has the following expressions in time/frequency domains [2]:

Time Domain Expression

$$s_1(t) = \text{rect}\left(\frac{t}{\tau}\right)e^{j2\pi\left(f_0 t + \frac{\mu}{2}t^2\right)} \tag{2.20}$$

where $\text{rect}(t/\tau)$ denotes a rectangular pulse of width τ. We can rewrite equation (2.20) as

$$s_1(t) = e^{j2\pi f_0 t}s(t) \tag{2.21}$$

where

$$s(t) = \text{rect}\left(\frac{t}{\tau}\right) e^{j\pi\mu t^2} \tag{2.22}$$

is the complex envelope of $s_1(t)$.

The Fourier integral of a rectangular chirp pulse is

$$S(\omega) = \int_{-\tau/2}^{\tau/2} \exp(j(\mu. t^2/2 - \omega t)) \, dt \tag{2.22a}$$

The exponent term factored out of the integral is

$$S(\omega) = \sqrt{\frac{\pi}{\mu}} . \exp\left(\frac{-j.\omega^2}{2.\mu}\right) \int_{-x_1}^{x_2} \exp(j.\pi x^2/2) \, dx \tag{2.22b}$$

where

$$x = \sqrt{\frac{\mu}{\pi}}\left(t - \frac{\omega}{\mu}\right), \, dx = \sqrt{\frac{\mu}{\pi}} \, dt$$

and μ is the chirp slope in radians per second and ω is the radian frequency, that is, $2.\pi.f$. This integral is recognized as

$$\left[\left(C(x_2) + C(x_1)\right) + j\left(S(x_2) + S(x_1)\right)\right]$$

where $C(x)$ and $S(x)$ are Fresnel integrals. We substitute

$$\mu = \frac{2.\pi.B}{\tau}$$

where B is the total chirp bandwidth and τ is the chirp pulse length.

Substitution yields:

$$\sqrt{\frac{\pi.\tau}{2.\pi.B}} . \exp\left(\frac{-j.\omega^2.\tau}{2.2.\pi.B}\right) = \tau.\sqrt{\frac{1}{\tau.B}} . \exp\left(\frac{-j.\omega^2.\tau}{4.\pi.B}\right) . \frac{1}{\sqrt{2}} . \left[\left(C(x_2) + C(x_1)\right) + j\left(S(x_2) + S(x_1)\right)\right] \tag{2.22c}$$

I am indebted to David Lynch[1] for this part of the derivation.

Frequency Domain Expression
From (2.22c) we obtain,

$$S(\omega) = \tau\sqrt{\frac{1}{B\tau}} e^{-j\omega^2/(4\pi B)} \left\{\frac{\left[C(x_2) + C(x_1)\right] + j\left[S(x_2) + S(x_1)\right]}{\sqrt{2}}\right\} \tag{2.23}$$

where

$$x_1 = \sqrt{\frac{B\tau}{2}}\left(1 + \frac{f}{B/2}\right) \tag{2.24}$$

---

1 Personal correspondence with the author.

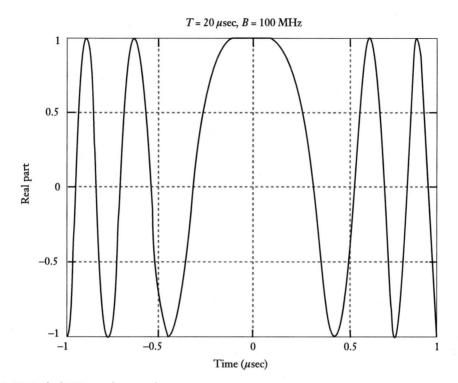

Figure 2–22 Typical LFM waveform, real part.

$$x_2 = \sqrt{\frac{B\tau}{2}}\left(1 - \frac{f}{B/2}\right) \qquad (2.25)$$

The Fresnel integrals, denoted by $C(x)$ and $S(x)$, are defined by

$$C(x) \approx \frac{1}{2} + \frac{1}{\pi x}\sin\left(\frac{\pi}{2}x^2\right); \quad x \gg 1 \qquad (2.26)$$

$$S(x) \approx \frac{1}{2} - \frac{1}{\pi x}\cos\left(\frac{\pi}{2}x^2\right); \quad x \gg 1 \qquad (2.27)$$

Figures 2–22 through 2–24 display the real part, imaginary part, and the spectrum of the LFM signal. These curves can be retrieved by using the program *LFM.m* supplied with this book. These curves are for an LFM bandwidth of 100 MHz and an uncompressed pulse width of 20 μsec.

2.4 MATCHED FILTER

The matched filter [1, 2, 6] is a filter which, for a specified signal waveform, will result in the maximum attainable signal-to-noise ratio at the filter output when both the signal and white noise are passed through it. Such filters are widely used in radars. This section is from [6] and reprinted with permission of John Wiley & Sons, Inc.

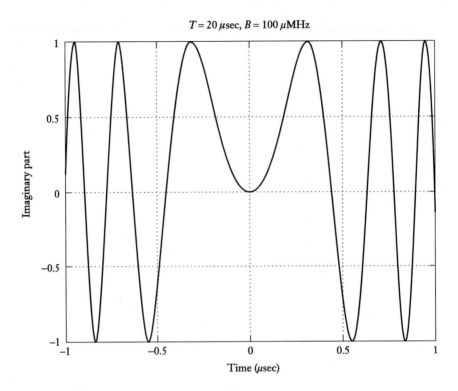

Figure 2–23 Typical LFM waveform, imaginary part.

Consider a signal $s(t)$ with additive white Gaussian noise which is two-sided with a spectral density $N_0/2$ which is passed through a linear filter with a frequency transfer function $H(\omega)$. The problem before us is as follows:

What is the filter response that will yield the highest SNR at the output, at a given observation time t_M?

We, therefore, need to search for such a transfer function $H(\omega)$ that will do this job for us and maximize the SNR given by

$$SNR = \frac{\left| s_0\left(t_M\right)\right|^2}{n_0^2\left(t\right)} \qquad (2.28)$$

Levanon [6] has proved that this maximum SNR holds when

$$H\left(\omega\right) = KS^*\left(\omega\right)e^{\left(-j\omega t_M\right)} \qquad (2.29)$$

where $S^*(\omega)$ is conjugate of the Fourier transform of $s(t)$.

Equation (2.29) when applied gives $2E/N_0$ as the highest attainable peak signal-to-noise ratio. The inverse Fourier transform of $H(\omega)$ will yield the impulse response of the desired filter

$$h\left(t\right) = Ks^*\left(t_M - 1\right) \qquad (2.30)$$

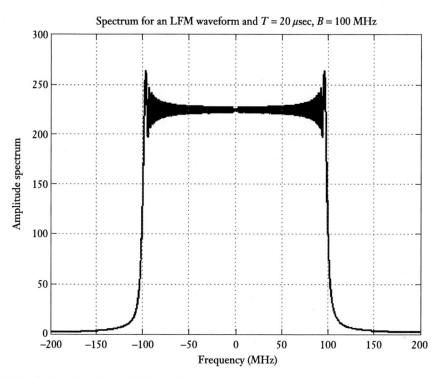

Figure 2–24 Typical spectrum for an LFM waveform.

Inspection of equation (2.29) implies that $H(\omega) = K|S(\omega)|$, which means that the filter frequency response depends upon the spectrum of the signal. The impulse response indicates that it is a delayed mirror image of the conjugate of the signal. For a *causal* filter $h(t)$ must be zero for $t < 0$. This can happen only if t_M is equal to or larger than the duration of the signal $s(t)$.

If we convolve the input signal with the impulse response of the filter matched to it, we obtain

$$s_0(t) = \int_{-\infty}^{\infty} s(\tau) h(t-\tau) d\tau = K \int_{-\infty}^{\infty} s(\tau) s^* \left[\tau - (t - t_M)\right] d\tau \qquad (2.31)$$

If we now put $t = t_M$, we obtain

$$s_0(t_M) = K \int_{-\infty}^{\infty} |s(\tau)|^2 d\tau = KE \qquad (2.32)$$

which says that at t_M the output signal is proportional to the *energy* of the input signal. This applies to all signals passing through their matched filters. The interested reader is referred to [6] for details on the proof.

Hence, to summarize [6], in the presence of white noise, the output SNR from a matched filter is the highest attainable one, viz. $2E/N_0$. This output SNR is a function of the signal energy E, but not of the signal form. The signal form will matter when the noise is nonwhite or when other considerations such as resolution, accuracy, and so forth, as well as detection are important [6].

We illustrate through an example. This example is attributed to Mahafza (From [2] © 2004. Reproduced by permission of Taylor & Francis, a division of Informa plc.).

Example

What is the maximum instantaneous SNR at the output of a linear filter whose impulse response is matched to the signal $s(t) = \exp(-t^2/2T)$?

Solution

The signal energy is

$$E = \int_{-\infty}^{\infty} |s(t)|^2 \, dt = \int_{-\infty}^{\infty} e^{(-t^2)/T} \, dt = \sqrt{\pi T} \ \ \text{J}$$

Hence, maximum instantaneous SNR is

$$SNR = \frac{\sqrt{\pi T}}{N_0/2}$$

where $N_0/2$ is the input noise power spectrum density.

2.4.1 Storing a Replica

Matched filter output can be computed from the cross-correlation between the radar received signal and a delayed replica of the transmitted waveform. Mathematically and structurally this is the same as is defined by the expression given in equation (2.29). If the input signal is the same as the transmitted signal, the output of the matched filter would be the autocorrelation function of the received (or transmitted) signal. This is a very popular method of implementing such filters and in practice, replicas of the transmitted waveforms are normally computed and stored in memory for use by the radar signal processor when needed.

References

1. Nathanson, F. E., *Radar Design Principles*, 2nd ed., McGraw-Hill, New York, 1991.
2. Mahafza, B. R. and Elsherbeni, A. Z., *MATLAB Simulations for Radar Systems Design*, Chapman and Hall/ CRC Press, Boca Raton, FL, 2004.
3. http://www.acfr.usyd.edu.au/teaching/4th-year/mech4721-Signals/material/lecture notes/HighRangeResolutionTechniques.pdf.
4. Skolnik, M. I., *Introduction to Radar Systems*, 3rd ed., McGraw-Hill, Boston, MA, 2001.
5. Caputi, W. J., Jr., "Stretch: A time-transformation technique," *IEEE Transactions on AES*-7, No. 2, March 1971.
6. Levanon, N., *Radar Principles*, John Wiley & Sons, New York, 1988.

3

The Radar Ambiguity Function

3.1 INTRODUCTION

The radar ambiguity function is defined as the absolute value of the envelope of the output of a matched filter when the input to the filter is a Doppler-shifted version of the original signal, to which the filter was matched [1]. If $s(t)$ is the complex envelope of the signal, then the ambiguity function is given by ([1], p. 120)

$$\left|\chi(\tau,\nu)\right|^2 = \left|\int_{-\infty}^{\infty} s(t) s^*(t-\tau) \exp\left(j2\pi\nu t\right) dt\right|^2 \tag{3.1}$$

The filter was originally matched to the signal at a nominal center frequency and a nominal delay. Thus, $|X(0,0)|$ is the output when the input signal is returned from a point target at the nominal delay and Doppler shift for which the filter was matched. The two parameters of the ambiguity function are an additional delay τ and an additional frequency shift ν. Therefore, any value of τ and/or ν other than zero, indicate a return from a target at some other range and/or velocity. The ambiguity function peaks at $\tau = 0$, $\nu = 0$ and is zero everywhere else. This will correspond to an ideal resolution between neighboring targets. However, we will see that such a shape of the ambiguity function is impossible to attain. Furthermore, even if we could, such a narrow function would not permit a radar to find a previously undetected target, because the probability of that target lying within the response region would be near zero. One requirement on a radar waveform is that it must be possible to search a large area of possible target locations (in both range and Doppler) with minimum losses, and a conflicting requirement is that it must be possible to resolve closely spaced targets and measure their positions with specified accuracy. Hence, there is no single "ideal" ambiguity function. Therefore, in the absence of an ideal ambiguity function that fits *all* requirements, we need to use waveforms which have ambiguity functions well suited to the task the radar is required to do. This is called waveform design and this is what motivates us to study ambiguity functions of various types of signals. The ideal ambiguity function is, however, not realizable. This is because the ambiguity function must have a finite peak value equal to (E) and a finite volume also equal to (E). Clearly, the ideal ambiguity function cannot meet these requirements.

Ambiguity functions are usually analyzed on a single pulse basis. Hence, in a work of this nature on continuous wave (CW) radars, there is no error as the results apply equally well to CW radar waveforms. However, in CW radars there is in advantage not shared by modulated pulsed radars, viz. the concept of periodic ambiguity function (PAF). The concept of PAF introduces the fact that in certain class of phase-coded signals employed in CW radars, one can obtain an autocorrelation function devoid of side lobes on the delay axis. This aspect will be examined in detail in Chapter 5.

The following list includes the properties of the radar ambiguity function:

1. The maximum value for the ambiguity function occurs at $(\tau,\nu)=(0,0)$ and is equal to E,

$$\max\left\{\left|\chi(\tau,\nu)\right|\right\}=\left|\chi(0,0)\right|=E \tag{3.2}$$

$$\left|\chi(\tau,\nu)\right|\le\left|\chi(0,0)\right|=E \tag{3.3}$$

2. The ambiguity function is symmetric

$$\left|\chi(\tau,\nu)\right|=\left|\chi(-\tau,-\nu)\right| \tag{3.4}$$

3. The total volume under the ambiguity function is a constant

$$\int_{-\infty}^{\infty}\int_{-\infty}^{\infty}\left|\chi(\tau,\nu)\right|^2 d\tau d\nu=E \tag{3.5}$$

4. If $s(t)\leftrightarrow\left|\chi(\tau,\nu)\right|$
   then

$$s(t)\exp\left(j\pi kt^2\right)\leftrightarrow\left|\chi(\tau,\nu+k\tau)\right| \tag{3.6}$$

Rule 1 says that for normalized signals ($E = 1$), the maximum value of the ambiguity function is one and it is achieved at the origin. Rule 2 says that the volume underneath the ambiguity function squared is a constant equal to one. The implication here is that if we squeeze the ambiguity function to a narrow peak near the origin, then that peak cannot exceed the value of one and the *volume squeezed out of that peak must reappear somewhere else*. This means that for an LFM (linear frequency modulation) pulse, if we try for a very narrow peak (for better range resolution), the side lobes increase and vice versa. Hence, when we try to weight the compressed LFM pulse, the side lobes decrease to the level desired, but the pulse widens. Rule 3 indicates that the ambiguity function is symmetrical with respect to the origin. Rule 4 says that multiplying the envelope of any signal by a quadratic phase (linear frequency) will shear the shape of the ambiguity function. We will apply this rule to LFM pulses further down in this chapter. Proofs for these rules are given in [1].

3.2 EXAMPLES OF AMBIGUITY FUNCTIONS

Ambiguity functions are usually discussed on a single pulse basis. Hence, in the following discussion, we shall retain the term "pulse," though this book is on CW radars. We shall now investigate the following basic types of radar signals:

- Single-frequency pulse
- LFM pulse
- Stepped frequency pulse train

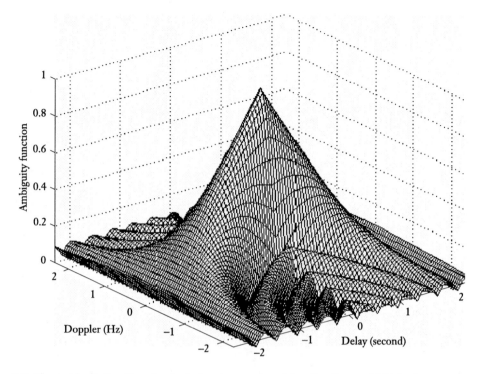

Figure 3–1 The ambiguity function of a single-frequency pulse of duration 2 second (3D view).

We will analyze the physical meaning of the various shapes of the functions and their importance in radar applications. The idea will be to convey more a physical idea rather than tedious mathematics.

We shall study the ambiguity functions of phase-coded signals in Chapter 5.

3.2.1 Single-Frequency Pulse

The single-frequency pulse is defined as

$$s(t) = \frac{1}{\sqrt{T}} \, rect \, \frac{t}{T} \tag{3.7}$$

The ambiguity function for this waveform has been derived by Levanon [1] and is given by

$$|\chi(\tau, \nu)| = \left| \left(1 - \frac{|\tau|}{T} \right) \frac{\sin \left[\pi T (1 - |\tau|/T) \nu \right]}{\pi T (1 - |\tau|/T) \nu} \right|, \quad |\tau \le T| \tag{3.8}$$

The program "*singlepulse.m*" in the accompanying software plots the ambiguity function and its contour plot, for a pulse duration of 2 second (see Figures 3–1 and 3–2).

The cut along the delay axis is obtained by setting $\nu = 0$. We obtain

$$|\chi(\tau, 0)| = 1 - \frac{|\tau|}{T} \qquad |\tau| \le T \tag{3.9}$$

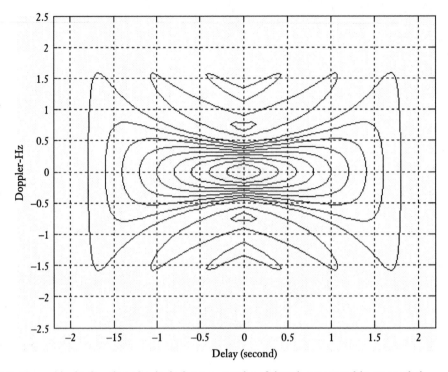

Figure 3–2 The ambiguity function of a single-frequency pulse of duration 2 second (Contour plot).

This shape is shown in Figure 3–3. Since the zero Doppler cut along the time delay axis extends between $-T$ and T, then close targets would be discriminated if they were to be at least T seconds apart.

The cut along the Doppler axis is obtained by setting $\tau = 0$. This yields

$$\left| \chi \left(0, \nu \right) \right| = \left| \frac{\sin \pi T \nu}{\pi T \nu} \right| \tag{3.10}$$

The shape is shown in Figure 3–4. It should be noted that the cut along the delay axis extends from $-T$ to T, while the cut along the Doppler axis extends from $-\infty$ to ∞. This is valid for any cuts along these two axes.

In Figure 3–4, we note that the first null occurs at $\pm 1/T$. In out case, it is at 0.5 Hz, since $T = 2$ second. This is a standard sinc function. Hence, the first side lobe level is at -13.3 dB, which indicates that the first null can be considered the practical end of the ambiguity function along the Doppler axis.

3.2.2 Linear FM Pulse

Consider the LFM complex envelope defined by [1]

$$s(t) = \frac{1}{T} \, rect \left(\frac{t}{T} \right) \exp \left(j \pi \mu t^2 \right) \tag{3.11}$$

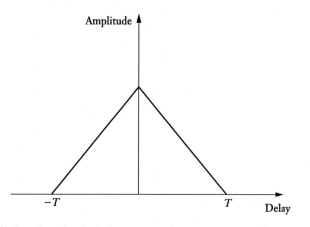

Figure 3–3 The ambiguity function of a single-frequency pulse cut at zero Doppler.

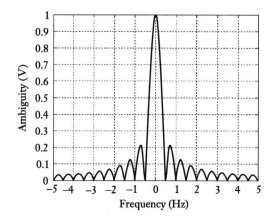

Figure 3–4 The ambiguity function of a single-frequency pulse cut at zero delay.

We differentiate the argument of the exponential and divide by 2π to obtain the instantaneous frequency $f(t)$ of $s(t)$. Hence

$$f(t) = \frac{1}{2\pi} \frac{d(\pi\mu t^2)}{dt} = \mu t \tag{3.12}$$

Equation (3.12) is a linear function. To obtain the ambiguity function of a signal with a complex envelope as given in equation (3.11), we apply Rule 4 defined in equation (3.6). Thus, we obtain the ambiguity function of equation (3.11) by replacing ν with $\nu + \mu\tau$ in equation (3.8) yielding

$$|\chi(\tau,\nu)| = \left| \left(1 - \frac{|\tau|}{T}\right) \frac{\sin\left[\pi T(1-|\tau|/T)(\nu+\mu\tau)\right]}{\pi T(1-|\tau|/T)(\nu+\mu\tau)} \right|, \quad |\tau \leq T| \tag{3.13}$$

Once again we look at the two cuts. The cut along the Doppler axis will not yield anything new, because we have added frequency modulation and not another amplitude modulation. The cut along the delay axis is obtained by setting $v = 0$, yielding

$$|\chi(\tau,0)| = \left\|\left(1 - \frac{|\tau|}{T}\right)\frac{\sin\left[\pi\mu\tau T(1-|\tau|/T)\right]}{\pi\mu\tau T(1-|\tau|/T)}\right\|, \quad |\tau| \le T \tag{3.14}$$

The plot of this curve is shown in Figure 3–7. The pulse width is 1 second and the bandwidth is 20 Hz. We find that it is radically different from that of the single pulse in Figure 3–3. The triangle is further multiplied by a sine function. To locate the first null, the argument of the sine should be equal to π. This occurs when [1]

$$\tau_{null1} = \frac{T}{2} - \left(\frac{T^2}{4} - \frac{1}{\Delta f}\right)^{1/2} \tag{3.15}$$

where Δf is the sweep bandwidth.

If $\Delta f T^2 >> 4$, (3.15) reduces to

$$\tau_{null1} \approx \frac{1}{\Delta f T} \tag{3.16}$$

or

$$\tau_{null1} \approx \frac{1}{\Delta f} \tag{3.17}$$

since $T << \Delta f$.

The pulse is much narrower compared to the unmodulated pulse cut in Figure 3–3, which indicates that the effective pulse width (compressed pulse width) of the matched filter output is completely determined by the radar bandwidth. In fact, it is narrower than the unmodulated pulse by a factor

$$\varepsilon = \frac{T}{(1/\Delta f)} = \Delta f T \tag{3.18}$$

ε is called the compression ratio, or time–bandwidth product, or compression gain.

If the radar bandwidth is increased, the compression ratio also increases. But the limitations here are the nonlinearities in large LFM sweeps as discussed in Section 2.3. The volume underneath the ambiguity function, in accordance with Rule 2, is a constant. Therefore, it has to reappear elsewhere when the pulse is compressed along the delay axis. It does so by stretching in the Doppler as far as Δf, the sweep bandwidth, as a diagonal ridge as shown in Figure 3–5.

There is another very important fact to be learned from the LFM ambiguity function. Normally, when there is no target Doppler, the peak of the ambiguity function is located at the origin $|\chi(0,0)|$. However, in the presence of target Doppler, this peak is shifted. This causes an error in delay. This is obvious from the contour plot in Figure 3–6, which shows a slope between range (delay) and Doppler, that is, a coupling. Hence, Doppler shifts will be reflected as an error in range. This means that a change in Doppler will yield a change in range and vice versa. This is the *range–Doppler coupling* and has been examined in the last chapter. It is only along the cardinal axes that there will be no coupling, that is, a case when the target Doppler is zero or the range is

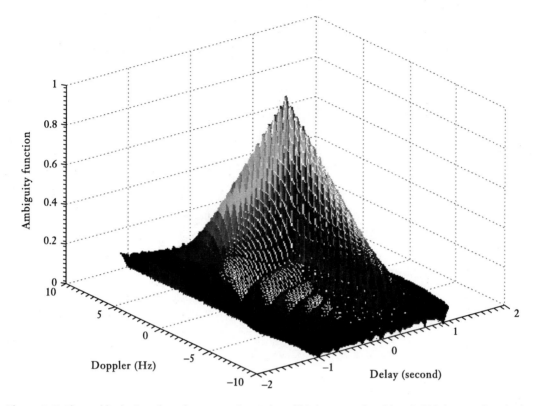

Figure 3–5 The ambiguity function of an LFM pulse: Pulse width is 1 second and bandwidth is 10 Hz (3D view).

zero. From (3.13), we note that if

$$\tau_{peak} = -\frac{\nu}{\Delta f} \tag{3.19}$$

the sinc function maximizes (its argument is zero). Hence, if the target has a radial velocity, its Doppler value will cause an error in the measurement of range, causing us difficulty in distinguishing between the two.

On the positive side, LFM exhibits a remarkable resilience to Doppler, unlike phase-coded signals. As a result the chirp pulse can tolerate considerable Doppler shift before it decorrelates. This aspect will be further examined in Part III of this book.

Figures 3–5 to 3–7 have been obtained using the program "*LFM.m*" in the accompanying software.

3.2.3 Stepped Frequency Waveform

The stepped frequency pulse train is shown in Figure 3–8.

The stepped frequency pulse train or stepped frequency waveform (SFW) is described in Figure 3–8. The burst consists of N pulses at frequencies of f_n and amplitudes of $A_n, n = 1, 2, ..., N$. The total coherence time is NT_R. T_R is the pulse repetition interval and T is the pulse width. In its basic linear form, the frequencies are equally spaced, that is, $f_n - f_{n-1} = \Delta f$ and the pulses are of a fixed amplitude, that is, $A_n = A$.

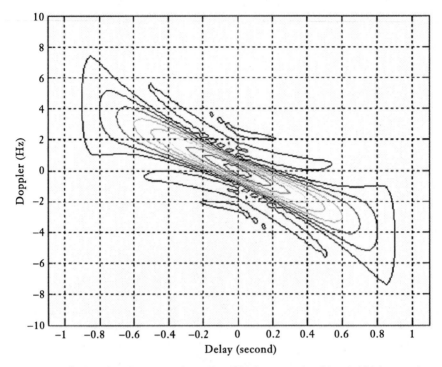

Figure 3–6 The ambiguity function of an LFM pulse: Pulse width is 1 second and bandwidth is 10 Hz (Contour plot).

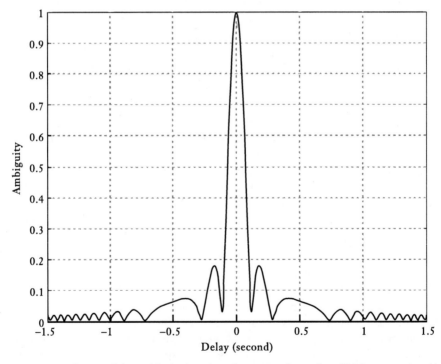

Figure 3–7 A zero-Doppler cut of the ambiguity function of an LFM pulse: Pulse width is 1 second and bandwidth is 20 Hz.

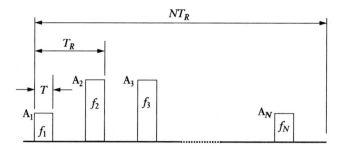

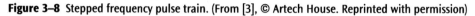

Figure 3–8 Stepped frequency pulse train. (From [3], © Artech House. Reprinted with permission)

Complex envelope can be stated as [2]:

$$g(t) = \frac{1}{\sqrt{N}} \sum_{n=0}^{N-1} g_n(t - nT_R) \exp(-j2\pi n\Delta ft) \qquad (3.20)$$

where

T_R is the pulse repetition period and

Δf is the frequency step size.

The division by $\sqrt{N}$ will maintain unit energy for the entire train.

Ambiguity function is defined as:

$$\chi(\tau, \omega_d) = \int_{-\infty}^{\infty} g(t) g^*(t - \tau) \exp(j2\pi\omega_d t) dt \qquad (3.21)$$

The reader should note that we have changed the variables for better readability, viz., $g(t)$ for $s(t)$ and ω_d for ν.

Substituting:

$$\chi(\tau, \omega_d) = \frac{1}{N} \sum_{n=0}^{N-1}\sum_{m=0}^{N-1} g_n(t - nT_R) g_m^*(t - mT_R - \tau) \exp(j2\pi\omega_d t) \exp(-j2\pi\Delta f(n - m - \tau)t)\, dt$$

$$= \frac{1}{N} \sum_{n=0}^{N-1}\sum_{m=0}^{N-1} g_n(t - nT_R) g_m^*(t - mT_R - \tau) \exp(j2\pi(\omega_d - (n - m - \tau)\Delta f)t)\, dt \qquad (3.22)$$

Substituting $t_1 = t - nT_R$

$$\chi(\tau, \omega_d) = \frac{1}{N} \sum_{n=0}^{N-1}\sum_{m=0}^{N-1} \int_{-\infty}^{\infty} g_n(t - nT_R) g_m^*(t - mT_R - \tau) \exp(j2\pi(\omega_d - (n - m - \tau)\Delta f)t)\, dt$$

$$= \frac{1}{N} \sum_{n=0}^{N-1}\sum_{m=0}^{N-1} \int_{-\infty}^{\infty} g_n(t_1 + nT_R - nT_R) g_m^*(t_1 + nT_R - mT_R - \tau) \exp(j2\pi(\omega_d - (n - m - \tau)\Delta f)\dots$$

$$\dots(t_1 + nT_R))dt_1$$

$$= \frac{1}{N} \sum_{n=0}^{N-1}\sum_{m=0}^{N-1} \int_{-\infty}^{\infty} g_n(t_1) \times g_m^*\{t_1 + nT_R - mT_R - \tau\} \exp(j2\pi(\omega_d - (n - m - \tau)\Delta f)(t_1 + nT_R))dt_1$$

$$= \frac{1}{N} \sum_{n=0}^{N-1}\sum_{m=0}^{N-1} \int_{-\infty}^{\infty} g_n(t_1) \times g_m^*\{t_1 - [\tau - (n - m)T_R]\} \exp(j2\pi(\omega_d - (n - m - \tau)\Delta f)(t_1 + nT_R))dt_1$$

Substituting $p = n - m$

$$= \frac{1}{N}\sum_{n=0}^{N-1}\sum_{m=0}^{N-1}\int_{-\infty}^{\infty} g_n(t_1) \times g_m^* \left\{ t_1 - \left[\tau - pT_R\right]\right\} \exp(j2\pi(\omega_d - p\Delta f - \tau\Delta f)(t_1 + nT_R))dt_1$$

$$= \frac{1}{N}\sum_{n=0}^{N-1}\exp(j2\pi(\omega_d - p\Delta f - \tau\Delta f)nT_R \sum_{m=0}^{N-1}\int_{-\infty}^{\infty} g_n(t_1) \times g_m^* \left\{ t_1 - \left[\tau - pT_R\right]\right\} \dots$$

$$\dots \exp(j2\pi(\omega_d - p\Delta f - \tau\Delta f)t_1 dt_1$$

$$= \frac{1}{N}\sum_{n=0}^{N-1}\exp(j2\pi(\omega_d - p\Delta f - \tau\Delta f)nT_R \sum_{m-0}^{N-1}\chi_p\left[\tau - pT_R, \omega_d - p\Delta f - \tau\Delta f\right] \qquad (3.23)$$

where χ_p is cross-ambiguity function.

If the pulse train is *uniform*, that is, if

$$g_n(t) = g_m(t) = g_c(t) \qquad (3.24)$$

Equation (3.23) reduces to

$$\chi(\tau, \omega_d) = \frac{1}{N}\sum_{n=0}^{N-1}\exp(j2\pi(\omega_d - p\Delta f - \tau\Delta f)nT_R \sum_{m-0}^{N-1}\chi_C\left[\tau - pT_R, \omega_d - p\Delta f - \tau\Delta f\right] \quad (3.25)$$

where $\chi_C(\tau, \omega_d)$ is the ambiguity function of one pulse in a uniform pulse train.

After Rihaczek [4], we note that

$$\sum_{n=0}^{N-1}\sum_{m=1}^{N-1} = \sum_{p=-(N-1)}^{0}\sum_{n=0}^{N-1-|p|}\Bigg|_{m=n-p} + \sum_{p=1}^{N-1}\sum_{m=0}^{N-1-|p|}\Bigg|_{n=m+p} \qquad (3.26)$$

Using equation (3.26) in equation (3.25)

$$\chi(\tau, \omega_d) = \frac{1}{N}\sum_{p=-(N-1)}^{0}\left[\chi_C(\tau - pT_R, \omega_d - p\Delta f - \tau\Delta f)\sum_{n=0}^{N-1-|p|}\exp(j2\pi(\omega_d - p\Delta f - \tau\Delta f)nT_R)\right]$$

$$= +\frac{1}{N}\sum_{p=1}^{N-1}\left[\exp(j2\pi(\omega_d - p\Delta f - \tau\Delta f)pT_R\chi_C(\tau - pT_R, \omega_d - p\Delta f - \tau\Delta f)\dots\right.$$

$$\left.\dots \sum_{m=0}^{N-1-|p|}\exp(j2\pi(\omega_d - p\Delta f - \tau\Delta f)mT_R\right]$$

$$\qquad (3.27)$$

Using,

$$\sum_{n=0}^{N-1-|p|}\exp(j2\pi(\omega_d - p\Delta f - \tau\Delta f)nT_R) = \exp\left[j\pi(\omega_d - p\Delta f - \tau\Delta f)(N - 1 - |p|T_R\right]\dots$$

$$\dots \frac{\sin\left[\pi(\omega_d - p\Delta f - \tau\Delta f)(N - |p|)T_R\right]}{\sin \pi(\omega_d - p\Delta f - \tau\Delta f)T_R}$$

$$\qquad (3.28)$$

in equation (3.27)
we obtain,

$$|\chi(\tau,\omega_d)| = \left| \frac{1}{N} \sum_{p=-(N-1)}^{N-1} \exp\left[j\pi(\omega_d - p\Delta f - \tau\Delta f)(N-1+p)T_R \right] \chi_C(\tau - pT_R, \omega_d - p\Delta f - \tau\Delta f) \dots \right.$$
$$\left. \dots \frac{\sin\left[\pi(\omega_d - p\Delta f - \tau\Delta f)(N-|p|)T_R \right]}{\sin \pi(\omega_d - p\Delta f - \tau\Delta f)T_R} \right|$$

(3.29)

Equation (3.29) is the ambiguity function of a train of N coherent stepped frequency pulses where $|\chi_C(\tau,\omega_d)|$ is the ambiguity function of an individual pulse. In cases where the *separation between pulses is larger than the duration of the individual pulses*, we can further simplify equation (3.29) to read as,

$$|\chi(\tau,\omega_d)| = \left| \frac{1}{N} \sum_{p=-(N-1)}^{N-1} |\chi_C(\tau - pT_R, \omega_d - p\Delta f - \tau\Delta f)| \frac{\sin\left[\pi(\omega_d - p\Delta f - \tau\Delta f)(N-|p|)T_R \right]}{\sin \pi(\omega_d - p\Delta f - \tau\Delta f)T_R} \right|$$

(3.30)

when

$$t_p < \frac{T_R}{2}$$

(3.31)

where t_p is the pulse width.

Expanding equation (3.30) we obtain,

$$|\chi(\tau,\omega_d)| = \left| \frac{1}{N} \sum_{p=-(N-1)}^{N-1} \left| \left(1 - \frac{|\tau - pT_R|}{t_p}\right) \frac{\sin\left[\pi(t_p - |\tau - pT_R|)(\omega_d - p\Delta f - \tau\Delta f) \right]}{\pi(t_p - |\tau - pT_R|)(\omega_d - p\Delta f - \tau\Delta f)} \right| \dots \right.$$
$$\left. \dots \left| \frac{\sin\left[\pi(\omega_d - p\Delta f - \tau\Delta f)(N-|p|)T_R \right]}{\sin \pi(\omega_d - p\Delta f - \tau\Delta f)T_R} \right| \right|$$

(3.32)

Equation (3.32) is plotted in Figure 3–9 for $T_R = 1$, $t_p = 0.2$, $N = 4$, and $\Delta f = 2$. The program "*sfw.m*" in the accompanying software is used to plot this ambiguity function.

The results obtained are similar to that obtained by Gill and Huang [5]. The cuts along the Doppler and delay axes are shown in Figure 3–10 (with Hamming weighting).

As a comparison the ambiguity function of a coherent pulse train is plotted in Figure 3–11 for $T_R = 1$, $t_p = 0.2$, and $N = 5$. We use the software "*cohopulsetrain.m*" supplied with this book.

The results obtained are exactly similar to that obtained by Levanon [1]. The cuts along the Doppler and delay axes are shown in Figure 3–12 (with Hamming weighting).

Comparison of both the above cases leads us to the following conclusions:

- Both the figures are spiky; however, the ambiguity diagram of the step frequency radar has a tilt in common with other linear frequency modulated waveforms.

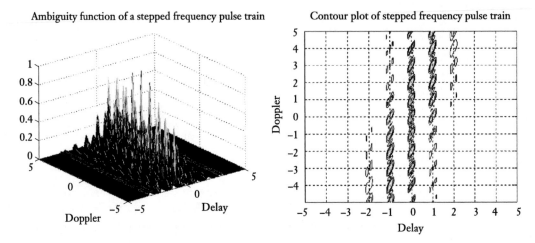

Figure 3–9 Mesh and contour plots of a stepped frequency pulse train.

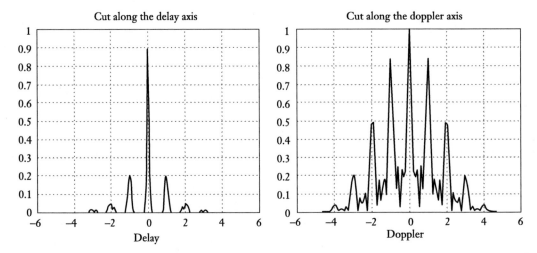

Figure 3–10 Cuts along the delay and Doppler axes.

- The central spike of the step frequency waveform is also tilted.
- Comparing both the cases, it can be seen that for the step frequency waveform, ambiguous spikes decrease very fast as compared to the constant frequency waveform.
- Along the delay axis, the null-to-null width of each spike for the step frequency radar is $2/N\Delta f$ as compared to $2t_p$ for the constant frequency pulses, thereby making it possible to decrease the effective pulse width of the step frequency radar by increasing N or Δf.

Range–Doppler Coupling: Examination of equation (3.32) shows that if the peak position of the signal output of the matched filter is the indication of the delay, then a return, shifted in Doppler, will cause an erroneously delayed peak. Near zero Doppler, that additional delay will be a linear function of the Doppler shift. From equation (3.32), we see that the peak of the ambiguity function

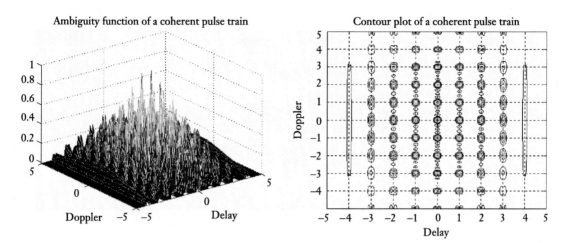

Figure 3–11 Mesh and contour plots of a coherent pulse train.

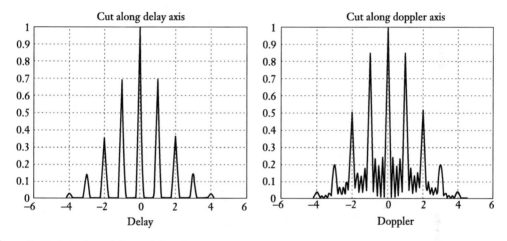

Figure 3–12 Cuts along delay and Doppler axes.

will correspond fairly well to the erroneous delay in which the argument of the sinc function is zero. This happens at

$$\tau = \frac{\omega_d - p\Delta f}{\Delta f} \tag{3.33}$$

$$\approx \frac{\omega_d}{\Delta f} \tag{3.34}$$

We find that in stepped frequency radar, the Doppler shift is coupled to a delay, making it difficult to distinguish between the two.

3.2.4 Digital-Coded Waveforms

Digital-coded waveforms can be classified into the following major groups:

- Phase-coded waveforms. This class of waveforms can be further grouped into two major groups:
- Binary phase codes, like Barker codes [1].
- Polyphase codes like Frank codes [1].
- Frequency-coded waveforms like Costas codes [1].

These waveforms and their ambiguity functions shall be examined in Chapters 5 and 6.

References

1. Levanon, N., *Radar Principles*, John Wiley & Sons, New York, 1988.
2. Jankiraman, M., de Jong, E. W., and van Genderen, P., "Ambiguity analysis of PANDORA multifrequency radar," in *Radar Conference 2000, Record of the IEEE 2000 International*, May 7–12, 2000, pp 35–41.
3. Wehner, D. R., *High Resolution Radar*, Artech House, Norwood, MA, 1987.
4. Rihaczek, A. W., *Princples of High Resolution Radar*, McGraw-Hill, New York, 1969.
5. Gill, G. S. and Huang, Jen-Chih, "The ambiguity function of the step frequency radar signal processor," in *Proceedings of the CIE International Conference Radar*, October 8–10, 1996, pp. 375–380.

FMCW Waveform

4.1 INTRODUCTION

In the previous chapters we had studied the various types of radar waveforms. These include the linear frequency modulated (LFM) waveform central to frequency modulated continuous wave (FMCW) radars and chirp pulse radars. We also studied an antitank missile, which uses continuous waveforms (CW) radar technology for reasons of low probability of intercept (LPI). We then studied the ambiguity functions of the various radar waveforms and their salient features. We noted that the LFM waveform, in particular, has a diagonal ridge. This ridge imparts to it Doppler tolerance essential in tracking fast targets, but the diagonal also imparts to it range–Doppler coupling which is not a good thing. In this chapter, we shall study waveform compression as applied to LFM waveforms. This technique is used in both pulse as well as CW radars. We shall then study the FMCW radar design in terms of range resolution and as to how it is affected by the sweep time, target spectral width, and receiver frequency resolution.

4.2 WAVEFORM COMPRESSION

In the preceding sections we have seen that in order to obtain high resolutions in radars, we need to increase the signal bandwidth. This is achieved in unmodulated pulsed radars by transmitting very short pulses. However, if we utilize short pulses, we also decrease the average transmitted power and hence, the radar detection range. We therefore need to look for a method which allows us to transmit at a large average power (by using long pulses) and at the same time achieve the same range resolution as given by short pulses. This paradoxical situation was resolved with the advent of the LFM pulse. The LFM pulse transmits a long pulse using a wide bandwidth and large average power and then the received pulse is compressed using pulse compression techniques to be discussed below, to achieve the desired range resolution. Hence, pulse compression allows us to achieve the average transmitted power of a long pulse while obtaining the range resolution corresponding to a short pulse. In CW radars, we use the term "waveform compression" instead of pulse compression. Henceforth, we shall use this term throughout this book.

There are two well-known techniques to achieve waveform compression:

- Correlation processing
- Stretch processing

We shall examine these aspects in the subsequent sections [1, 2].

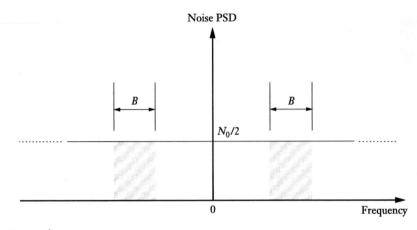

Figure 4–1 Input noise power.

4.2.1 Time–Bandwidth Product

We examine a matched filter radar receiver. The filter has a white noise bandwidth, which has a two-sided spectrum as discussed earlier. This noise power is given by

$$N_{wn} = 2 \frac{N_0}{2} B \tag{4.1}$$

where B is the matched filter bandwidth and the factor of two is used to account for both the negative and positive frequency bands as shown in Figure 4–1.

The average input signal power over a signal duration T is [2]

$$S_{sp} = \frac{E}{T} \tag{4.2}$$

where E is the signal energy. Hence, matched filter input SNR (signal to noise ratio) is given by

$$\left(SNR_{input}\right) = \frac{S_{sp}}{N_{wn}} = \frac{E}{N_0 BT} \tag{4.3}$$

The ideal output peak instantaneous SNR to the input SNR ratio is

$$\frac{SNR_{output}}{SNR_{input}} = 2BT \tag{4.4}$$

The quantity BT is referred to as the "time–bandwidth" product for a given waveform or its corresponding matched filter. The factor BT by which the output SNR is increased over the input SNR is called the matched filter gain, or compression gain.

The time–bandwidth product of an unmodulated signal approaches unity. We can increase the time–bandwidth product of a signal to a value greater than unity by using frequency or phase modulation. If the radar receiver matched filter is perfectly matched to the incoming waveform,

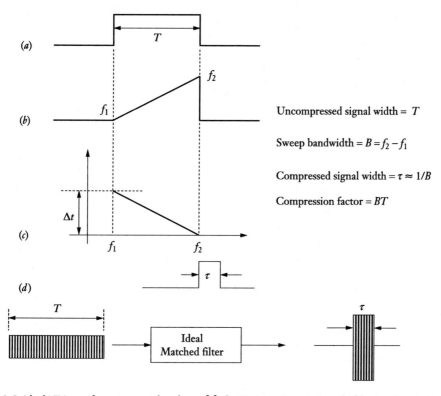

(a)

(b)

Uncompressed signal width $= T$

Sweep bandwidth $= B = f_2 - f_1$

Compressed signal width $= \tau \approx 1/B$

Compression factor $= BT$

(c)

(d)

Ideal
Matched filter

Figure 4–2 Ideal LFM waveform compression. (From [2], © CRC Press, Boca Raton, Florida. Reprinted with permission)

the compression gain is equal to BT. If the matched filter spectrum deviates from that of the input signal, the compression gain proportionally falls.

4.2.2 LFM Waveform Compression

This section is based on the work done by Mahafza (From [2] © 2004. Reproduced by permission of Taylor & Francis, a division of Informa plc.). LFM waveform compression is achieved by adding frequency modulation to a long pulsed chirp signal at transmission and by using a matched filter receiver in order to compress the received signal. Hence, the matched filter output is compressed by a factor BT where T is the uncompressed signal width and B is the bandwidth of the LFM signal. Hence, we can use long chirp signals and LFM modulation to achieve large compression ratios.

Figure 4–2 shows the LFM waveform compression process.

In Figure 4–2(a)–(d), (a) shows the envelope of the single step signal; (b) shows the frequency modulation (up-chirp in this case) with bandwidth $B = f_2 - f_1$; (c) shows the time-delay characteristic of the matched filter; while (d) shows the compressed signal. This picture is conceptual as in reality the output is a *sinc* function.

We now examine the effect of waveform compression bandwidth. We assume two targets of RCS 1 m² and 2 m² located at 15 m and 25 m. The sweep bandwidth in the initial case is 10 MHz. The range resolution in this case is given by

$$\Delta R = \frac{c}{2\Delta f} = \frac{3 \times 10^8}{2 \times 10 \times 10^6} = 15 \text{ m} \tag{4.5}$$

This is also called the "Rayleigh Resolution." Since the bandwidth is insufficient, the radar is unable to resolve the targets. If we now increase the bandwidth to 50 MHz, then

$$\Delta R = \frac{c}{2\Delta f} = \frac{3 \times 10^8}{2 \times 50 \times 10^6} = 3 \text{ m} \tag{4.6}$$

Clearly, the radar can now resolve the targets. The program is given in the accompanying software and entitled "LFM_resolve.m". The result for the two cases is shown in Figure 4–3(a) and 4–3(b).

4.2.2.1 Correlation Processor   We define the radar range window as the difference between the radar maximum and minimum range. This is also called a receive window. All target returns within the receive window are collected and passed through a matched filter to perform waveform compression. This matched filter is implemented in many ways. One is to use a Surface Acoustic Wave (SAW) device [3]. Alternately, we can perform the correlation process digitally using the fast Fourier Transform (FFT). This method is called Fast Convolution Processing (FCP) and has been already mentioned in Section 2.3. We now examine the math behind this process with reference to FCP as shown in Figure 4–13.

Consider a receive window defined by

$$R_{rec} = R_{max} - R_{min} \tag{4.7}$$

where R_{max} and R_{min}, respectively, define the maximum and minimum radar detection ranges. The normalized complex transmitted signal has the form [2]

$$s(t) = \exp\left(j2\pi\left(f_0 t + \frac{\mu}{2}t^2\right)\right) \quad 0 \leq t \leq T \tag{4.8}$$

where T is the signal width, $\mu = B/T$, and B is the bandwidth.

The radar return is the same as the transmitted signal, but with a time delay and an amplitude change that corresponds to the target RCS. We assume that the target is located at range R_1. The echo received by the radar is then

$$s_{rec}(t) = a_1 \exp\left(j2\pi\left(f_0(t - \tau_1) + \frac{\mu}{2}(t - \tau_1)^2\right)\right) \tag{4.9}$$

where a_1 is proportional to target RCS, antenna gain, and range attenuation. The time delay τ_1 is given by

$$\tau_1 = 2R_1/c \tag{4.10}$$

Initially, we remove the frequency f_0. This is achieved by mixing the received signal $s_{rec}(t)$ with a reference signal whose phase is $2\pi f_0 t$. The phase of the resultant signal, after low pass filtering is then given by

$$\psi(t) = 2\pi\left(-f_0 \tau_1 + \frac{\mu}{2}(t - \tau_1)^2\right) \tag{4.11}$$

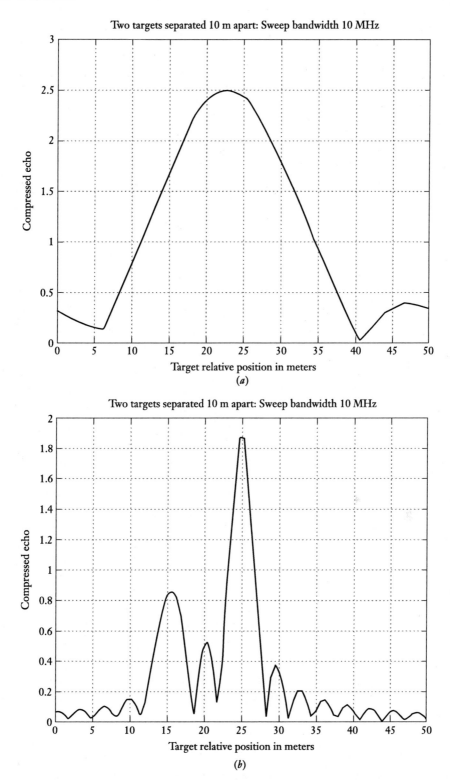

Figure 4–3 Resolution bandwidth (*a*) Sweep bandwidth 10 MHz (top) and (*b*) Sweep bandwidth 50 MHz (bottom).

and the instantaneous frequency is

$$f_i(t) = \frac{1}{2\pi}\frac{d}{dt}\psi(t) = \mu(t-\tau_1) = \frac{B}{T}\left(t - \frac{2R_1}{c}\right) \qquad (4.12)$$

The quadrature components are

$$\begin{pmatrix} x_I(t) \\ x_Q(t) \end{pmatrix} = \begin{pmatrix} \cos\psi(t) \\ \sin\psi(t) \end{pmatrix} \qquad (4.13)$$

We next sample the quadrature components by selecting a sampling frequency $f_s > 2B$ (to satisfy the Nyquist criterion, so as to avoid ambiguity in the spectrum). The sampling interval is then $\Delta t \le 1/2B$. Using equation (4.12) it can be shown that the frequency resolution of the FFT is

$$\Delta f = 1/T \qquad (4.14)$$

The minimum required number of samples is

$$N = \frac{1}{\Delta f \Delta t} = \frac{T}{\Delta t} \qquad (4.15)$$

Using $\Delta t \le 1/2B$ and substituting in equation (4.15) we obtain

$$N \ge 2BT \qquad (4.16)$$

Hence, we require a total of $2BT$ real samples or BT complex samples to completely describe an LFM waveform of duration T and bandwidth B. For example, an LFM signal of duration $T = 10\ \mu\text{sec}$ and bandwidth $B = 4$ MHz requires 80 real samples to determine the input signal (40 samples for the I-channel and 40 samples for the Q-channel).

If we assume that there are I targets at ranges R_1, R_2, and so on, within the receive window, then from the superposition theorem, the phase of the down-converted signal is given by [2]

$$\psi(t) = \sum_{i=1}^{I} 2\pi\left(-f_0\tau_i + \frac{\mu}{2}(t-\tau_i)^2\right) \qquad (4.17)$$

The times $\{\tau_i = (2R_i/c); i = 1,2,...,I\}$ represent the two-way time delays, where τ_i coincides with the start of the receive window.

4.2.2.2 Stretch Processing   Stretch processing is also called "*active correlation*" and is used in order to process extremely wide bandwidth LFM waveforms. The reader was introduced to this type of signal processing in Section 2.3. We now examine this important technique in more detail as it is extremely popular in FMCW radars [1, 2, 4, 5].

The salient steps towards stretch processing are shown in Figure 4–4 [2]. The radar returns are mixed with a longer replica (reference signal) of the transmitted waveform. We then pass the signal through a low pass filter to eliminate harmonics at the output of the mixer (it also acts as an anti-aliasing filter for the ADC) and then to a coherent detector and Analog to Digital converter

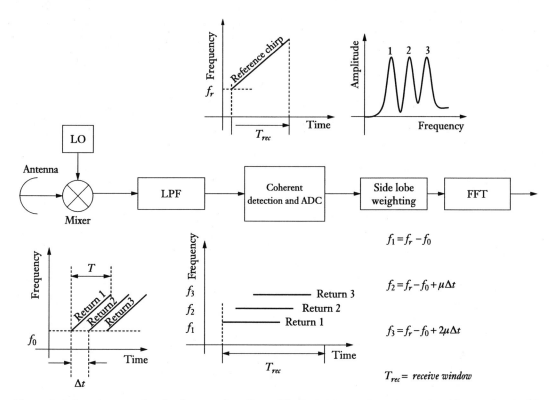

Figure 4–4 Stretch processing implementation. (From [2], © CRC Press, Boca Raton, FLorida. Reprinted with permission)

(ADC). If the chirp slopes of the transmitted and receiver waveforms are identical, then it will be appreciated that as the transmitted chirp progresses from some initial frequency f_r (in Figure 4–4, f_0 is the initial frequency of the radar return; if Δt were zero then $f_r = f_0$) through $f_1, f_2 \ldots$ and so on up to the end of its sweep bandwidth, so does the received radar return, but after a time delay Δt, proportional to the propagation time. Therefore, the reader can visualize that there is always a steady frequency gap between the frequency of the transmitted waveform at any instant and the frequency of the received waveform. Furthermore, this frequency gap is constant and its value is directly dependent upon the time delay Δt. This is the beat frequency and it is proportional to range. The equations in Figure 4–4 reflect this view point.

The coherent detector is implemented as an I–Q demodulator. The signal is then sampled by the ADC and then weighted to reduce the time side lobes. Finally, the signal is given to a bank of narrow band filters, implemented as an FFT. The output of the FFT is a tone proportional to the target range, since the beat frequencies are proportional to range. All returns from the same range bin produce the same constant frequency. This FFT is sometimes also called the "range FFT" as it deals with target ranges. Earlier in this discussion, we had assumed that the transmitted chirp and the received chirp have the same slope. This is true only if the target is static. If it has a Doppler value, then the received return will have an incremental frequency which will give the beat frequency a steady additional frequency difference proportional to the target Doppler. This will cause an error in range. This phenomenon is called range-Doppler coupling and is prominent in LFM waveforms. This aspect will be examined in more detail in the next chapter. The other contributory cause for this is the fact that the received pulse width is expanded (or compressed) by the time dilation factor due to the target radial velocity. This phenomenon can be corrected in two popular ways:

1. We can take repeated measurements of the target returns and then determine the Doppler value. We then adjust the chirp slope and pulse width of the next transmitted pulse to account for the estimated Doppler frequency and time dilation.
2. The second method is to so choose the width of the range bin such that the signal does not change range bins due to target Doppler.

If we wish to measure the target Doppler directly, then one way is to have multiple and identical channels in the radar. Suppose we have seven additional channels. In such a case, we route the outputs of all the eight range bins (for the same target) to an eight-point Doppler FFT. The output of this FFT will be the target Doppler. This aspect will be examined in Part III of this book. Alternately, we can store eight target returns from a single channel and then route these signals to an eight-point Doppler FFT to extract the Doppler value.

We now prove the stretch signal processing mathematically. This is derived by Mahafza [2] and is reproduced with permission. The normalized transmitted signal can be expressed as

$$s_{tr}(t) = \cos\left(2\pi\left(f_0 t + \frac{\mu}{2}t^2\right)\right) \quad 0 \le t \le T \tag{4.18}$$

where $\mu = B/T$ is the LFM coefficient and f_0 is the chirp start frequency. If we assume a point scatterer at range R, the signal received by the radar is

$$s_{rec}(t) = a \operatorname{rect}\left(\frac{t}{T} - \Delta\tau\right)\cos\left[2\pi\left(f_0(t - \Delta\tau) + \frac{\mu}{2}(t - \Delta\tau)^2\right)\right] \tag{4.19}$$

where a is proportional to target RCS, antenna gain, and range attenuation. The time delay $\Delta\tau$ is $\Delta\tau = 2R/c$. The reference signal is

$$s_{ref}(t) = 2\cos\left(2\pi\left(f_r t + \frac{\mu}{2}t^2\right)\right) \quad 0 \le t \le T_{rec} \tag{4.20}$$

The receive window is

$$T_{rec} = \frac{2(R_{max} - R_{min})}{c} - T = \frac{2R_{rec}}{c} - T \tag{4.21}$$

The reader will recall that earlier we had stated that $f_r = f_0$ if there were no propagation delay, i.e., $\Delta t = 0$. Putting it in other words, f_r and f_0 are the same frequencies, the only difference being that the former pertains to the transmitted signal and the latter to the received signal. Hence, we can, for the purposes of this derivation, state that $f_r = f_0$. The output of the mixer is the product of the received and reference signals. After low pass filtering,

$$s_0(t) = a \operatorname{rect}\left(\frac{t}{T} - \Delta\tau\right)\cos\left(2\pi f_0 \Delta\tau + 2\pi\mu\Delta\tau t - \pi\mu(\Delta\tau)^2\right) \tag{4.22}$$

Using $\Delta\tau = 2R/c$ and substituting in equation (4.22), we obtain

$$s_0(t) = a \operatorname{rect}\left(\frac{t}{T} - \Delta\tau\right)\cos\left[\left(\frac{4\pi BR}{cT}\right)t + \frac{2R}{c}\left(2\pi f_0 - \frac{2\pi BR}{cT}\right)\right] \quad (4.23)$$

Since, $T >> 2R/c$, we can approximate equation (4.23) as

$$s_0(t) = a \operatorname{rect}\left(\frac{t}{T} - \Delta\tau\right)\cos\left[\left(\frac{4\pi BR}{cT}\right)t + \frac{4\pi R}{c}f_0\right] \quad (4.24)$$

The instantaneous frequency is

$$f_{inst} = \frac{1}{2\pi}\frac{d}{dt}\left(\frac{4\pi BR}{cT}t + \frac{4\pi R}{c}f_0\right) = \frac{2BR}{cT} \quad (4.25)$$

which clearly indicates that the target range is proportional to the instantaneous frequency. Therefore, after sampling the LPF output and taking FFT, we obtain

$$R_1 = f_1 cT/2B \quad (4.26)$$

for a target located at R_1 with a beat frequency f_1.

If there are I close targets at ranges R_1, R_2, and so forth ($R_1 < R_2 < ... < R_I$), we obtain by superposition, the total signal as

$$s_{rec}(t) = \sum_{i=1}^{I} a_i(t)\cos\left[2\pi\left(f_0(t - \tau_i) + \frac{\mu}{2}(t - \tau_i)^2\right)\right] \quad (4.27)$$

where $\{a_i(t); i = 1, 2, ..., I\}$ are proportional to the targets' cross-sections, antenna gain, and range. The times $\{\tau_i = (2R_i/c); i = 1, 2, ..., I\}$ represent the two-way time delays, where τ_1 coincides with the start of the receive window. Using equation (4.23) the overall signal at the output of the LPF can then be described as

$$s_0(t) = \sum_{i=1}^{I} a_i \operatorname{rect}\left(\frac{t}{T} - \frac{2R_i}{c}\right)\cos\left[\left(\frac{4\pi BR_i}{cT}\right)t + \frac{2R_i}{c}\left(2\pi f_0 - \frac{2\pi BR_i}{cT}\right)\right] \quad (4.28)$$

Hence, the target returns appear as constant frequency tones that can be resolved using FFT. Consequently, determining the proper sampling rate and FFT size is critical. It is proved in [2] that the number of samples N is given by

$$N \geq 2BT_{rec} \quad (4.29)$$

There is a program "*stretch_processing.m*" in the accompanying software that carries out this exercise. We have four targets with an RCS of 1, 2, 1, 2 m^2 and located at ranges of 15, 20, 23, and 25 m, respectively. A Hamming window is assumed. The transmitted pulse width T is assumed as 20 msec with a bandwidth B of 1 GHz. Figures 4–5 and 4–6 demonstrate the effect. The initial frequency is 5.6 GHz and the receive window is 60 m.

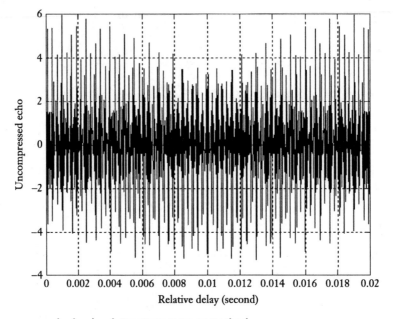

Figure 4–5 Uncompressed echo signal. Four targets are unresolved.

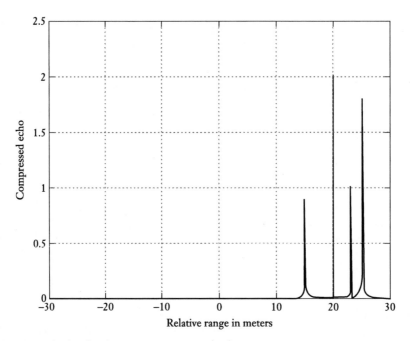

Figure 4–6 Compressed echo signal. Four targets are resolved.

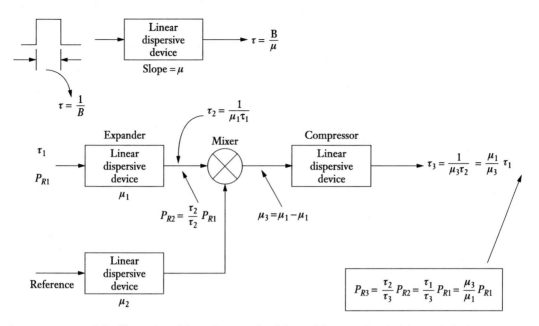

Figure 4–7 Toward the illustration of "stretch processing." (From [4], © Reprinted with permission)

The Rayleigh resolution for this bandwidth of 1 GHz is 15 cms. The reader is invited to try targets less than 15 cm apart. The radar will then be unable to resolve the targets, unless the sweep bandwidth is suitably increased. This is left to the reader as an exercise.

It will be interesting to determine the origin of the word "stretch." Toward this end we examine the paper by Caputi [4, 5]. In Figure 4–7, slope signifies chirp rate and P_{R1} corresponds to the energy of the pulsed input signal. The expander transforms the pulse of length τ_1 to a pulse of length τ_2 with slope μ_1 and energy P_{R2}. In our context, this is the transmitted signal, i.e., an LFM pulse. The target return is also P_{R2}, neglecting losses. This is then correlated with a signal whose slope is μ_2 which is equal to or *less* than μ_1. Therefore, the slope of the output signal is

$$\mu_3 = \mu_1 - \mu_2 \tag{4.30}$$

If we now send this output signal with slope μ_3 through a dispersive delay line having a characteristic which is an inverse of slope μ_3, the pulse is compressed to a width of τ_3. However, μ_3 is a very small value (in our context it varies exclusively because of target Doppler, which causes a very small change in slope, due to the Doppler frequency). Therefore, the bandwidth given by $(1/\tau_3)$ is very small, i.e., drastically reduced (see Figure 4–8(c)). Hence, the pulse compression gain is proportionately less, since the time–bandwidth product is reduced. We note that τ_3 is much larger than τ_1, the original input pulse width, i.e., the pulse gets *stretched*! This is the origin of the term "stretch processing."

We note that the output pulse width and amplitude is proportional to the input pulse width and amplitude, with proportionality constant determined by the design of the input and output devices.

$$P_{R3} = \frac{\tau_2}{\tau_3} P_{R2} = \frac{\tau_1}{\tau_3} P_{R1} = \frac{\mu_3}{\mu_1} P_{R1} \tag{4.31}$$

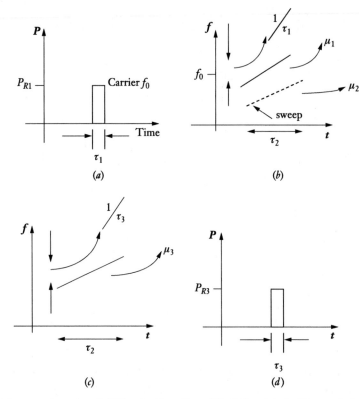

Figure 4–8 Stretch processing—bandwidth reduction. (From [4], © Reprinted with permission)

Specifically, if we assume that the reference slope μ_2 is half μ_1,

$$\mu_3 = \mu_1 - \frac{\mu_1}{2} = \frac{\mu_1}{2}, \text{ i.e., the bandwidth is halved.}$$

Hence,
$P_{R3} = \frac{\mu_1}{\mu_1} P_{R1} = \frac{P_{R1}}{2}$, i.e., the amplitude is halved, since the time-bandwidth product is halved.

$$\therefore \tau_3 = \tau_1 \times \frac{P_{R1}}{P_{R3}} = 2\tau_1, \text{ i.e., pulse width is doubled, since the signal energy is a constant.}$$

Since the same thing happens to noise, hence, the SNR remains the same. It must be clarified that the SNR is the same at the output of the mixer and *not* after the range FFT as we shall see. Furthermore, since the output signal energy is always the same as the input signal energy, there will be no change in the SNR of a noisy input signal (the noise power is also reduced by the ratio of the input-to-output bandwidth). Even though pulse-compression is taking place at the final stage, the signal strength with respect to noise at the final stage *even after the pulse compression*, is the same as the signal strength with respect to noise at the initial stage, with the only exception that due to the bandwidth being halved, the pulse width is doubled. It can be shown mathematically [4] that if

$$m(t) = \exp i\omega t \text{ is the input signal}$$

the output signal after stretch processing is given by

$$M(t) = \sqrt{\frac{1}{S}} m\left(\frac{t}{S}\right)$$

where S is the stretch factor.

This means that the input signal is stretched in time by the factor S and scaled in voltage amplitude by the factor $\sqrt{\frac{1}{S}}$ (or $\frac{1}{S}$ in terms of power).

It can be seen that in the special case, if μ_3 is very small (i.e., $\mu_3 = 0$ since $\mu_1 = \mu_2$) as happens in the case of radars, where the slope of the transmitted reference is the same as the slope of the radar return (neglecting Doppler, which in any case is very small), the bandwidth will be practically zero. In such cases, there is no point in carrying out pulse compression in the conventional sense. We can instead take advantage of the reduced bandwidth conditions and use a low bandwidth device like a spectrum analyzer. The stretch approach allows the full range resolution of a wide bandwidth waveform to be realized with restricted bandwidth processor like a spectrum analyzer. If $\mu_3 = 0$, the output will be a beat frequency signal proportional to range. A range FFT will put this signal into its corresponding range bin. The resolution of the range FFT is dictated by the sampling time, the upper ceiling being the sweep bandwidth Δf. The advantage here is that we can achieve as high a range bin resolution as we desire by simply increasing the resolution of the range FFT by increasing the sampling rate. Strictly speaking, we are not *directly* compressing the pulse, but by narrowing the range bin width we are in effect increasing the range bin resolution. This is the benefit of using a range FFT as a spectrum analyzer.

Therefore, since the input and output signal energies are a constant, *there is no change in SNR*. This is because both the signal and noise are subjected to the same processing, i.e., bandwidth reduction.

Figure 4–8 illustrates the process.

If we have two pulses separated by Δt_1, it can be shown that

$$\Delta t_2 = \frac{\mu_1 \Delta t_1}{\mu_2} = \frac{\mu_1}{\mu_3} \Delta t_1 \qquad (4.32)$$

Since μ_3 is much smaller than μ_1, time "slows" down. Therefore, in stretch processing both the pulse widths and the separation between them gets "*stretched*". Hence, the term "*stretch processing*".

The output of the mixer is now given to a range FFT via an ADC. The signal processing gain at the output of the FFT (all values in dBs) is given by ([1], p. 607)

$$SNR_{output} = FFT\ length + SNR_{input} - SPL \qquad (4.33)$$

where SPL is the signal processing loss due to the mixer, due to weighting, and due to ADC quantization. SNR_{input} is the SNR at the input of the mixer. BT is the time–bandwidth product of the LFM signal and (BT) dBs = (FFT length) dBs. It is interesting to note that there is a sliding range window effect, in that the T in BT product changes with target range, the beat frequency also changes with range causing the FFT length to change correspondingly. Finally, it is pointed out to the reader that whereas techniques like SAW filters yield a processing gain of around 1,000 typically, stretch processing quite commonly yields gains of 50,000! This type of processing is particularly suited for high bandwidth signals and utilize a receive window of up to 100 m though windows as high as 10 km are common.

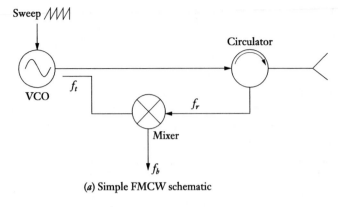

(*a*) Simple FMCW schematic

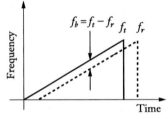

(*b*) Sweep characteristic

Figure 4–9 FMCW principle.

4.3 FMCW SYSTEM

The fundamental principle of FMCW radar is illustrated in Figure 4–9. A CW transmitter is swept in frequency linearly with time and the received signals are mixed with a portion of the signal currently being transmitted. Because of the time delay between the transmitted and received signals, there is a frequency difference between these two signals and a beat frequency is produced. These aspects have already been examined in earlier chapters. The beat frequency is directly proportional to the time delay and hence to the range to the target.

Since the radar will receive returns from a multitude of targets at many different ranges, the received signal will be a composite of many components of the type described above. It is, therefore, necessary to perform a frequency analysis on the received signals in order to resolve the individual frequency components and hence, the individual targets. The chosen frequency analysis method is the Fast Fourier Transform (**FFT**).

It is fundamental for a coherent radar system that the ability of the radar to resolve targets at only slightly different ranges is proportional to the total frequency deviation in the sweep and hence, for a fixed total dwell time on the target, is proportional to the sweep rate. The mathematics of this has already been examined, when we studied the LFM pulse. *It is a feature of FMCW radars that almost any range resolution can easily be obtained by varying the frequency sweep.* The equivalent unmodulated pulse radar could require extremely short transmitted pulses posing severe technological difficulties.

Until now the reader has been presented a more physical idea of this technology. We now pursue a more rigorous mathematical approach, investigating aspects like range resolution and non-linearities in the waveform.

4.4 ADVANTAGES OF AN FMCW SYSTEM

FMCW modulation has the following advantages:

1. It is well matched to simple solid state transmitters which lead to systems with low initial cost, high reliability, and low maintenance costs. This is because LPI radars are usually low power radars [6].
2. The technique allows a wide transmitted spectrum to be used giving a very good range resolution without the need to process very short pulses. This applies to magnetron-based radars. However, chirp pulse or phase-coded pulse radars do not have this problem.
3. The radar has LPI capability.
4. The power spectrum is nearly rectangular over the modulation bandwidth, making interception difficult [6, 7]. However, this is common to chirp pulse and phase coded radars also.
5. Easier to implement as compared to phase-coded modulation.

4.5 BASIC EQUATIONS OF FMCW RADARS

We revisit equation (2.12) reproduced here for convenience.

$$R = \frac{f_b T_s c}{2\Delta f} = \frac{Nc}{4\Delta f} \tag{4.34}$$

where R is target range, f_b is beat frequency, T_s is sweep time, Δf is sweep bandwidth or frequency deviation, and c is the velocity of light.

We introduce a new variable f_{max}, which is the maximum beat frequency corresponding to maximum range. Let $T_s = \Delta t$ the sweep time. We then write for maximum detection range of FMCW radar as

$$R_{max} = \frac{CT_s}{2\Delta f} f_{max} = \frac{NC}{4\Delta f} \tag{4.35}$$

where N is the number of samples in one sweep time, $f_{max} = \frac{f_s}{2}$ where f_s is the sampling frequency satisfying the Nyquist sampling criterion, and R_{max} is usually taken as the *maximum unambiguous range* of the radar.

We also know from earlier chapters that

$$\Delta R = \frac{c}{2\Delta f} \tag{4.36}$$

where ΔR is the range resolution.

If we now use a 64-point FFT and a sweep time $T_s = 330 \ \mu\text{sec}$ and $\Delta f = 5 \ \text{MHz}$, we obtain

$$R_{max} = \frac{Nc}{4\Delta f} = \frac{64 \times 3 \times 10^8}{4 \times 5 \times 10^6} = 960 \approx 1 \ \text{km} \tag{4.37}$$

The range resolution at this maximum indicated range is given by

$$\Delta R = \frac{c}{2\Delta f} = \frac{3 \times 10^8}{2 \times 5 \times 10^6} = 30 \ \text{m}$$

The range resolution of 30 m at a range of 1 km has been achieved with a sweep bandwidth of 5 MHz extended over a sweep time of 330 μsec. Compare this with a modulated pulse radar employing LFM pulses which require a compressed pulse width of 200 nanosecs to achieve the same resolution. This means that in terms of modulated pulse radars the compression ratio is 1,650, which is very high. This is made possible because of stretch processing, common to FMCW radars. The reader should also note that in equation (4.36), the sweep bandwidth is in the denominator. This implies that the range resolution is constant throughout the radar receiver window, because all targets in the receiver window will experience the same sweep bandwidth. Hence, we can expect that like in pulse radars, the range-resolution should always be a constant regardless of target range. However, in reality this is not true, since in FMCW radars, the range resolution is a function of sweep time. We shall investigate this aspect further in this chapter.

In order to yield 64 samples in one sweep, i.e., to carry out a 64-point FFT, the sampling frequency

$$f_s = \frac{64}{330 \times 10^{-6}} \approx 200 \text{ kHz}$$

$$\therefore f_{max} = \frac{f_s}{2} = 100 \text{ kHz} \leftrightarrow \text{Maximum beat frequency}$$

Since we have 64 samples/sweep, we use a 64-point FFT for range determination. But the spectrum is symmetrical for real samples. Therefore, we need to use only half the spectrum, i.e., $N/2$ which in our example is 32. Hence, the number of range cells (or bins) is 32. However, we will need to process both the halves of the spectrum as otherwise we will lose half the power, i.e., we carry out complex processing.

4.5.1 The FMCW Equation

We notice that there is a direct relationship between the frequency deviation, modulation period, beat frequency, and transit time. This relationship is called *FMCW equation*.

$$\frac{f_b}{t_d} = \frac{\Delta f}{T_s} \tag{4.38}$$

where f_b is beat frequency, t_d is round trip propagation time delay = $\frac{2R}{c}$ where R is the range to target, Δf is the sweep bandwidth or frequency deviation, and T_s is the modulation period (sweep time).

In view of the range–Doppler coupling inherent in LFM waveforms as discussed earlier, the beat frequency for upsweep depends upon both range and velocity (see Appendix C).

$$f_b = \frac{\Delta f 2R}{T_s c} + \frac{2Vf}{c} \tag{4.39}$$

where V is the target velocity relative to the radar, i.e., radial velocity and f is the nominal radar frequency.

The second expression in equation (4.39) constitutes the Doppler frequency shift of the target. In order to resolve this coupling, we need to have two frequency slew rates or slopes. Alternately, we live with it and control the range–Doppler coupling to within one range cell, as discussed in Chapter 2. The dual slew rate technique is called triangular waveform generation as against the one previously discussed, which is called sawtooth generation.

4.6 TRIANGULAR WAVEFORM

The frequency of the transmitted waveform in the first half is given by [7–9]

$$f_1(t) = f_0 - \frac{\Delta f}{2} + \frac{\Delta f}{T_s} t \qquad (4.40)$$

for $0 < t < T_s$ and zero elsewhere. Here T_s is the modulation period or sweep time, f_0 is the RF carrier, and Δf is the sweep bandwidth.

The frequency of the transmitted waveform in the second half is given by

$$f_1(t) = f_0 + \frac{\Delta f}{2} - \frac{\Delta f}{T_s} t \qquad (4.41)$$

for $0 < t < T_s$ and zero elsewhere. Here T_s is the modulation period or sweep time, f_0 is the RF carrier, and Δf is the sweep bandwidth.

It is easy to see from Figure 4–10 that the Doppler frequency

$$f_d = \frac{f_{1b} + f_{2b}}{2}. \qquad (4.42)$$

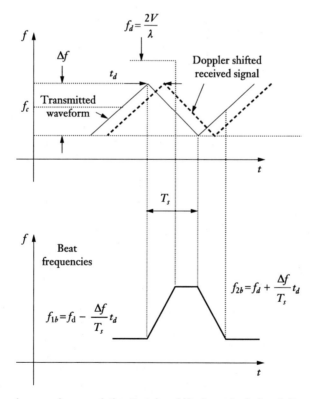

Figure 4–10 LFM triangular waveform and the Doppler shifted received signal (From [6], © Artech House, Reprinted with permission)

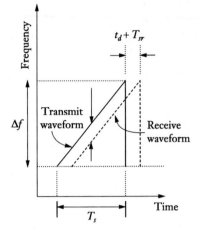

Figure 4–11 Toward the explanation of the sweep time effect on range resolution.

4.7 EFFECT OF SWEEP TIME ON RANGE RESOLUTION

The total time delay to the target and back is caused due to two factors [7–9]:

(a) Round trip propagation delay to the maximum range.
(b) Sweep recovery time T_{sr}. This is the time elapsed with the end of one sweep and the beginning of the next sweep. Therefore, if T_s is the sweep period:

$$T_{mod} = T_s - t_d - T_{sr} \tag{4.43}$$

where

$$t_d = \frac{2R_{max}}{c} \tag{4.44}$$

where R_{max} is the maximum range.

Hence, till we get our return from the maximum range, we cannot process the range FFTs. This causes a reduction in the effective processed bandwidth Δf_{eff} given by

$$\Delta f_{eff} = \Delta f \left(1 - \frac{t_d}{T_s}\right) \tag{4.45}$$

Therefore, the degraded range resolution ΔR_{deg} is given by

$$\Delta R_{deg} = \frac{c}{2\Delta f \left(1 - \dfrac{t_d}{T_s}\right)} \tag{4.46}$$

It can be seen from the previous discussion, that mere sweep bandwidth does not guarantee the range resolution. We have a lot more tedious spadework ahead before we can realize our entitled range resolution. These issues will be examined below.

Example 1

We now take an example to better explain the process. Suppose we have the following parameters:

$$\Delta f = 5 \text{ MHz}, \; T_{sr} = 3.3 \; \mu\text{sec}, \; T_s = 1 \text{ ms}, \; R_{max} = 1 \text{ km, then}$$

$$t_d = \frac{2R_{max}}{c} = \frac{2 \times 1000}{3 \times 10^8} = 6.7 \; \mu\text{sec}$$

$$\therefore T_{mod} = T_s - t_d - T_{sr} = 990 \; \mu\text{sec}$$

The 6.7 μsec transit time plus 3.3 μsec the sweep recovery time reduces the sweep bandwidth by 0.05 MHz. This makes the effective transmitting bandwidth 4.95 MHz.

The ideal range resolution was

$$\Delta R = \frac{c}{2\Delta f} = \frac{3 \times 10^8}{2 \times 5 \times 10^6} = 30 \text{ m}$$

The degraded range resolution now becomes

$$\Delta R_{deg} = \frac{c}{2\Delta f} = \frac{3 \times 10^8}{2 \times 4.95 \times 10^6} = 30.3 \text{ m}$$

This is the worst case, i.e., *range resolution is minimal at maximum range*. It gets progressively better at near ranges, since the t_d value decreases as we come closer to the radar. Hence, normally the modulation period T_s is kept at least 5 times the transit time for maximum range, so that the effective processed bandwidth is at least 80% of the total bandwidth [8, 9].

The important conclusion from all this discussion is that the range resolution decreases with range and is worst at the maximum range. This is fundamentally because of the increase in the value of t_d with range.

4.8 EFFECT OF RECEIVER FREQUENCY RESOLUTION AND TARGET SPECTRAL WIDTH ON RANGE RESOLUTION

Already the sweep time problem has degraded the range resolution to 30.3 m. Now, there are two more problems that need to be addressed so that the range resolution does not degrade any further! These are:

- *Receiver frequency resolution*: The name indicates that it is the frequency resolution of the receiver. This means range-bin resolution, since beat frequency resolution in FMCW systems corresponds to range-bin resolution. The receiver frequency resolution is given by

$$\Delta f_{rec} = 1/T_{mod} \tag{4.47}$$

In range profiling systems, we need to so design the radar that it yields the best range-bin resolution. This makes for better target profiling. However, there are other considerations, which we need to investigate.

- *Minimum target spectral width*: Even a point target will have a certain spectral width. This is dictated by the non-linearities in the FMCW waveform. It is given by

$$\Delta f_{tar} = \frac{NL}{100} \times \Delta f \qquad (4.48)$$

where NL is the non-linearity in percentage. In the limiting case, if $NL = 0$, $\Delta f_{tar} = \Delta f_{rec}$.

In our example, for an in-band time $T_{mod} = 990 \ \mu sec$, the receiver frequency resolution is 1,010 Hz [8], i.e. $(1/T_{mod})$.

We define scale factor as

$$SF = \frac{T_s \times c}{2 \times \Delta f} \ \text{m/Hz}$$

A 5 MHz frequency deviation and a 1 msec modulation period yield a scale factor of 30 m/kHz. Thus, the receiver frequency resolution of 1,010 Hz corresponds to a range width of 30.3 m. This is fine, as the range width has not deteriorated further.

In view of the interaction of Δf_{rec} and Δf_{tar} on Δf_b (where Δf_b is the beat frequency resolution, which determines in the final analysis, the range-bin width), obviously they have a combined relationship on Δf_b. The beat frequency resolution is the convolution of the target beat frequency spectral width and the receiver frequency resolution. This resolution can be estimated [8] by the square root of the sum of the squares (RSS) of the target beat frequency spectral width and the receiver filter bandwidth as

$$\Delta f_b = \sqrt{\Delta f_{tar}^2 + \Delta f_{rec}^2} \qquad (4.49)$$

where Δf_{tar} is the target beat frequency spectral width. For a point target, it is $(1/T_{mod})$, i.e., it equals the receiver frequency resolution (assuming perfect sweep linearity), and Δf_{rec} is the receiver frequency resolution.

It is pointed out that Δf_{tar} has a width even if there is no non-linearity and this width equals the receiver frequency resolution. But in the presence of non-linearities, it widens further and brings down the quality of the beat frequency resolution. Hence, in order to minimize it, i.e., to make it ideally equal to the receiver frequency resolution, we need to control the non-linearities.

If $\Delta f_{tar} = 1,010$ Hz, i.e., it equals the receiver frequency resolution in the ideal case, then

$$\Delta f_b = \sqrt{1010^2 + 1010^2} = 1,428 \ \text{Hz}$$

Therefore, each 1.428 kHz frequency bin corresponds to

$$\Delta R_{deg} = SF \times 2.09 = 30 \times 1.428 = 42.84 \ \text{m}$$

In the event we use Hamming weighting to reduce the side lobes, the bin width increases by 1.81 (say, for a 6 dB resolution, for Hamming weighting), corresponding to 1,830 Hz. This means that

$$\Delta f_b = \sqrt{1010^2 + 1830^2} = 2.09 \ \text{KHz}$$

is the overall frequency resolution. This is approximately 2.1 times greater than the 30 m ideal range resolution for a 5 MHz frequency deviation that we started out with. This is the ideal case, i.e., when we neglect non-linearities in the sweep. If we take into account sinusoidal non-linearities, then the target beat frequency spectral width of 1.10 kHz, in this example, further broadens, decreasing the range resolution even further.

We started out with wanting 30 m range resolution due to a 5 MHz sweep bandwidth and we now end up with 42.84 m, neglecting Hamming weighting. Very obviously we need to evolve a strategy to avoid getting cheated like this! A suggested approach is discussed below.

The first thing that comes to mind is to make T_{mod} as high as possible by increasing the sweep time. This will straightaway take Δf_{rec} out of the reckoning. The side effect is that the instrumentation range (see equation 4.9) will be high. But that need not deter us. This leaves us to concentrate only on the one remaining parameter, viz reducing target spectral width (meaning bringing non-linearities to zero). If we succeed in doing this, then $\Delta f_{tar} = \Delta f_{rec}$ (do not forget that Δf_{rec} is already now a negligible value) and (if the sweep time is long enough) we can get the full benefit of the sweep bandwidth towards range resolution. Therefore, it now becomes a two-pronged approach, viz controlling the non-linearities and utilizing a large sweep time.

From the FMCW equation,

$$f_b = \frac{t_d \Delta f}{T_s} = \frac{2R\Delta f}{cT_s} \tag{4.50}$$

$$\text{or} \quad R = \frac{T_s c}{2\Delta f} f_b$$

$$\therefore \Delta R = \frac{T_s c}{2\Delta f} \Delta f_b, \text{ where } \Delta f_b \text{ is the beat frequency resolution} \tag{4.51}$$

Equation (4.51) only says that the range resolution is a direct function of the beat frequency resolution.

Now, from equation (4.50)

$$T_s = \frac{\Delta f \, 2R}{f_b c} \tag{4.52}$$

Therefore, substituting equation (4.52) in equation (4.51), we obtain

$$\Delta R = \frac{\Delta f \, 2R}{f_b c} \frac{c \Delta f_b}{2 \Delta f} = R \frac{\Delta f_b}{f_b} \tag{4.53}$$

We know that,

$$\Delta f_b = \sqrt{\Delta f_{tar}^2 + \Delta f_{rec}^2} \tag{4.54}$$

Since beat frequency resolution is limited by the convolution of the target spectral width with the receiver frequency resolution, substituting equation (4.54) in equation (4.53) we obtain

$$\Delta R_{conv} = \frac{R}{f_b} \sqrt{\Delta f_{tar}^2 + \Delta f_{rec}^2} \tag{4.55}$$

Equation (4.55) tells us that it is possible to control the deterioration of range resolution due to the convolution of target spectral width and the receiver frequency resolution by varying the beat frequency f_b. This means if we hike the beat frequency the range resolution will improve. However, f_b depends upon Δf as given by (4.38). We know this already and therefore it does not help us as we do not wish to increase the sweep bandwidth, but rather we wish to attain whatever resolution the present sweep bandwidth can give us. But equation (4.55) tells us something more. Given that f_b is *not* planned to be increased, we can still improve ΔR_{conv} be reducing Δf_{rec} and Δf_{tar} to the extent possible and by increasing f_b (see equation 4.55).

Our strategy will, therefore, be to *initially* increase the sampling frequency (with a view to increasing f_b as defined by $f_b = f_s/2$ where f_s is the sampling frequency) to the extent possible (based on the speed of the ADCs in the market) and then *subsequently* improve the range resolution by reducing the non-linearities in the sweep, as this is more difficult.

Finally, to summarize, it should be noted that the range resolution depends upon two factors:

1. Range resolution due to decrease in bandwidth caused due to the increase in t_d.
2. Range resolution due to the convolution of the target spectral width and the receiver frequency resolution.

This can be expressed mathematically as

$$\Delta R = \max\left\{\frac{R}{f_b}\sqrt{\Delta f_{tar}^2 + \Delta f_{rec}^2}, \quad \frac{c}{2\Delta f\left(1 - \dfrac{t_d}{T_s}\right)}\right\} \tag{4.56}$$

Figure 4–12 illustrates this.

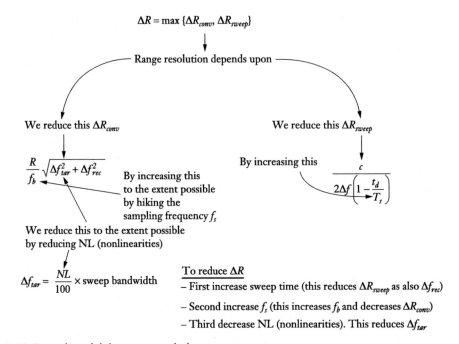

Figure 4–12 Towards explaining range resolution management.

This concept has been implemented in Appendix B and in the accompanying software. The question now is which of the two arms in Figure 4–12 is greater?

In order to answer this, let us take an example.

Example 2

Assume R_{max} = 3 kms, sampling frequency f_s = 1.5 MHz, sweep time T_s = 2 ms, sweep recovery time T_{sr} = 3.3 μsec, and sweep bandwidth Δf = 25 MHz. We also assume Hamming weighting.

$$t_d = \frac{2R}{c} = \frac{2 \times 3000}{3 \times 10^8} = 20 \ \mu\text{sec}$$

$$T_{mod} = T_s - t_d - T_{sr} = 2000 - 20 - 3.3 = 1976.7 \ \mu\text{sec}$$

$$\Delta f_{tar} = \frac{1}{T_{mod}} = \frac{1}{1976.7 \times 10^{-6}} = 500 \ \text{Hz}$$

This is for a point target, i.e., an ideal case with perfect linearity.

$$f_b = f_{max} = \frac{f_s}{2} = \frac{1.5 \times 10^6}{2} = 750 \ \text{KHz}$$

$$\Delta f_{rec} = \Delta f_{tar} = 500 \ \text{Hz}$$

Correcting for Hamming weighting

$$\Delta f_{rec} = 1.81 \times \Delta f_{rec} = 1.81 \times 500 = 905 \ \text{Hz}$$

In such a case, the first expression becomes

$$\frac{R}{f_b}\sqrt{\Delta f_{tar}^2 + \Delta f_{rec}^2} = \frac{3000}{750 \times 10^3}\sqrt{500^2 + 905^2} = 4.13 \approx 4 \ \text{m}$$

The second expression becomes

$$\frac{c}{2\Delta f\left(1 - \dfrac{t_d}{T_s}\right)} = \frac{3 \times 10^8}{2 \times 25 \times 10^6\left(1 - \dfrac{20 \times 10^{-6}}{2000 \times 10^{-6}}\right)} = 6 \ \text{m}$$

Both the above values are comparable. If anything, the first value is an ideal one, i.e., it does not take onto account non-linearities. If we do so, then it will easily equal or even exceed 6 m. This implies that loss of resolution due to round trip propagation delay is not all that critical as receiver frequency resolution. The loss of resolution due to round trip delay can always be brought under control, by having large sweep times in excess of 1 msec thereby ensuring that the modulation time T_s is at least five times the transit time for the maximum range so that the effective processed bandwidth is at least 80% of the total bandwidth. This will also keep the loss in effective processed transmit power less than 1 dB ([10], p. 304). The beat frequency resolution problem, however, needs to be brought under control by controlling the non-linearities (this is more difficult) and having a high sampling rate. However, we can also control the loss of resolution due to the round trip delay by utilizing *delayed sweeps*. This concept is hardware intensive and is discussed in Part III.

4.9 CONCEPT OF INSTRUMENTED RANGE

Normally, the radar range equation like (1.22) gives us the energetic range of the radar. This is the range one can achieve, given the transmitter power levels and the other radar parameters, as well as the type of target and the propagation conditions. However, in an effort to control receiver frequency resolution we have resorted to increasing the sampling rate to a large value. This causes the unambiguous range R_{max} to be high as is clearly seen from equation (4.35), due to the high number of samples/sweep, even exceeding the energetic range of the radar as given by the range equation. This makes the sweep time cover ranges from zero range to something even exceeding the energetic range. This range is called *radar instrumented range*. This is the range the radar is designed to cover, i.e., it is instrumented for. It can happen that a radar has an energetic range of, say, 2 km, but an instrumented range of 3 km. Ideally, the designer should strive to make the instrumented range as close as possible to the energetic range, as otherwise it is pointless, as the radar will not detect much beyond the energetic range. The radar receiver window, in other words, becomes excessive. This should be avoided to the extent possible. In Part III, we shall examine these trade-offs.

4.10 NON-LINEARITY IN FM WAVEFORMS

In order to illustrate the problem, we utilize equation (1.27) to determine the maximum range (also called energetic range) based on the following parameters.

The desired range resolution:

$$\Delta R = \frac{c}{2\Delta f} = \frac{3 \times 10^8}{2 \times 500 \times 10^6} = 0.3 \text{ m}$$

The radar parameters are:

$P_{CW} = 1 \text{ W},$

$T_s = 1 \text{ ms},$

$\Delta f = 500 \text{ MHz},$

$SNR_{output} = 10 \text{ dB for a } P_D = 0.25 \text{ and } P_{FA} = 10^{-6} \text{ for a Swerling 0 target}$

$\sigma_T = 2 \text{ m}^2$

$\lambda = 0.032 \text{ m}$

$G_t = 15 \text{ dB}$

$G_r = 15 \text{ dB}$

$SRF \text{ (sweep repetition frequency)} = 1 \text{ kHz}$

System losses $= 10 \text{ dB}$

Noise figure $= 3 \text{ dB}$

System noise temperature $T_0 = 400° \text{ K}$

Using equation (1.27), we achieve an energetic range of 993 m for a 2 m² target.

For the purposes of our calculation we shall assume that the energetic range is equal to the unambiguous range R_{max}.

The beat frequency for such a radar is given by

$$f_b = \frac{\Delta f \times \tau}{T_s}$$

where t is the round trip propagation time.

The round trip propagation time for this R_{max} is 6.62 μsec.

Hence,

$$f_b = f_{max} = \frac{500 \times 10^6 \times 6.62 \times 10^{-6}}{1000 \times 10^{-6}} = 3.3 \text{ MHz}$$

This yields a 3.3 MHz/993 m or 3323.3 Hz/m beat frequency to range ratio (scale factor, *SF*).

The ideal range resolution for such a radar is

$$\Delta R = \frac{c}{2\Delta f} = \frac{3 \times 10^8}{2 \times 500 \times 10^6} = 0.3 \text{ m}$$

Thus, a 0.3 m range resolution requires a 996.98 Hz (using scale factor) receiver frequency resolution and consequently the frequency sweep linearity is 0.0002% of the 500 MHz frequency deviation. This is, therefore, the linearity required to achieve our desired range resolution of 0.3 m. The reader can verify that a higher non-linearity of, say, 0.04% does not satisfy our requirement.

The logic behind this is as follows. Let us assume that the sweep recovery time is negligible. Then

$$T_{mod} = T_s - t_d = 1000 - 6.62 \approx 993 \text{ } \mu\text{sec}$$

This means that

$$\Delta f_{tar} = \frac{1}{T_{mod}} = \frac{1}{993 \times 10^{-6}} = 1007 \text{ Hz}$$

This is for a point target, i.e., an ideal case with perfect linearity.

If the frequency sweep linearity exceeds 1.007 kHz (in our example), then the target spectral width increases beyond that of an ideal point target given by $(1/T_{mod})$ that we calculated above. This will lead to loss of beat frequency resolution due to the excessive widening of the target spectral width. The better method will be to control the non-linearities. Piper [11] has shown that the non-linearities can be reduced by time gating the minimum and maximum points of the frequency sweep, since the greatest non-linearities occur in these regions. For example, blocking 5% at each end of the sweep period reduces the effective signal power by less than 0.5 dB. But this may yield a 2.1 dB decrease in the standard deviation of the non-linearity. This makes a case for employing a digital FMCW generator, like a DDS discussed in Part III of this book. Such generators will not have discontinuities at the ends of the sweep, as the ends of the sweep are gated.

Modern closed-loop compensation techniques are achieving good frequency sweep linearity [12]. Many linearizer designs divide the frequency tuning curve for the RF source into segments and provide compensation for each segment. After this compensation, the residual non-linearity

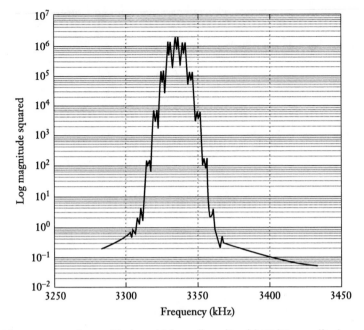

Figure 4–13 Beat frequency spectrum with sinusoidal non-linearity with 30 kHz amplitude and 5 kHz frequency. (From [9], © IEEE 1995)

error has a periodicity equal to the frequency sweep period divided by the number of linearizer segments [9]. Thus, the linearizer will reduce the peak non-linearity error and increase the frequency of the error. The spectral width of the target return depends upon both the bandwidth of the frequency sweep non-linearity and the correlation time of the non-linearity relative to the target transit time [3]. The target return spectral width will be less for a given non-linearity bandwidth when the correlation time is long relative to the transmit time. The frequency of the non-linearity determines where the frequency side lobes appear and the amplitude of the non-linearity determines the side lobe amplitude. Thus, a linearizing compensator should decrease the non-linearity amplitude as the non-linearity frequency is increased to avoid frequency and range side lobes of the target return [9]. Figure 4–13 shows the beat frequency spectrum for 30 kHz non-linearity in amplitude and 5 kHz non-linearity spacing in frequency. The first side lobes are comparable with the main lobe. Non-linearity amplitude of 30 kHz is only 0.006% of the 500 MHz frequency deviation and 5 kHz non-linearity frequency corresponds to five times the modulation frequency (1 kHz). Figure 4–14 shows the spectrum with 10 kHz non-linearity amplitude and 10 kHz non-linearity frequency. A 10-segment linearizer and 1 kHz modulation frequency may have residual error with 10 kHz non-linearity frequency. The side lobes have moved out to ±10 kHz corresponding to ±3 m in range. The side lobe amplitude dropped to approximately 15 dB below the peak. Decreasing the peak non-linearity reduces the side lobe levels and increasing non-linearity frequency increases the side lobe separation [9].

4.10.1 Coherent Processing Interval

We need to determine the size of the range FFT. The samples processed should be matched to the modulation period, with the condition that the number of samples be a power of two for

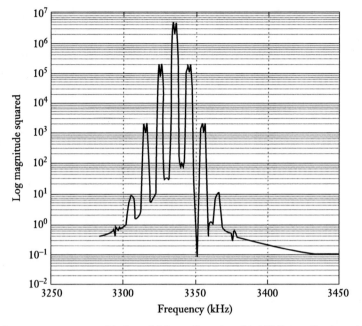

Figure 4–14 Beat frequency spectrum with sinusoidal non-linearity with 10 kHz amplitude and 10 kHz frequency. (From [9], © IEEE 1995)

convenience. This yields

$$T_s = \frac{N}{f_s} = \frac{2^n}{f_s} \tag{4.57}$$

where $N = 2^n$ is the number of samples, so that

$$f_s = \frac{2^n}{T_s} \tag{4.58}$$

Using the FMCW equation (4.38), we can express range in terms of beat frequency by rearranging the terms.

$$R = \frac{T_s c}{2\Delta f} f_b \tag{4.59}$$

The maximum beat frequency should be

$$f_{max} \leq \frac{f_s}{2} \tag{4.60}$$

Substituting for f_{max} from equation (4.59) we obtain

$$2\frac{\Delta f}{T_s} \frac{2R_{max}}{c} \leq f_s \tag{4.61}$$

Using results of example in Section 4.6 above, $f_{max} = 3.3\,\text{MHz}$. Hence, f_s must be at least 6.7 MHz.

We now substitute equation (4.61) into equation (4.58) to obtain

$$2\frac{\Delta f}{T_s}\frac{2R_{max}}{c} \le \frac{2^n}{T_s} \qquad (4.62)$$

so that

$$\frac{4\Delta f R_{max}}{c} \le 2^n \qquad (4.63)$$

Once again, using the example in Section 4.6, the left side of equation (4.63) is 6,620, so the right side must be 8,192 or 2^{13}. A total of 8,192 samples in 1 msec corresponds to 8.192 MHz sample frequency. The 8,192-point FFT will cover beat frequencies up to 4.096 MHz, corresponding to ranges up to 1,232 m. The beat frequency sample spacing will be 1 kHz, corresponding to range spacing of 0.3 m. The frequency resolution with Hamming window equals the 6 dB bandwidth of 1.81 times the 1 kHz frequency sample spacing or 1.81 kHz and that corresponds to 0.545 m in range.

In this example, the energetic range is 993 m, while the instrumented range is 1,232 m. If we increase the sampling frequency, the receiver frequency resolution will improve, but the range resolution will still be limited by the target spectral width, which again depends upon the non-linearities. Extending the FFT input sequence with zero padding is preferred [8]. Finally, if we increase the sampling, the instrumented range will also increase, unnecessarily.

In surveillance applications, the radar would need to process most of the receiver window. But for an application with a limited slant range swath, only a portion of the FFT output is needed. Thus, the beat frequency spectrum of interest can be offset to lower frequency and low pass filtered so that an FFT with fewer points can cover the frequency band and the corresponding range swath of interest. For example, a 128-point FFT would cover a 38.4 m range swath [8].

We now examine curves obtained from [8] in order to consolidate our ideas. This was based on the simulation done by Piper for a radar designed on the same lines as discussed by us earlier in the preceding paragraphs.

The radar parameters are [8]:

Table 4–1 Radar Parameters

Radar Parameter	Value
RF center frequency	35 GHz
RF wavelength	8.6 mm
FMCW waveform	Sawtooth
Frequency deviation Δf	500 MHz
Ideal range resolution ΔR	0.3 m
Modulation frequency SRF	1 kHz
Modulation period T_s	1 msec
Sweep recovery time T_{sr}	3.3 μsec
Beat frequency/range ratio (scale factor)	3.3 kHz/m
Maximum range R_{max}	1 km

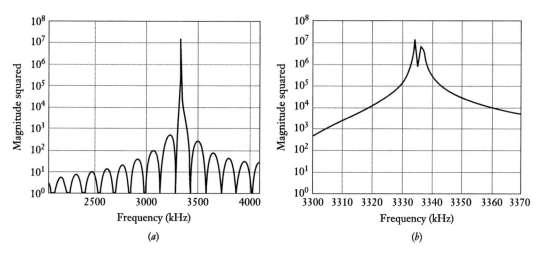

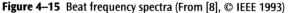

Figure 4–15 Beat frequency spectra (From [8], © IEEE 1993)

Maximum transit time τ	6.7 μsec
Maximum beat frequency f_{max}	3.3 MHz
Minimum beat frequency interval	993 μsec
Minimum beat frequency spectral width	1.007 kHz
Range resolution limit	0.302 m
ADC sample rate	8.192 MHz
FFT length	8,192 points
FFT frequency sample spacing	1 kHz
FFT range sample spacing	0.3 m
Window	Hamming
Window range resolution	0.543 m

Figure 4–15(a) shows the FFT output from 2.048 MHz to 4.096 MHz for one target at 1,000 m range and another at 0.6 m greater range with equal power. Note the peak at spectral sample 3,333 corresponding to 3.3 MHz. The two targets are not resolvable in this plot. Since the input sequence is real, the output sequence is symmetric about sample 4,096.

Figure 4–15(b) shows, for the same targets as in Figure 4–7(a), the expanded beat frequency spectrum 3.30 to 3.37 MHz, corresponding to ranges from 990 to 1,011 m.

Figure 4–16(a) shows the beat frequency spectrum for the same two target ranges as in Figure 4–15(a) and 4–15(b) with Hamming windows. Note that the frequency side lobes are much lower. The two targets are barely resolvable because of the convolution of the 0.302 m target range extent and 0.543 m receiver range resolution.

Figure 4–16(c) shows that the targets are more resolvable, because the separation was increased to 0.602 m. This additional 2 mm range separation corresponds to approximately one-quarter of the RF wavelength, illustrating the sensitivity of the combined spectrum to the interference between spectral components of each target.

Figure 4–16(b) shows the spectrum for a 16,384-point FFT. Here, 8,192 zeros were appended to the 8,192 input samples. The spacing between spectral samples in this spectrum is 0.5 kHz, so that the

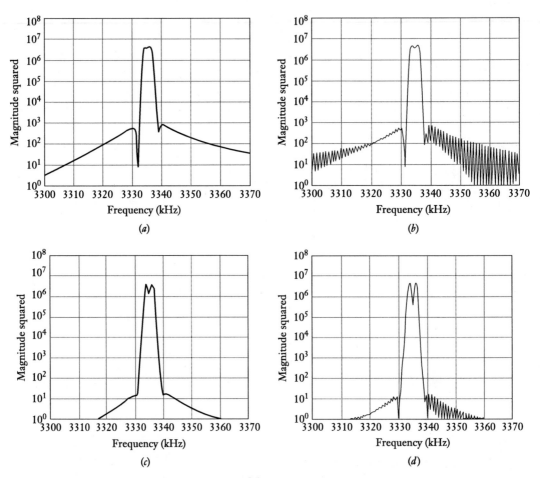

Figure 4–16 Beat frequency spectra (contd.) (From [8], © IEEE 1993)

receiver range resolution is 0.272 m. Here, the two target peaks are clearly resolvable. Figure 4–16(*d*) shows the beat frequency spectrum with zero padding as in Figure 4–16(*b*) and with 0.602 m range separation as in Figure 4–16(*c*).

4.11 IT ALL COMES TOGETHER. APPLICATION: BRIMSTONE ANTITANK MISSILE

The Brimstone Missile is one of the guided missiles developed for the Longbow Apache AH-64D attack helicopter [13] (see Figure 4–17).

4.11.1 System Specifications

- Length: 1.8 m, diameter: 178 mm
- Mass: 50 kg
- Operation: 24 h, day/night, all weather
- Mode: Totally autonomous, fire-and-forget, lock-on after launch (LOAL)
- Resistant to camouflage, smoke, flares, chaff, decoys, jamming
- Operational range: 8 km

Figure 4–17 The Brimstone FMCW seeker (From [13], © Reprinted with permission)

Figure 4–18 Processor for an FMCW seeker and a monkey (From [13], © Reprinted with permission)

- Designation: Accepts any or no target information
- Motor: Boost/coast, burns for 2.75 sec with a thrust of 7.5 kN
- Guidance: Digital autopilot, 2 gyros (25°/h drift), 3 accelerometers

4.11.2 Seeker Specifications (Known)

- 94 GHz active radar
- Low power, narrow beam
- Dual polar, dual look
- Fast 96002 processor
- Detection/classification software (see Figure 4–18)

4.11.3 Operational Procedure

The operational procedure for lock-on after launch is as follows (see Figure 4–19):

- Rough target designations including, range bearings and rates downloaded to missile
- Missile fired in general direction of target
- Updates designation from initial positions and rates

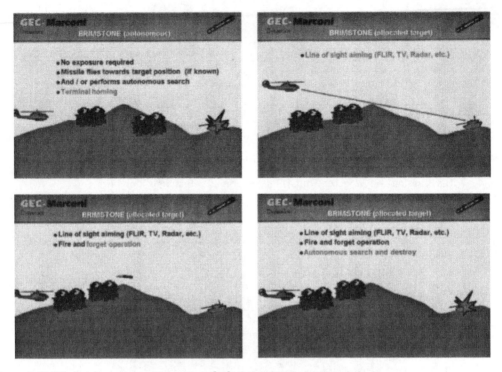

Figure 4–19 Missile engagement options (From [13], © Reprinted with permission)

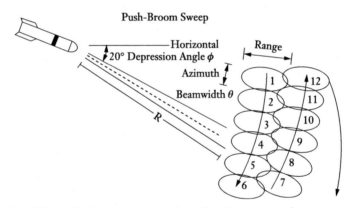

As the vehicle travels along its trajectory, the on-board sensor scans for its target from side-to-side in a "push-broom search."

Figure 4–20 Push broom search (From [13], © Reprinted with permission)

- Flies up to 7 km toward target using INS guidance only
- In the last 1 km it activates the radar seeker and searches for target
- Search footprint scans search box in 200 msec
- Acquisition algorithms map all targets in box (exclude trucks)
- Track-while-scan enables optimum decision on target priority
- Algorithm selects MBT (main battle tank)
- Moving armor given the highest priority (see Figure 4–20)

4.11.4 System Performance (Speculated)

This missile is classified. Hence, the radar community outside this project can only speculate as to how this missile might actually work. Therefore, the reader is cautioned against concluding that the missile actually works as indicated in this section. The idea here is to convey to the reader that based on available radar information and technology, how the various aspects in the missile design come together. Hence, this is purely speculative. This analysis is based on the work done by Brooker [13].

Target Detection and Identification: Target identification is based on a combination of the high range-resolution and polarization characteristics of the radar echo. The system transmits horizontal polarization (H) and receives vertical (V) and horizontal (H) returns and the range gate size is matched to the radar bandwidth for high resolution ≈ 0.5 m. This puts between 6 and 10 range cells on a typical MBT (3 m $\times$ 5 m).

Doppler processing is used to distinguish moving targets.

Radar Front End: To make the radar **LPI**, the transmit power will be low and spread spectrum. This almost certainly implies FMCW operation.

FMCW operation through a single antenna generally limits the transmit power to less than 50 mW. However, with good matching and active leakage compensation, transmit powers can be as high as 1 W. We believe that the Brimstone transmit power $P_{tx} \approx (100 \text{ mW})20 \text{ dBm}$ as it includes an injection-locked amplifier stage.

Transmitter swept bandwidth $\Delta f = 300$ MHz to meet the 0.5 m range resolution requirement,

$$\Delta R_{chirp} = \frac{c}{2\Delta f} = \frac{3 \times 10^8}{2 \times 300 \times 10^6} = 0.5 \text{ m}$$

To allow for Doppler processing a triangular waveform may be used as shown below. For an operational range of 1 km with a 0.5 m bin size, 2,000 gates are required. It is speculated that a 4,096-point FFT will produce 2,048 bins for both the co- and cross-polar receive channels.

Because the time available to perform a search is limited, the data rate will be as high as possible. However, there is a limit to the speed that the loop linearization and the ADC can operate. We will assume a total sweep time of 1 msec (500 μsec for each the up and down sweeps) (see Figure 4–21).

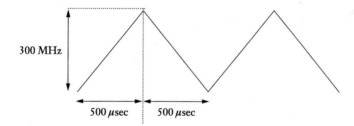

300 MHz

500 μsec   500 μsec

Figure 4–21 Doppler processing a triangular waveform.

The beat frequency for an FMCW radar is given by the following equation

$$f_b = \frac{\Delta f}{\Delta t} T_r = \frac{\Delta f}{\Delta t} \frac{2R}{c} = \frac{300 \times 10^6 \times 2 \times 1000}{500 \times 10^{-6} \times 3 \times 10^8} = 4 \text{ MHz}$$

Using the Nyquist criterion, the minimum sample rate required to digitize a signal with a 4 MHz bandwidth is 8 MHz. Because of limited filter skirt selectivity anti-aliasing filter characteristics, the sample rate is generally 2.5$\times$ making the sample rate 10 MHz.

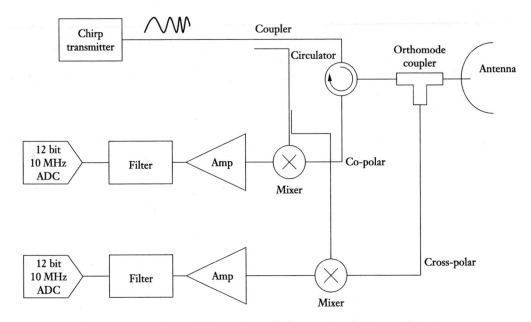

Figure 4–22 Brimstone seeker schematic diagram (From [13], © Reprinted with permission)

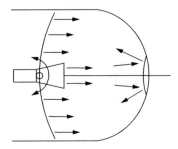

Figure 4–23 Brimstone antenna schematic (From [13], © Reprinted with permission)

To ensure sufficient dynamic range, an ADC with at least 12 bits of resolution is required.

A total of 5,000 samples can be taken over each of the up and the down sweep. This is just about perfect for the 4,096-point FFT because the sweep linearity is generally not good at the start and the end (see Figure 4–22).

Antenna and Scanner: For a missile diameter of 178 mm, the antenna cannot be much more than 160 mm in diameter (see Figure 4–23).

For $\lambda = 3.2$ mm at 94 GHz, the 3 dB beamwidth will be

$$\theta_{3\,dB} = \frac{70\lambda}{D} = \frac{70 \times 3.2}{160} = 1.4°$$

The antenna uses an interesting Cassegrain configuration [4] with a scanned parabolic mirror as shown in Figure 4–27.

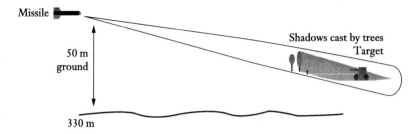

Figure 4–24 Shadow effects due to low grazing angle (From [13], © Reprinted with permission)

Table 4–2 Relationship Between Radar Height and Beam Footprint Length

Height (m)	Angle 1 (deg)	Angle 2 (deg)	×2 (m)	Footprint (m)
10	0.57	1.97	290.29	709.71
20	1.15	2.55	449.83	550.17
30	1.72	3.12	550.67	449.33
40	2.29	3.69	620.13	379.87
50	2.86	4.26	670.87	329.13
100	5.71	7.11	801.64	198.36

From [13], © Reprinted with permission.

The gain of the pencil beam antenna will be approximately (assuming antenna efficiency η of 0.6).

$$G = \frac{4\pi\eta A}{\lambda^2} = \frac{4\pi \times 0.6 \times \pi \times 0.08^2}{0.00319^2} = 14897 \ \left(41.7 \ \text{dB}\right)$$

The critical aspect is the sub-reflector beam shaping that allows a limited scan using the parabolic prime reflector without generating large side lobes.

At a range of 1 km, the width of the footprint will be 24.5 m and the length of the footprint will be a function of the operational height at an operational range of 1 km (see Table 4–2).

To limit the amount of potential shadowing of the target area due to trees and undulating terrain, while maintaining a reasonable size footprint on the ground, an operational height of 50 m would be reasonable. This results in a footprint length of approximately 330 m.

It can be assumed that a single mechanical scan takes place in the 200 msec search time (see Figure 4–24).

Because the missile is coasting, it will have limited lateral acceleration capability, and so it is pointless searching beyond the boundaries that the missile can reach.

It is reasonable to assume that a square search area of 330 × 330 m will be covered. At a range of 1,000 m, this equates to an angular scan of about 18° if the antenna beamwidth is considered. To scan 18° in 200 msec requires an angular rate of 90°/s

Signal Processing: The time-on-target for a beamwidth of 1.4° and an angular rate of 90°/s is 15.5 msec. For a total sweep time of 1 msec, a total of nearly 16 hits per scan occur.

This allows for 16 pulse integration to improve the SNR if it is required, it also gives the processor more information to identify the target type.

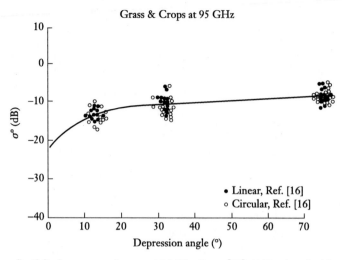

Figure 4–25 Clutter reflectivity for grass and crops at 94 GHz (From [13], © Reprinted with permission)

Each target can be identified using the following information:

- 5–10 gates that span it in range
- 16 time slices
- Two orthogonal polarizations

This may be sufficient information to discriminate between a truck and an MBT.

Signal to Clutter Ratio: Clutter Levels. Single look signal to clutter ratio (SCR) is determined from the target RCS, the clutter reflectivity σ^0 and the area of a range gate.

The following graphs show measured clutter reflectivity data at 94 GHz for grass and crops (see Figure 4–25).

At a grazing (depression) angle of between 3° and 4°, the mean reflectivity of grass will be about −20 dBm²/m² (reduces to dB).

The clutter cross-section is the product of the clutter reflectivity σ^0 and the area of the gate footprint $\tau R \theta_{3\,dB}$ on the ground for flat terrain (the beamwidth must be in radians) [13].

$$\sigma_{clut} = \sigma^0 \tau R \theta_{3\,dB} = -20 + 10 \log_{10}\left(0.5 \times 1000 \times 1.4 \times \frac{\pi}{180}\right) = -9 \text{ dBm}^2$$

Because tank commanders are aware that they are vulnerable when out in the open, they tend to make use of the available local cover, and will position themselves on the borders of lines of trees (see Figure 4–26).

The reflectivity of lines of trees observed broadside is much higher than that of the canopy, as shown in the following image which shows rows of pine trees between orchards, and a double line of eucalyptus straddling a railway line (see Figure 4–27).

Measurements indicate that the mean reflectivity of deciduous trees is typically −10 dBm²/m². The clutter RCS in this case is product of the area of trees illuminated by the radar and the reflectivity. Because of the narrow gate, there will be areas where the tree reflectivity is very strong and areas where it is very low. There will also be areas where a tank is sticking out from under a tree, in which case the clutter level is determined by the ground clutter only (see Figure 4–28).

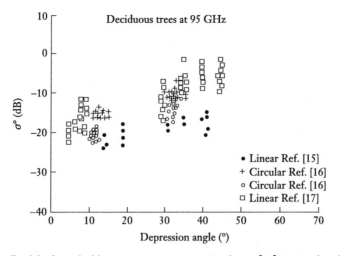

Figure 4–26 Clutter reflectivity for a deciduous tree canopy at 94 GHz (From [13], © Reprinted with permission)

Figure 4–27 94 GHz radar image of trees and scrub (From [13], © Reprinted with permission)

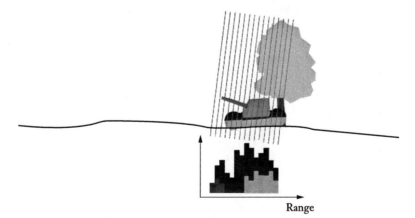
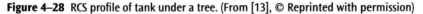

Figure 4–28 RCS profile of tank under a tree. (From [13], © Reprinted with permission)

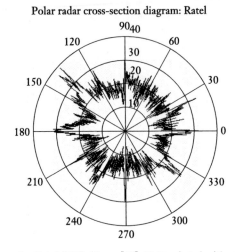

Polar radar cross-section diagram: Ratel

Figure 4–29 Radar cross-section of a Ratel (APC). (From [13], © Reprinted with permission)

If in a 4-m high hedge of trees the width of the range gate is illuminated, then the RCS will be calculated as

$$\sigma_{dut} = \sigma^0 hR\theta_{3\,dB} = -10 + 10\log_{10}\left(4 \times 1000 \times 1.4 \times \frac{\pi}{180}\right) = +10 \text{ dBm}^2$$

In general, however, a much smaller section of the tree will be illuminated, within a single gate. For a tree of 4-m tall and 3-m wide, roughly elliptical in shape, a maximum area of 8 m² will be illuminated

$$\sigma_{dut} = \sigma^0 A = -10 + 10\log_{10}(8) = -1 \text{ dBm}^2$$

Target Levels: The RCS of a tank depends on the observation angle as shown in Figure 4–29.
The maximum RCS can reach 40 dBm² and the minimum seldom falls below 10 dBm². Hence, to ensure that the vehicle is always detected irrespective of the angle the 10 dBm² threshold must be selected.
Signal to Clutter Ratio: In open ground, the SCR is then

$$SCR = \sigma_{tan} - \sigma_{dut} = 10 - (-10) = 20 \text{ dB}$$

For the tank under a tree, the worst case will be

$$SCR = \sigma_{tan} - \sigma_{dut} = 10 - 10 = 0 \text{ dB}$$

Typical SCR will be more reasonable

$$SCR = \sigma_{tan} - \sigma_{dut} = 10 - 1 = 9 \text{ dB}$$

Without resorting to the statistics of the variation in tank RCS and that of trees, it can be seen that if the range bin is sufficiently narrow, parts of the tank will be visible if it is parked on the border of a row of trees.

When the radar is looking for a moving target, the clutter signals (because they are static) are suppressed.

Signal to Noise Ratio: The SNR is determined using the characteristics of the radar and the target as they are related in the radar range equation. The total noise at the output of the receiver N can be considered to be equal to the noise power output from an ideal receiver multiplied by a factor called the noise figure, NF ($NF_{dB} \approx 1.5$ dB for an FMCW radar). It was discussed in Chapter 1 on the Calypso radar that noise figures are typically 3 dB for FMCW radars. However, in the case of this missile, it is 15 dB because of the absence of an LNA at frequencies of 94 GHz. Furthermore, in a missile, the noise floor increases due to leakage from the transmitter to the receiver as the missile uses a common aperture both for transmission as well as reception. These aspects will be investigated elsewhere in this book.

In this case β is the bandwidth of a single bin output by the FFT and widened by the window function 1.3×5 MHz/2,048 ≈ 3 kHz

$$N_{dB} = 10 \log_{10} P_N NF = 10 \log_{10} kT_{sys}\beta + NF_{dB} = -154 \text{ dBW}$$

Because the transmitter power is in mW, this value is generally converted from dBW to dBm by adding 30 dB.

$$N_{dB} = -154 + 30 = -124 \text{ dBm}$$

Writing the range equation for a monostatic radar system in dB

$$10 \log_{10}(P_r) = 10 \log_{10}(P_t) + 10 \log_{10}\left(\frac{\lambda^2}{(4\pi)^3}\right) + 20 \log_{10}(G) + 10 \log_{10}(\sigma)$$
$$- 10 \log_{10}(L) - 40 \log_{10}(R) - 2\alpha R_{km}$$

This is best tackled in MATLAB$^{\circledR}$ as the attenuation α is a function of the weather conditions. Figure 4–30 shows the missile performance in adverse weather.

The SNR is sufficient for detection up to a rain rate of about 10 mm/h.

Target Identification: Doppler Processing. The bandwidth of each bin output by the FFT is about 3 kHz. This is equivalent to a Doppler velocity of

$$v_r = \frac{f_d \lambda}{2} = \frac{3 \times 10^3 \times 0.00319}{2} = 4.8 \text{ m/s}$$

Because a Doppler shift causes an upward shift for half the sweep and a downward shift for the other, the range profiles generated by the up and down sweeps will diverge. For a target with a radial velocity of 4.8 m/s, this will be two bins, and will increase to six bins at a speed of 50 km/h, which is reasonable for a tank on the move.

A simple form of moving target discrimination is obtained by taking the difference between the up-sweep and the down-sweep range profiles. Static targets will cancel if the correct shift to compensate for the missile velocity is applied, but moving targets will appear as two large peaks as shown in Figure 4–31.

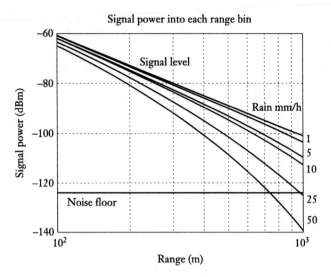

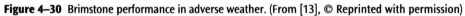

Figure 4–30 Brimstone performance in adverse weather. (From [13], © Reprinted with permission)

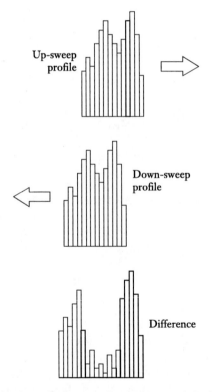

Figure 4–31 Moving target detection (From [13], © Reprinted with permission)

Target Identification: Other Techniques. Different target types are identified by the differences in their co- and cross-polar signatures. Targets with lots of corners and attachments tend to reflect signals after more than one bounce, and that rotates the polarization. Because there are lots of scatterers each rotating the polarization by a different amount, the overall return will have a random polarization i.e. uniformly spread. The signal is said to be depolarized.

Smooth targets reflect with a single bounce, so the polarization is not rotated (see Figure 4–32).

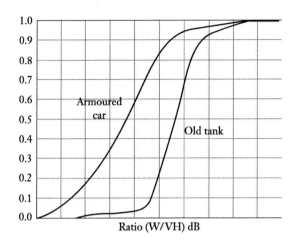

Ratio (W/VH) dB

Figure 4–32 Polarization ratio used to identify vehicles. (From [13], © Reprinted with permission)

The Results

Figure 4–33 Antitank missile scoring a direct hit. (From [13], © Reprinted with permission)

4.12 SUMMARY

In this chapter, we have studied LFM waveform compression using correlation and stretch techniques. We then studied the basic FMCW radar theory and equations. We have also studied the effect of target Doppler on radar performance and how to measure it. We have then investigated the factors affecting range resolution like sweep times and beat frequency resolution. In particular, we have investigated problems like target return spectral width and receiver frequency resolution, which play such a key role in determining the beat frequency resolution which lead to our final receiver range bin resolution. Finally, through worked examples we investigated the problems pertaining to non-linearities and its control. In this process, we have investigated the trade-offs one needs to make between controlling the level (percentage) of non-linearities and the receiver frequency resolution leading to the final beat frequency resolution. This needs to be done without an excessive instrumented range in the radar. We shall use all this knowledge in Part III when we design the Pandora radar. We concluded this chapter by studying an interesting example of what can be achieved with FMCW technology in the area of antitank missiles.

References

1. Nathanson, F. E., *Radar Design Principles*, 2nd edn., McGraw-Hill, New York, 1991.
2. Mahafza, B. R. and Elsherbeni, A. Z., *MATLAB Simulations for Radar Systems Design*, Chapman & Hall/ CRC Press, Boca Raton, FL, 2004.
3. Skolnik, M. I., *Introduction to Radar Systems*, 3rd edn., McGraw-Hill, Boston, MA, 2001.
4. Caputi, W. J., Jr., "Stretch: A time-transformation technique," *IEEE Transactions on AES* -7, No. 2, March 1971.
5. Jankiraman, M., *Pandora Multifrequency Radar—Project Report*, IRCTR-S-014–99, Delft, The Netherlands, April 1999.
6. Pace, P. E., *Detecting and Classifying Low Probability of Intercept Radar*, Artech House, Norwood, MA, 2004.
7. Barrett, M., Beasley, P. D. L., and Stove, A. G., *An Advanced FMCW Radar*, PRL Tech. Report 3344, 1986.
8. Piper, S. O., "Receiver frequency resolution for range resolution in Homodyne FMCW radar," in *IEEE Radar Conference Proceedings*, 1993.
9. Piper, S. O., "Homodyne FMCW radar range resolution effects with sinusoidal non-linearities in the frequency sweep," in *IEEE Radar Conference Proceedings*, 1995.
10. Scheer, J. A., et al., *Coherent Radar Performance Estimation*, Artech House, Norwood, MA, 1993.
11. Piper, S. O., "FMCW radar linearizer bandwidth requirements," in *IEEE Radar Conference Proceedings*, 1991.
12. Stove, A.G. "Linear FMCW radar techniques," *IEE Proceedings-F*, Vol. 139, No. 5, October 1992.
13. http://www.acfr.usyd.edu.au/teaching/4th-year/mech4721-Signals/material/lecture notes/High RangeResolutionTechniques.pdf

5

Phase-Coded Waveform

5.1 INTRODUCTION

In the last chapter, we had examined the performance and design aspects of frequency modulated CW signals, the so-called FMCW signal. This signal has found wide application in LPI radars. In this chapter, we shall, investigate the design and performance of phase-coded signals that have also found popularity in LPI radar waveform compression design. In the class of *phase-coded signals*, there is a wide variety. This class of signals transmits at one frequency, but changes the phase as it transmits in a predetermined order. This is the exact reverse concept as compared to frequency-coded signals that will be studied in Chapter 6, wherein the phase is kept constant and the frequency is varied at each discrete step. Usually, this class comprises codes like Barker, Frank, complementary pairs or P1, P2, P3, and P4 codes. Barker codes fall into a class called *binary phase codes*, in that they can take only one of two phase values, zero, or one. Frank code and codes like P1 through P4 fall into a class called *polyphase codes*, in that they can take values other than one. There are also a class of codes called *polytime codes* T1, T2, T3, and T4 [1]. These code sequences use fixed phase states with varying time periods at each phase state. Due to lack of space we cannot examine polytime codes. The reader is referred to [1] for information on polytime codes and the references listed therein. We shall examine for each of these codes, the phase characteristics as well as the generated spectrum. The reader is then introduced to the concept of periodic ambiguity function (PAF) and periodic auto correlation function (PACF) that are peculiar to CW signals.

5.2 PHASE-CODED RADAR SIGNALS

5.2.1 Barker Coding

Barker codes belong to a class of binary phase codes [1–5]. We consider such a code, wherein a long signal of duration T is divided into N smaller signals, called segments, each of width $\Delta \tau = T/N$. The phase of each segment is then randomly chosen as either 0 or π radians relative to some CW reference signal. Usually, we characterize a segment having 0 phase as "1" or "+" and π phase as "0" or "−". The compression ratio associated with this class of codes is given by $\varepsilon = T/\Delta \tau$ and the bandwidth of such codes is defined by the segment bandwidth. The peak value of the compressed signal is N times larger than that of the long signal. The degree of compressibility of such codes depends upon the random sequence of the phases of the individual segments.

One such class of binary phase codes, which produce compressed waveforms with constant side lobe levels equal to unity is the Barker code. Figure 5–1 illustrates a Barker code of length 7. A Barker code of length n is depicted as B_n. There are only seven known Barker codes that share this unique property. These are listed in Table 5–1. An important point to note here is that, since there are only seven such codes, there is no radar security.

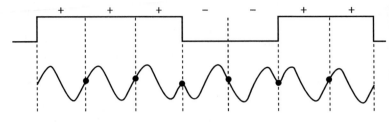

Figure 5–1 Binary phase code of length 7.

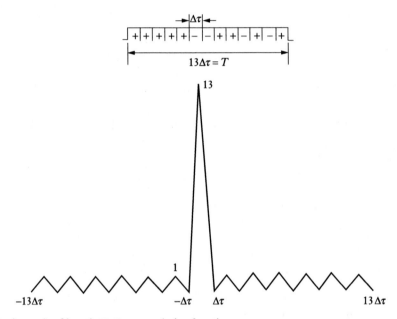

Figure 5–2 Barker code of length 13: Autocorrelation function.

Table 5–1 Barker Codes

Code Symbol	Code Length	Code Element	Side Lobe (dB)
B_2	2	+ −/+ +	6.0
B_3	3	+ + −	9.5
B_4	4	+ + − +/+ + + −	12.0
B_5	5	+ + + − +	14.0
B_7	7	+ + + − − + −	16.9
B_{11}	11	+ + + − − − + − − + −	20.8
B_{13}	13	+ + + + + − − + + − + − +	22.3

    The autocorrelation function for this type of code is $2N\Delta\tau$ wide ($\Delta\tau$ is the segment width) for a code of length N. The main lobe is $2\Delta\tau$ wide; the peak value is equal to N. There are $(N-1)/2$ side lobes on either side of the main lobe. Figure 5–2 shows the autocorrelation function for a Barker code of length 13. We note that the main lobe is equal to 13 and all the side lobes are unified.

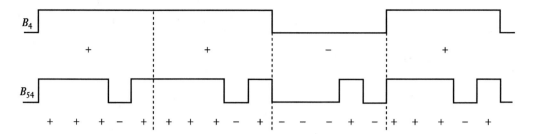

Figure 5–3 A B_{54} Barker code. (From [3], © Reproduced with permission)

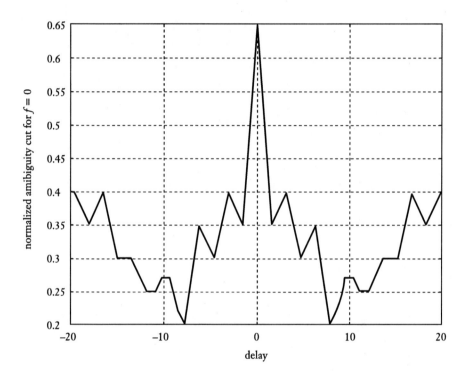

Figure 5–4 Zero Doppler cut for the B_{54} Barker code ambiguity function.

The most side lobe reduction offered by this type of code is -22.3 dB, which is not enough for radar applications. However, we can combine Barker codes to generate much longer codes. For example, a B_m code can be used within a B_n code to generate a code of length mn. The compression ratio for the combined B_{mn} code is equal to mn. For example, a B_{54} code is given by [3]

$$B_{54} = \{11101, 11101, 00010, 11101\} \tag{5.1}$$

This is illustrated in Figure 5–3.

The side lobes of such a waveform are no longer equal to unity. Figure 5–4 shows the zero Doppler cut for the B_{54} Barker code ambiguity function.

The ambiguity function for a Barker code of length 7 is shown in Figure 5–5.

The ambiguity function of a B_7 Barker code shows that at higher Doppler shifts, the side lobe level deteriorates rapidly. In fact, *phase-coded signals exhibit poor Doppler resilience* as compared to

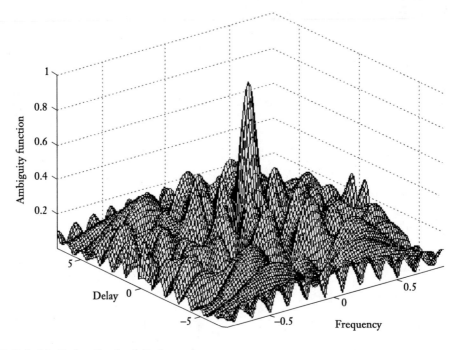

Figure 5–5 Ambiguity function for B_7 Barker code.

LFM signals, because with rising target Doppler, they decorrelate rapidly. Hence, for the radar designer, this becomes a design problem. These aspects will be discussed further in Part III. The figures for Barker code have been generated using the software "*barker.m*" accompanying this book (see Figures 5–6 and 5–7).

5.2.2 Frank Codes

A Frank code [6] has a length of $N = M^2$, where M is an integer [2]. This is like a set of M sequences of length M. The phase of the qth element in the pth sequence is given by

$$\varphi_{p,q} = \frac{2\pi k}{M}(p-1)(q-1), \quad p=1,2,...,M, \quad q=1,2,...,M \tag{5.2}$$

where k is a prime integer to M and is usually 1. When listed one underneath the other, they form an $M \times M$ matrix of phases whose elements are defined by equation (5.2). This is shown at equation (5.3), where the numbers represent multiplying coefficients of the basic phase angle $2\pi/M$.

The peak side lobe level (PSL) of the Frank code is $PSL = 20\log_{10}(1/(M\pi))$ [2, 6]. For $M = 4$ and $N = 16$, the $PSL = -22$ dB.

$$\begin{bmatrix} 0 & 0 & 0 & \cdots & 0 \\ 0 & 1 & 2 & \cdots & (M-1) \\ 0 & 2 & 4 & \cdots & 2(M-1) \\ \vdots & \vdots & \vdots & & \vdots \\ 0 & (M-1) & 2(M-1) & \cdots & (M-1)^2 \end{bmatrix} \tag{5.3}$$

The phase matrix for a Frank code of length 16 is shown in Table 5–2.

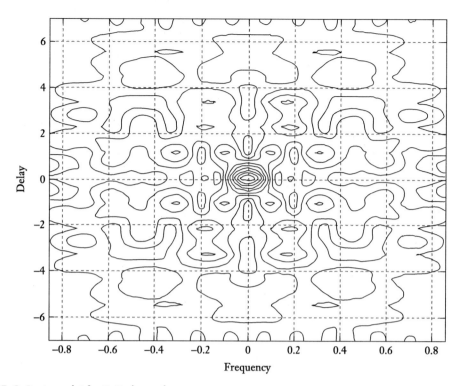

Figure 5–6 Contour plot for B_7 Barker code.

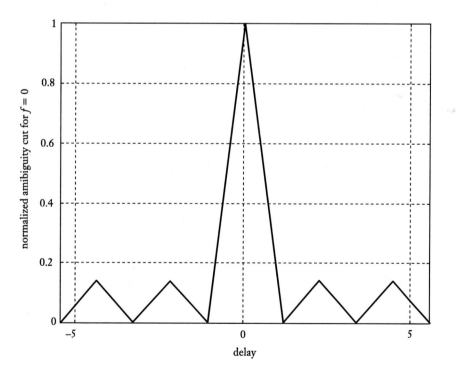

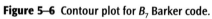

Figure 5–7 Zero Doppler cut for the B_7 ambiguity function.

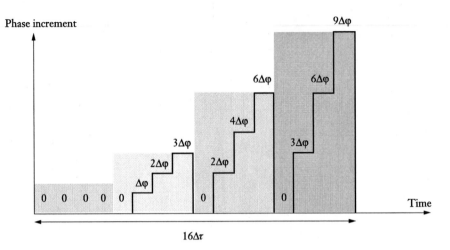

Figure 5–8 Phase increments for a Frank code of length 16.

Table 5–2 Phase Matrix for a Frank Code of Length 16

$$\begin{bmatrix} 0 & 0 & 0 & 0 \\ 0 & 90° & 180° & 270° \\ 0 & 180° & 0 & 180° \\ 0 & 270° & 180° & 90° \end{bmatrix} \leftrightarrow \begin{bmatrix} 1 & 1 & 1 & 1 \\ 1 & j & -1 & -j \\ 1 & -1 & 1 & -1 \\ 1 & -j & -1 & j \end{bmatrix}$$

$$F_{16} = \{1,1,1,1,1, j,-1,-j,1,-1,1,-1,1,-j,-1, j\}$$

The phase increments are shown in Figure 5–8.

The phase increments for a code of length 16 are calculated using equation (5.2) with $M = 16$, yielding $\Delta\varphi = 90°$. The ambiguity function for a Frank code of length 16 is shown in Figure 5–9 and its contour plot in Figure 5–10.

A diagonal ridge can clearly be seen, but it is limited in Doppler to $\pm 1/T$ and it is broken into discrete peaks. There are also additional ridges on each side of the main ridge. This is not the case with other polyphase codes, which we shall discuss later in this chapter. Digital LFM also has these ridges [2].

Figure 5–11 shows the cut of the ambiguity function along the delay axis. *This figure is not related to Figure 5–9 and Figure 5–10*, but has been taken from [2] and reproduced with permission. Unlike the Barker code, the normalized side lobe peaks at zero Doppler for the code of length $N = 16$ and reaches a level of $\sqrt{2}/16$ as against a $1/N$ level for Barker codes. Frank codes are harder to implement since they require generation of many phases (polyphase signals) rather than only phase reversals required by Barker codes. The big advantage is that they can be used for large lengths [2].

Digital LFM Signal: Frank codes are the digital variants of LFM signals. This can be proved as follows [2]. We add a 2π phase to all the elements of row 3 in Table 5–2 and 6π to row 4 (adding a phase that is a multiple of 2π does not affect the signal). Figure 5–12 demonstrates this analogy, wherein a discrete LFM signal with four frequencies is plotted together with that of a Frank-coded signal of length 16 and whose phase matrix has been modified as discussed. Indeed, if we denote the duration of each of the 16-signal elements of the Frank-coded signal as T_{Frank} and the duration of each of the four steps of the discrete LFM as T_{LFM} and the frequency spacing between

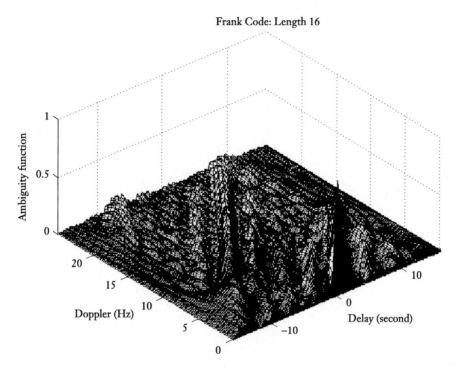

Figure 5–9 Ambiguity function for a Frank code of length 16.

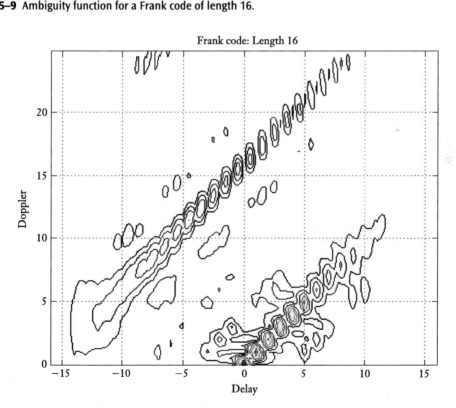

Figure 5–10 Contour plot for a Frank code of length 16.

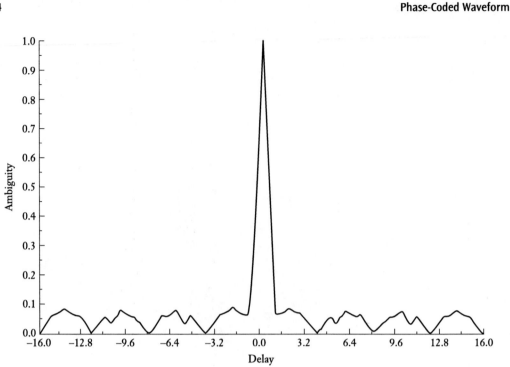

Figure 5–11 Zero Doppler cut of the ambiguity function for the Frank code of length 16. (From [2], © John Wiley & Sons. Reproduced with permission)

steps as Δf, we obtain

$$T_{LFM} = MT_{Frank} \tag{5.4}$$

and

$$\Delta f = \frac{1}{T_{LFM}} \tag{5.5}$$

The phases of both the signals are identical at all multiples of T_{Frank}. Furthermore, comparing the ambiguity function as depicted in Figures 5–9 and 5–10, we note that it is similar to LFM but is broken into discrete peaks. There are also additional parallel ridges on either side of the main ridge. Similar parallel ridges can be found in the digital LFM ambiguity function [7]. This implies that Frank codes exhibit range–Doppler coupling like LFM signals (as can be seen from the contour plot which shows that the ambiguity function has a distinctive slope giving rise to the coupling effect). In fact, it was the quest to find Doppler tolerant codes that led to the development of Frank codes, which are the digital variants of FMCW.

5.3 PERIODIC AMBIGUITY FUNCTION

Periodically modulated CW signals can be considered as the ultimate in waveform compression [8]. In fact, in Chapter 4, we had investigated the performance of CW radars using stretch processing, which enables them to achieve very high resolutions, that are difficult to attain using SAW filters.

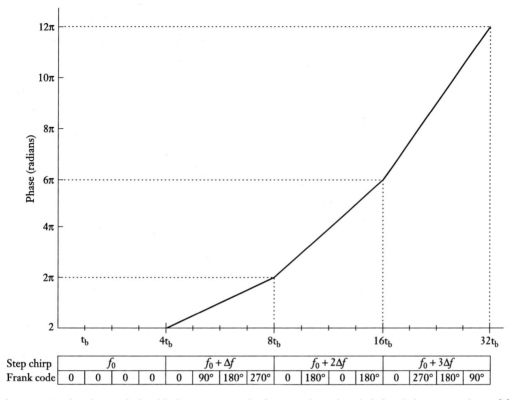

The table below the figure:

Step chirp	f_0				$f_0 + \Delta f$				$f_0 + 2\Delta f$				$f_0 + 3\Delta f$			
Frank code	0	0	0	0	0	90°	180°	270°	0	180°	0	180°	0	270°	180°	90°

Figure 5–12 The phase relationship between quantized LFM and Frank-coded signals for $M = 4$. (From [1], © Reproduced with permission)

This makes CW radars extremely attractive for high resolution range profiling. This, however, is not the only reason. We already know that CW signals achieve a unity peak-to-average power ratio. This makes them very useful from the point of view as LPI radars. There is also another fact that makes them advantageous. They can also exhibit an ideal (zero) range side lobes on the zero–Doppler axis. This property is not shared by pulse modulated radars.

The ideal (zero) range side lobes property of CW signals can be reached when the signal exhibits perfect periodic autocorrelation and the correlation receiver is matched to an integral number of modulation periods [8]. The delay–Doppler response of such a system can be predicted by the PAF. The PAF describes the response of a correlation receiver to a continuous signal modulated by a periodic waveform with period T, when the reference signal is constructed from an integral number N periods of the transmitted signal (we shall examine the methodology of implementing this for phase-coded signals below in this chapter) [8]. Thus, the reference signal is of duration NT. This is illustrated in Figure 5–13.

The response is a function of both delay and Doppler shift. The PAF is a two-dimensional generalization of the periodic autocorrelation function (PACF) by including the effect of Doppler shift. CW signals with periodic modulation are important because only they can yield a perfect autocorrelation. This means that it takes a value 1 when $\tau = 0 \ (\text{mod } T)$ and zero elsewhere. A finite length signal inherently cannot achieve such an ideal autocorrelation, since as the first

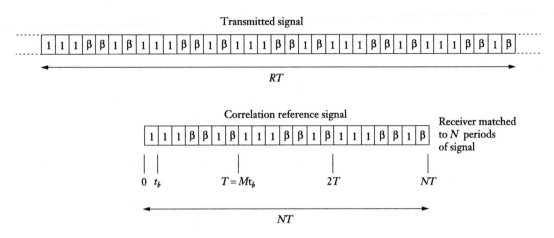

Figure 5–13 Typical transmitted signal and its reference signal. (From [8], © IEEE 1992)

sample (or last sample) enters (or leaves) the correlator, there is no sample that can cancel the product to yield a zero output. The target illumination time (dwell time RT) must be longer than NT. If the delay τ is shorter than the difference between the dwell time and the length of the reference signal $0 \leq \tau \leq (R - N)T$, the illumination time can be considered infinitely long and the receiver response can be described by the PAF given by [8]

$$\left| \chi_{NT}(\tau, \nu) \right| = \left| \frac{1}{NT} \int_{0}^{NT} s(t - \tau) s^*(t) \exp(j2\pi\nu t) dt \right| \tag{5.6}$$

where

$$s(t) = s(t + nT), \quad n = 0, \pm 1, \pm 2, \ldots \tag{5.7}$$

where τ is assumed to be a constant and the delay rate of change is represented by the Doppler shift ν. The PAF for N periods is related to the single-period ambiguity function by the relationship [8]

$$\left| \chi_{NT}(\tau, \nu) \right| = \left| \chi_{T}(\tau, \nu) \right| \left| \frac{\sin(N\pi\nu T)}{N \sin(\pi\nu T)} \right| \tag{5.8}$$

where

$$\left| \chi_{T}(\tau, \nu) \right| = \frac{1}{T} \left| \int_{0}^{T} s(t - \tau) s^*(t) \exp(j2\pi\nu t) dt \right| \tag{5.9}$$

is the single period ambiguity function. The single period ambiguity function is multiplied by a *universal function* of N and T, which is independent of the complex envelope of the signal and which does not change with τ. Equation (5.8) tells us that if N is large, the PAF gets increasingly attenuated for all values of ν except at multiples of $1/T$. It also has main lobes at $\nu T = 0, \pm 1, \pm 2, \ldots$ and relatively strong Doppler side lobes. In the limiting case, for large N, the PAF of a sequence *exhibiting perfect periodic autocorrelation* will strongly resemble the ambiguity function of a coherent pulse train (see Figure 3–11). The PAF serves CW radar signals in a similar manner as the traditional ambiguity function serves finite duration signals.

There is an important point to note regarding perfect periodic autocorrelation in CW signals. It only applies to phase-coded signals and not to signals like LFM, unless the frequency modulated signals can be approximated by phase-coded signals [9]. Examples of such discrete LFM signals include Frank codes [6] and Kretschmer and Lewis P3 and P4 signals [10]. Even in phase-coded signals, not all yield the desired perfect PACF. For example, the Barker code of length 7 does yield a perfect PACF, but Barker codes of lengths 5 and 13 do not [8]. How does one identify such special phase-coded signals? This problem was addressed by Golomb [11] to be discussed below.

5.4 PERIODIC AUTOCORRELATION FUNCTION

CW waveforms yield a PACF when they are modulated by a periodic function such as a phase-coded sequence. Each period in such a waveform is constructed from a sequence of M segments of duration t_b.

The complex envelope for one period is given by [8]

$$s(t) = \sum_{m=1}^{M} s_m \left[t - (m-1) t_b \right], \quad 0 \le t \le M t_b \tag{5.10}$$

where

$$s_m(t) = \begin{cases} \exp(j\varphi_m) & 0 \le t \le t_b \\ 0, & \text{elsewhere} \end{cases} \tag{5.11}$$

Equation (5.11) is valid with the relation

$$T = M t_b \tag{5.12}$$

The periodic autocorrelation at delays which are multiples of t_b are given by

$$C(r) = C(r t_b) = \frac{1}{M} \sum_{m=1}^{M} s(m) s^*(m+r) \tag{5.13}$$

Ideally, we would like a perfect PACF given by

$$C(r) = \begin{cases} 1, & r = 0 \,(\text{mod}\, M) \\ 0, & r \ne 0 \,(\text{mod}\, M) \\ & i.e.\ r = 1, 2, \ldots, M-1 \end{cases} \tag{5.14}$$

In the absence of any restriction on the values φ_m can take, we arrive at what are called polyphase codes. A number of polyphase sequences yield perfect PACF [12, 13]. In order for the phase sequence $\{\varphi_m\}$ to yield perfect PACF, it needs to satisfy the following relation [8]

$$\frac{1}{\sqrt{M}} \begin{bmatrix} 1 & 1 & 1 & \cdots & 1 \\ 1 & \omega & \omega^2 & \cdots & \omega^{(M-1)} \\ 1 & \omega^2 & \omega^4 & \cdots & \omega^{2(M-1)} \\ \vdots & \vdots & \vdots & & \vdots \\ 1 & \omega^{(M-1)} & \omega^{2(M-1)} & \cdots & \omega^{(M-1)^2} \end{bmatrix} \begin{bmatrix} 1 \\ \exp(j\varphi_1) \\ \exp(j\varphi_2) \\ \vdots \\ \exp(j\varphi_{(M-1)}) \end{bmatrix} = \begin{bmatrix} 1 \\ \exp(j\alpha_1) \\ \exp(j\alpha_2) \\ \vdots \\ \exp(j\alpha_{(M-1)}) \end{bmatrix} \tag{5.15}$$

where

$$\omega = \exp\left(-j\frac{2\pi}{M}\right) \tag{5.16}$$

and there are no restrictions on the resulting $\{\alpha_m\}$.

Assume that φ_m can take one of two values, 0 or φ. This will cover cases of binary as well as polyphase codes. Hence, s_m can be either 1 or a value of say, β (see equation (5.11)), where

$$\beta = \exp(j\varphi) \tag{5.17}$$

We can now conclude that the complex envelope of such a signal can be described by a periodic two-valued sequence $S = \{s_m\}$ having a period M, where the two values are the complex numbers 1 and β.

We now seek a sequence S, which has a value of phase φ, such that it will yield a perfect PACF. Golomb [11] shows that the sequence S must correspond to a (M, k, λ) difference set D, where $\alpha_m = 1$ for $m \in D$ and $s_m = \beta$ for $m \notin D$. The sequence S corresponds to a (M, k, λ) difference set [14], if M is the length of the sequence, k is the number of 1s and λ is a constant number of times that 1s in the original sequence coincide with 1s in the shifted sequence, for any cyclic shift. It should be noted that all "Maximal length linear shift register sequences" correspond to such difference sets [14].

Example

In the sequence $S = \{1 \quad 1 \quad 1 \quad \beta \quad \beta \quad 1 \quad \beta\}$ the 1s correspond to the set $D : \{0 \quad 1 \quad 2 \quad 5\} \bmod 7$, which is a (7, 4, 2) difference set [8].

For such a sequence, Golomb shows that the autocorrelation function is given by

$$C(0) = 1 \tag{5.18}$$

$$C(r \neq 0) = \frac{M - 2k + 2\lambda + 2(k - \lambda)\cos\varphi}{M} \tag{5.19}$$

To meet the ideal requirement in equation (5.14) we equate equation (5.19) to zero and obtain

$$\varphi = \arccos\left[-\frac{M - 2k + 2\lambda}{2(k - \lambda)}\right] \tag{5.20}$$

In the example for the sequence of length 7,

$$\varphi = \arccos\left(-\frac{3}{4}\right) = 138.59° $$

This is a Barker code of length 7, where -1 elements were replaced by β. However, Barker sequences of lengths 5 and 13 produce an absolute value of the argument of the arccos that is greater than 1. Hence, there is no φ that will provide a perfect PACF for either of these sequences. However, there are sequences other than Barker, which do produce a perfect PACF, for example, Hadamard sequences [14].

5.5 CUTS OF PAF ALONG DELAY AND DOPPLER AXES

The cut along the delay axis (zero Doppler) will be the perfect PACF as given in equation (5.14). The delay response will be triangles of width $2t_b$, spaced T periods apart. The cut along the Doppler axis is obtained by setting $\tau = 0$ in (5.6) and is given by

$$\chi_{NT}(0,\nu) = \frac{1}{NT}\int_0^{NT} |s(t)|^2 \exp(j2\pi\nu t)\, dt \tag{5.21}$$

If we assume a constant amplitude signal as in phase-coded CW signals, $|s(t)| = 1$, then

$$|\chi(0,\nu)| = \left|\frac{\sin(\pi\nu NT)}{\pi\nu NT}\right| \tag{5.22}$$

and

$$|\chi(0,0)| = 1 \tag{5.23}$$

The periodicity of the delay axis [8] for any integer n is given by

$$|\chi_{NT}(nT,\nu)| = |\chi_{NT}(0,\nu)| \tag{5.24}$$

For the Doppler axis $\nu = m/T$, for $m = 0, \pm1, \pm2,\ldots$

$$|\chi_{NT}(\tau, m/T)| = |\chi_{NT}(\tau + nT), m/T| \tag{5.25}$$

The symmetry cuts are a function of the three parameters: the code period T, the number of phase modulation bits in the sequence M, and the number of periods used in the correlation receiver, N. Further details are given in [8].

5.6 PAF AND PACF OF FRANK CODES

We investigate the PAF and PACF of Frank codes discussed earlier. These are shown in Figures 5–14, 5–15, and 5–16.

When we compare Figure 5–14 to Figure 5–9, we note the absence of side lobes along the delay axis. The number of reference waveforms $N = 1$. If N were very large, the PAF will tend to look like that of a coherent pulse train as in Figure 3–11. Figure 5–15 is the contour plot. We again note the total absence of side lobes along the delay axis. This aspect is clearer in Figure 5–16. In Figure 5–16(a), we see the ACF for a Frank code of length 16. The peak side lobe level is –21 dB as discussed earlier. Figure 5–16(b) shows the complete absence of side lobes in the PACF. Figure 5–16(c) is the spectrum of the Frank code. Note the broad spectrum, reflecting the wideband nature of the modulation. However, as can be seen, the PAF of the Frank code exhibits strong Doppler side lobes. The *PSL* for these lobes is given by [2]

$$PSL = 20\log_{10}\left(\frac{1}{NM\pi}\right) \tag{5.26}$$

Comparing equation (5.26) to that of *PSL* pertaining to finite signals, discussed earlier, we note the addition of the factor N in the denominator, where N is the number of periods of the

Frank code: Length 16: Periodic ambiguity function

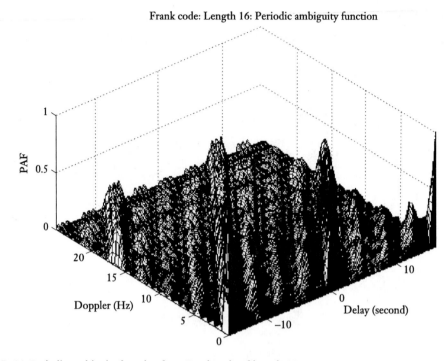

Figure 5–14 Periodic ambiguity function for a Frank code of length 16.

reference signal. This makes the *PSL* along the Doppler axis (there are no side lobes along the delay axis) as –21 dB (since $N = 1$ in these figures). If N were, say, equal to 4, then $PSL = -34$ dB. This proves that the higher the number of references, the better for side lobe control along the Doppler axis. We can reduce the Doppler side lobes by weighting. Interested readers are referred to [9]. It should be noted that weighting covering N periods of the signal, affects only the Doppler behavior. It has no influence on the PACF. In phase-coded signals, the delay response remains a triangle with base $2t_b$, regardless of any amplitude taper along the N periods of the signal [1].

The figures pertaining to Frank code were obtained using the software "*ambfn7.m*" provided by Nadav Levanon. The software is readily available at his website [15].

5.7 MATCHED FILTER FOR PHASE-CODED SIGNALS

We learned that the LPI receiver for phase-coded radars must correlate (or compress) the received signal from the target using their stored reference signal, in order to perform target detection. The correlation receiver is a "matched" receiver if the reference signal is exactly the same duration as the finite duration return signal.

The PAF describes the response of a correlation receiver to a continuous signal modulated by a periodic waveform with period T, when the reference signal is constructed from an integral number N of periods of the transmitted signal [8] (see Figure 5–13). Thus, the reference signal is of the duration NT. Increasing the number of receive reference waveforms N improves the target detection capabilities by increasing the resolution of the receiver response.

The LPI radar Doppler matrix receiver can be modeled as a coherent correlation processor of finite duration NT as shown in Figure 5–17. This figure shows a transmitted waveform

Frank code: Length 16: PAF contour plot

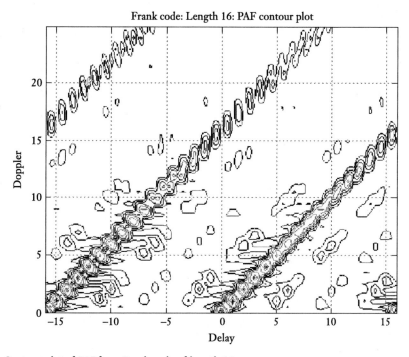

Figure 5–15 Contour plot of PAF for a Frank code of length 16.

(a phase-coded signal) of length RT, where T is the code period, t_b is the subcode period and $R >> 1$. This particular signal is of length $M = 7$. The receiver is matched to N periods. The receiver contains a bandpass filter matched to a rectangular RF (or IF) pulse of duration t_b. The target return is received by the correlation receiver containing the reference signal which is the conjugate of N periods of the transmitted signal with $N < R$. Correlation is now performed between these two signals. This is achieved by sending this signal (a binary-coded signal in this example) through a filter matched to a rectangular subcode of length t_b, followed by a detector that sends forward a one or a zero. The detected output signal is then passed to a tapped delay line where each delay D is t_b second. The signal in the tapped delay line (of length MNt_b) is first multiplied by the reference signal. The output of each of these multiplications is then summed separately for each of the N code periods. The output of the sum block can then be weighted as C_1 through C_N. If we use uniform weights, the first stage represents the response of the receiver for a zero Doppler shift signal ($v = 0$) and is identical to the ideal autocorrelation function [16]. The response of the receiver to a Doppler-shifted signal Δv is obtained from the second stage by first multiplying the output (before addition) from the first stage with q^0 through q^{MN-1} where $q = e^{j2\pi\Delta v t_b}$.

If the reference signal is weighted in order to reduce Doppler side lobes, such filters are called *mismatched* filters as was already discussed in Chapter 2. Finally, if the number of references N is very large, the PAF of a *sequence exhibiting perfect periodic autocorrelation* will strongly resemble the ambiguity function of a coherent pulse train [16]. This means that since the sequence exhibits perfect PACF, there are no delay side lobes. But there are Doppler side lobes. These are controlled by weighting and increasing the number of reference periods N to a large value. Having done this,

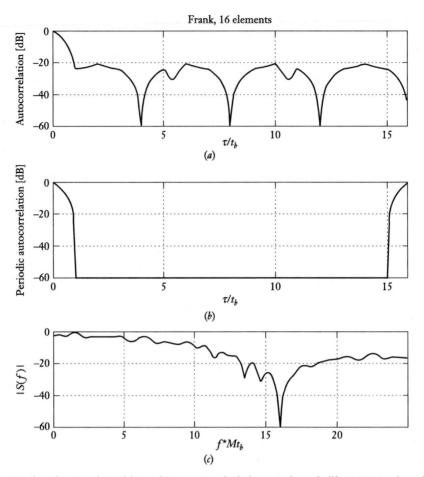

Figure 5–16 Frank code: Length 16 (*a*) ACF (*PSL* = −21 dB below maximum), (*b*) PACF: Number of reference waveforms $N = 1$ (*PSL* = −21 dB below maximum), and (*c*) Spectrum.

then what remains is a bed of "nails," that is, the ambiguity function of a coherent pulse train. Correspondingly, the target dwell time must be at least $N + 2$ [16], so that the processor is filled up. It is important to note that Doppler compensation Δv in Figure 5–17 must be performed for each bit, prior to adding them up. Hence, it is necessary to deal with MN samples. A processor for the corresponding CPT has to deal with only N samples. The resulting additional processing is a major penalty for using a CW phase-coded signal [16].

5.8 POLYPHASE CODES

Frank code and P1- through P4-coded signals belong to this class. These signals were developed by approximating a stepped frequency or LFM waveform, where the phase steps vary as needed to approximate the underlying waveform and the time spent at any phase state is a constant. The idea here was to emulate the LFM waveform, the positive aspects being its Doppler tolerance and pulse compression, while the negative aspect being its range–Doppler coupling. Frank code has already been discussed by us. We shall now study P1 through P4 polyphase codes.

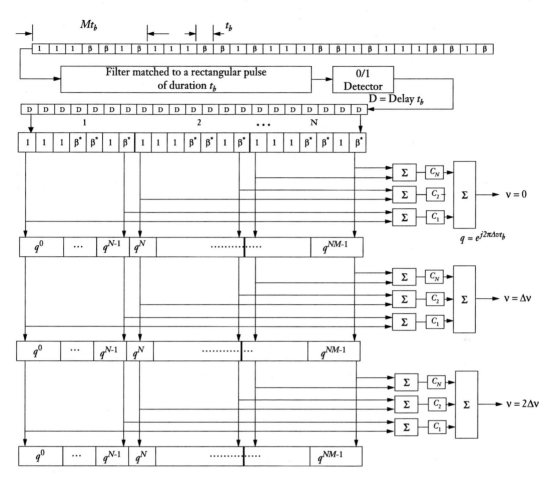

Figure 5–17 Doppler matrix correlation receiver matched to N periods of a two-valued phase-coded signal of length $M = 7$ including weighting C_i for Doppler side lobe reduction. (From [16], © IEEE 1993)

5.8.1 P1 Codes

The P1 code is a step approximation of the LFM waveform. This was born out of the desire to digitize the well-tested and reliable LFM signal. However, there were limitations, which make this class of phase-coded signals inferior to LFM as we shall see. This code comprises M frequency steps and M samples per frequency, which are obtained using double sideband detection with the local oscillator at band center [10]. The compression ratio of this code is $N_s = M^2$. If i is the number of the sample in a given frequency and j is the number of the frequency, the phase of the ith sample of the jth frequency is

$$\phi_{i,j} = \frac{-\pi}{M}\big[M-(2j-1)\big]\big[(j-1)M+(i-1)\big] \tag{5.27}$$

where $i = 1, 2, \ldots, M$ and $j = 1, 2, \ldots, M$ and $M = 1, 2, 3, \ldots$. The *PSL* for this code is given by $PSL = 20\log_{10}(1/NM\pi)$ where N is the number of reference periods. This is the same as Frank code and the result is not surprising since the Frank code also is similar to digital LFM.

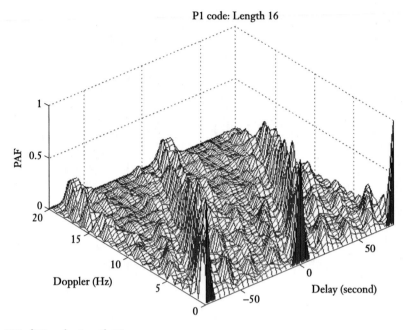

Figure 5–18 PAF of P1 code: Length 16.

Figure 5–18 shows the PAF of P1 code for $N_s = 16$ and $N = 1$. The sampling frequency f_s is 10 KHz and the carrier frequency f_c is 2 KHz. The number of cycles/period cpp = 1. Note that the PAF repeats at $N_s(cppf_s/f_c) = 80$ samples. The contour plot is shown in Figure 5–19. We first generate the basic signal using the accompanying software as detailed in Appendix K. This basic signal is stored in a .mat file. We then use the software of Nadav Levanon [15] "*ambn7.m*" in the "user defined" mode as outlined in Appendix K to plot the PAF, ACF and PACF. This same procedure is adopted for all codes in this chapter.

We note in Figure 5–20 that this code has a perfect PACF making it suitable for LPI radars since we can reap the advantage of zero time side lobes. This is for $N_s = 16$ and $N = 1$. The *PSL* is –22 dB down as per the formula given above.

5.8.2 P2 Codes

During the generation of the P2 code, we take M as even, since this gives rise to low autocorrelation side lobes [10]. This code has a compression ratio of $N_s = M^2$. The P2 code is given by [17]

$$\phi_{i,j} = \frac{-\pi}{2M}[2i-1-M][2j-1-M] \tag{5.28}$$

where $i = 1,2,3,...,M$ and $j = 1,2,3,...,M$ and $M = 2,4,6,....$ The PSL for this code is given by $PSL = 20\log_{10}(1/NM\pi)$ where N is the number of reference periods. This is the same as for Frank code and P1 code.

Figure 5–21 shows the PAF of P2 code for $N_s = 16$ and $N = 1$. The sampling frequency is 10 kHz and the carrier frequency is 2 kHz. The number of cycles/period *cpp* = 1. Note that the PAF repeats at $N_s(cppf_s/f_c) = 80$ samples.

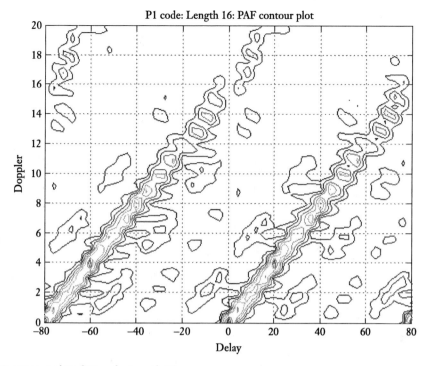

Figure 5–19 Contour plot of P1 code: Length 16.

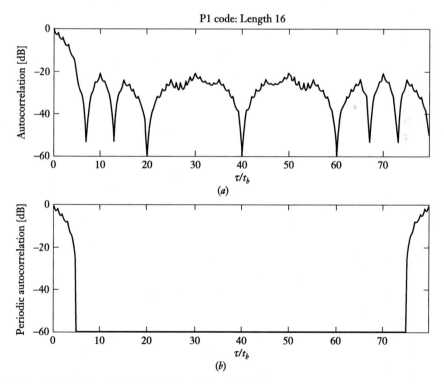

Figure 5–20 P1 code (*a*) ACF and (*b*) PACF for $N_s = 16$ and $N = 1$.

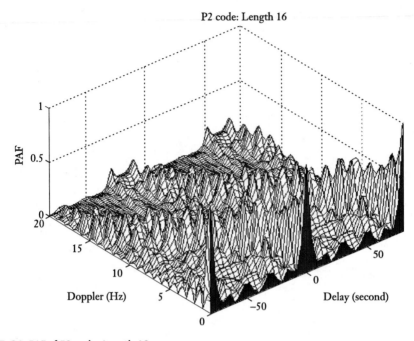

Figure 5–21 PAF of P2 code: Length 16.

The contour plot is shown in Figure 5–22. Note that the PAF of P2 code has an opposite slope compared to P1. This is evident in Figure 5–21 also.

The ACF and PACF are shown in Figure 5–23. We note that P2 code does not have a perfect PACF. It is, in fact, identical to ACF.

5.8.3 P3 Codes

An LFM waveform is converted to baseband by using a synchronous oscillator on one end of the frequency sweep (single sideband detection) and sampling the I and Q channels at the Nyquist rate (first sample of I and Q is taken at the leading edge of the waveform) [18]. The phase of the ith sample of the P3 code is given by

$$\phi_i = \frac{\pi}{N_s}(i-1)^2 \tag{5.29}$$

where $i = 1, 2, \dots N_s$ and N_s is the compression ratio.

Figure 5–24 shows the PAF of P3 code for $N_s = 16$ and $N = 1$. The sampling frequency is 10 kHz and the carrier frequency is 2 kHz. The number of cycles/period $cpp = 1$. Note that the PAF repeats at $N_s(cppf_s/f_c) = 80$ samples. We note that the slope of the PAF for P3 code is opposite to that for the P2 code. This is verifiable from the contour plot at Figure 5–25.

We note from Figure 5–26 that the ACF has a peak side lobe ratio a bit larger than the Frank, P1, and P2 codes. The *PSL* is given by $PSL = 20\log_{10}(\sqrt{2/(N_s\pi^2)})$ dB. In our case for $N_s = 16$ and $N = 1$ we obtain –19 dB down from the peak. For example, for the P2 code the PSL

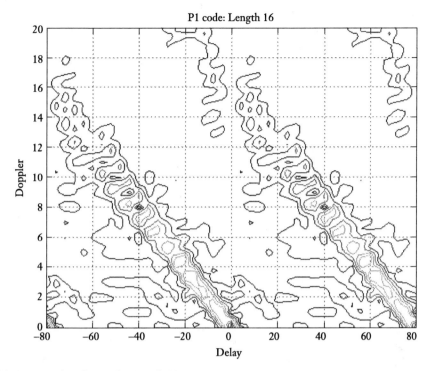

Figure 5–22 Contour plot of P2 code: Length 16.

was −22 dB down. Clearly, P3 is inferior from this point of view. But it scores over P2 code in the PACF structure. It is a perfect one.

5.8.4 P4 Codes

The P4 code is derived from LFM in the same manner as P3 code, except that the local oscillator frequency is offset in the I and Q detectors, resulting in coherent double sideband detection. We then sample this at Nyquist rate [17, 18]. The P4 code consists of discrete phases of the linear chirp waveform taken at specific intervals and it consequently exhibits the same range–Doppler coupling associated with the chirp waveform. Unlike LFM, however, it exhibits multiple ridges. It shares this trait with P1, P2, P3, and Frank codes which also (because their ambiguity functions are skewed to the cardinal axes) exhibit multiple ridges and range–Doppler coupling. The implications of this will be examined in greater detail toward the end of this chapter. However, P4 code is more Doppler resilient than the other codes. The phase sequence of a P4 signal is given by

$$\phi_i = \frac{\pi(i-1)^2}{N_s} - \pi(i-1) \tag{5.30}$$

for $i = 1, 2, \ldots, N_s$ where N_s is the compression ratio. Figure 5–27 shows the discrete phase values that result for the P4 code for $N_s = 16$. Compare this to the discrete phase values of P3 code shown in Figure 5–28.

Figure 5–29 shows the PAF for P4 code for $N_s = 16$ and $N = 1$. The sampling frequency is 10 kHz and the carrier frequency is 2 kHz. The number of cycles/period $cpp = 1$. Note that the PAF repeats at $N_s(cppf_s/f_c) = 80$ samples.

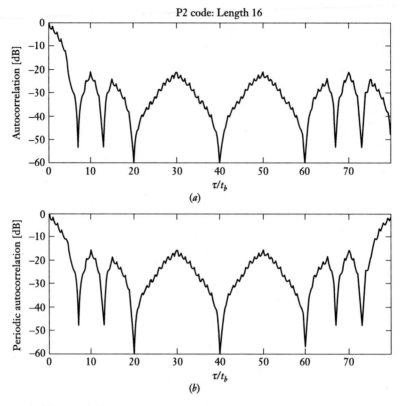

Figure 5–23 P2 code (*a*) ACF and (*b*) PACF for $N_s = 16$ and $N = 1$.

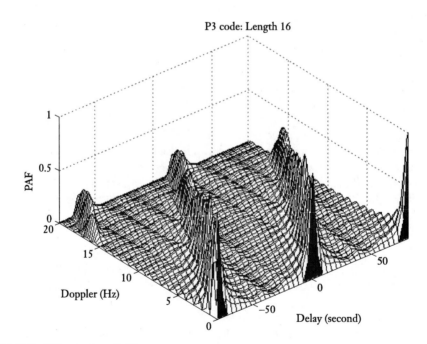

Figure 5–24 PAF of P3 code: Length 16.

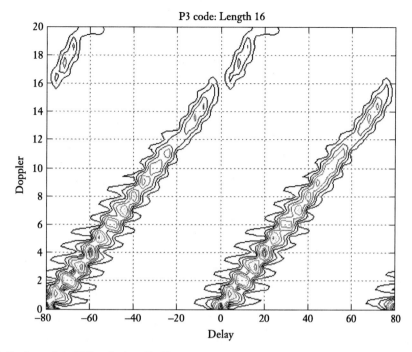

Figure 5–25 Contour plot of P3 code: Length 16.

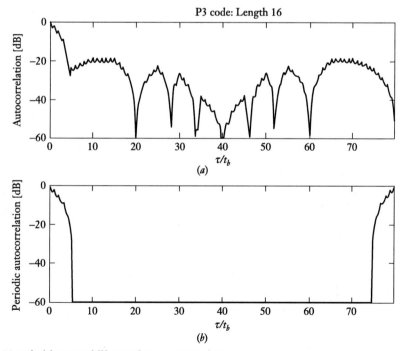

Figure 5–26 P3 code (*a*) ACF and (*b*) PACF for $N_s = 16$ and $N = 1$.

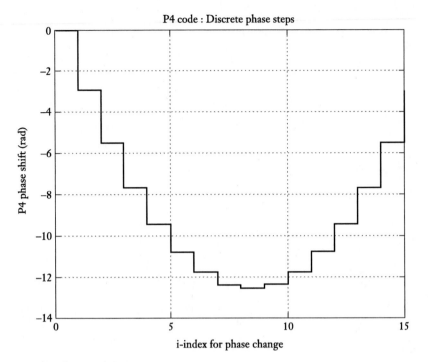

Figure 5–27 P4 code: Discrete phase steps.

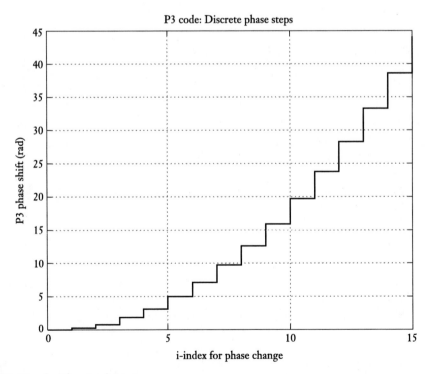

Figure 5–28 P3 code: Discrete phase steps.

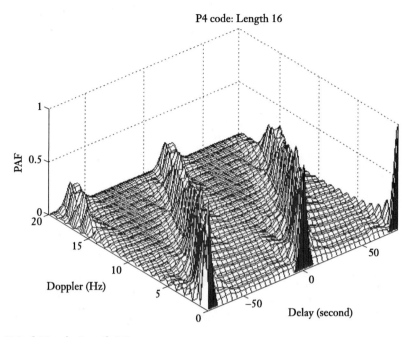

Figure 5–29 PAF of P4 code: Length 16.

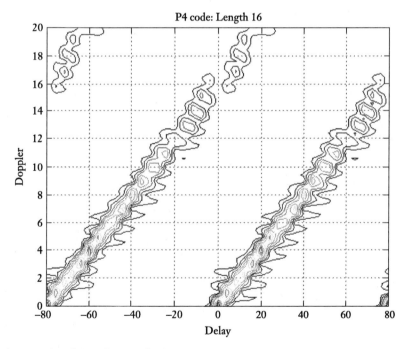

Figure 5–30 Contour plot of P4 code: Length 16.

The contour plot is shown in Figure 5–30.

The ACF and PACF are shown in Figure 5–31.

The peak side lobe level is given by $PSL = 20\log_{10}(\sqrt{2/(N_s \pi^2)})$. This is the same as for P3 code. The PACF shows us that P4 is a perfect code, viz., zero PACF side lobes.

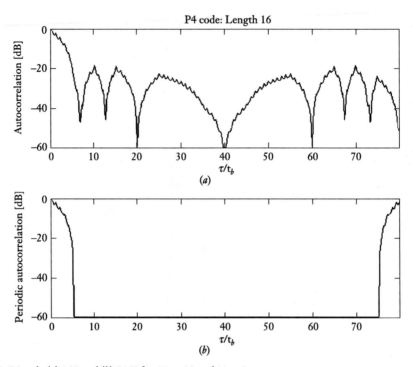

Figure 5–31 P4 code (*a*) ACF and (*b*) PACF for $N_s = 16$ and $N = 1$.

5.9 PERFORMANCE ANALYSIS OF PHASE-CODED SIGNALS

We have now come to the end of our study of phase-coded signals. During this study, we investigated:

1. Binary phase codes exemplified by the Barker code and
2. Polyphase codes like Frank and P1 through P4 codes.

Before proceeding with out analysis, it is essential to revisit the coherent pulse train (CPT), which we studied earlier in Chapter 3. The CPT is one of the most important radar signals. It provides independent control of both delay and Doppler resolution. It also exhibits a range window which is inherently free of side lobes. The ambiguity function of a CPT is shown in Figure 5–32. The ambiguity function of a CPT represents the magnitude of the matched receiver output in the delay–Doppler domain. The ambiguity function of the CPT also indicates that the Doppler resolution is the inverse of the total duration of the signal – NT, while the delay resolution is the pulse duration of the signal $t_p = T/M$ where T is the pulse interval, N is the number of pulses processed coherently and M is the inverse of the duty cycle. Figure 5–32 pertains to $N = 16$ and $M = T/t_p = 16$. This plot has been obtained using "*cohopulsetrain.m*" in the accompanying software supplied with Chapter 3.

The receiver response in Figure 5–32 approaches the ideal response, if we ignore the unavoidable ambiguity of T in delay and $1/T$ in Doppler. It is this ideal response that makes the CPT such an important radar signal [16]. The main drawback to the CPT signal is that it requires a high ratio of peak to average power. The average power is what determines the detection performance and estimation accuracy of the parameters of the target. To maintain sufficient average power, the CPT signal *usually* requires high peak power, based on vacuum tubes, high voltages, etc. [16]. LPI radars depend upon CW signals with peak-to-average ratio as unity. But they do not provide the

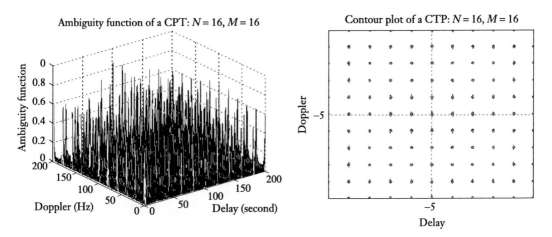

Figure 5–32 Ambiguity function and contour plot of a CPT for $N = 16$ and $M = T/t_p = 16$.

detection range or the Doppler resolution as CPT signals. We would, therefore, like our CW signal to provide us an ambiguity function like the one in Figure 5–32. We do this by using the special family of signals in the phase-coded class, which yield us a perfect PACF along with the signal processing suggested in Section 5.3.

Guided by our CPT ambiguity function, we note that there are three main qualities that the reader should bear in mind when analyzing suitable signals:

1. *Range Resolution*: In phase-coded signals, it is possible to achieve HRR by narrowing the segment width. However, excessive narrowing of this width will also lead to the problem of designing ultra wideband radar, which is still a research topic. We can get around this by adopting the proposal given by Levanon of using multi-carrier phase-coded signals [19].

2. *Range–Doppler Coupling*: This should be preferably nonexistent like in CPT signals. This can happen if the ambiguity function is parallel to the cardinal axes like for the single pulse (see Chapter 3). Unfortunately, if we look at perfect PACF signals like Frank code, P1, P3, and P4 codes, we note that their ambiguity functions are skewed to the delay axis. This is what imparts range–Doppler coupling just like an LFM signal from which these are derived. If we use the receiver technique discussed in Figure 5.17, and match the target return to a very large number of reference periods N, the ambiguity function can become like the one for CPT. The stipulation is that we need sufficient dwell time to achieve this, at least $N + 2$ [16]. This is shown in Figure 5.33 for a P4 code of length 16, $cpp = 1$, but with $N = 6$ instead of $N = 1$ as in Figure 5.29. The remaining parameters are the same. Notice the bed of "nails." The Doppler side lobes can be controlled by Hamming weighting or any other suitable weighting. In this figure, we had used uniform weighting. Already the Doppler side lobes have decreased due to more copies of the reference signal (see Fig 5.29 for $N = 1$ for comparison).

   The contour plot is shown in Figure 5–34. The bed of "nails" is evident.

   The ACF and PACF plots are shown in Figure 5–35.

   Notice in Figure 5–35, that the time side lobes of the ACF have decreased due to the higher number of reference periods $N = 6$. This means that delay–Doppler side lobe performance improves with more reference copies. It is important to note that Doppler compensation $\Delta\nu$ in Figure 5–17 must be performed for each bit, prior to adding them up. Hence, it is necessary to deal with MN samples. A processor for the corresponding

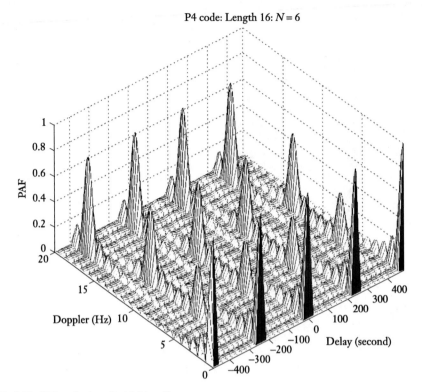

Figure 5–33 PAF of P4 code: Length 16 (N = 6).

CPT has to deal with only N samples. The resulting additional processing is a major penalty for using a CW phase-coded signal [16].

3. *Doppler Tolerance*: The issue of Doppler tolerance is poorly defined, and can be interpreted in many ways. When most radar experts talk about Doppler tolerance they mean that the single-pulse ambiguity function (AF) exhibits a ridge and this ridge must be diagonal, and which extends relatively far in the Doppler dimension. In the AF of a Barker code, the main AF peak reaches a value of zero at Doppler equal to the inverse of the pulse duration T (in Barker 13, $T = 13t_b$, where t_b is the bit duration). On the other hand, in phase codes like P3 and P4, the ridge remains high up to much higher Doppler values. For example, in a P4 of length 25, the ridge drops to a value of 0.5 (compared to a peak of 1 at the origin) at Doppler that is approximately 10 times the inverse of the pulse duration. Yet, due to the range–Doppler coupling, that point occurs at a delay of about $10t_b$ (out of $T = 25t_b$) (N. Levanon and M. Jankiraman, personal communication, August 2005). Frank code is different from the P3 and P4 family because its AF exhibits two more ridges parallel to the main ridge. Doppler tolerance applies only to a single pulse. When we process coherently many pulses (i.e., periods), we can create many filters, each one matched to a different Doppler shift, by simply performing FFT after pulse compression (N. Levanon and M. Jankiraman, personal communication, August 2005). In phase-coded CW radars, the target return signals do not correlate perfectly because of the target Doppler shift, which changes the phase of the code across its period. This

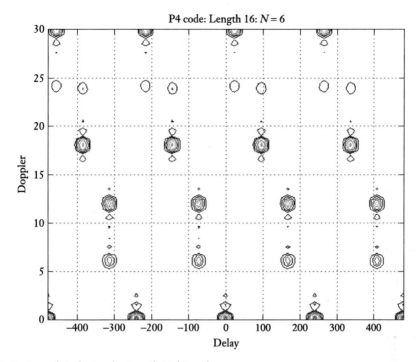

Figure 5–34 Contour plot of P4 code: Length 16 (*N* = 6).

causes imperfect compression. The poor Doppler tolerance of Barker code, for example, makes the signal decorrelate when the phase shift due to Doppler exceeds 90°. So far as LFM-derived signals are concerned the most popular is the Frank code, which has been known for its Doppler constancy. However, it has been proven in a study that even the Frank code has relatively poor Doppler tolerance [20]. Though phase-coded waveforms perform well when the Doppler is minimal, it was found that their performance degrades dramatically with increased frequency shifts due to Doppler effects. Ideally, we would like the Doppler tolerance exhibited by LFM. Figure 5–36 shows us the LFM AF cut along the Doppler axis, where T is the pulse width. This gradual decline of signal energy from the signal maximum (at zero Doppler) as the target Doppler increases ensures Doppler tolerance or resilience.

This sort of performance cannot be expected in phase-coded signals, even if they are derived from the LFM waveform. This makes a case for using the receiver schematic suggested in Figure 5–17 and to transform the PAF to that of a CPT, which is even better. Once again, the reader is reminded that this is hardware intensive. In HRR radars, the target Doppler is one cause for problems in phase-coded signal decorrelation. The other cause is its own platform motion, if the radar is not static, for example, it is mounted on an aircraft. In such a case even if the target were static, that is, no target Doppler, the own platform Doppler needs to be nulled so as to fool the signal processor into believing that the radar platform is static. The own Doppler nulling (ODN) needs to be well matched to the maximum range resolution of the radar. This means that when we go in for HRR, we cannot afford large time side lobes, as this will ruin the quality of target resolution. It will be recalled that phase-coded signals are sensitive to target Doppler. In the presence of target Doppler,

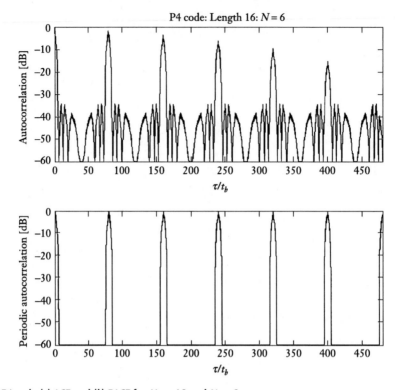

Figure 5–35 P4 code (*a*) ACF and (*b*) PACF for $N_s = 16$ and $N = 6$.

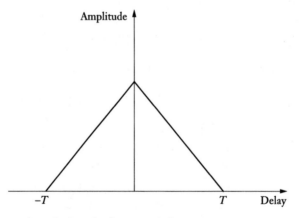

Figure 5–36 AF for LFM: Cut along the Doppler frequency axis.

they decorrelate leading to large time side lobes. Suppose the target were static and we are observing it from a moving platform like an aircraft. The same argument will apply here also but in reverse.

The above-mentioned viewpoints make phase-coded waveforms suitable only for low Doppler or static target applications from slow moving or static radar platforms. In spite of this inconvenience, these waveforms have shown significant promise, because of the fact that polyphase codes exhibit low side lobe levels without having to resort to weighting unlike LFM signals and are also directly compatible with digital generation and compression. This makes them very attractive to radar designers. Before concluding we need to clarify that despite all these arguments, phase-coded radars are sometimes used to track fast targets, but the hardware is intensive and such radars are relatively costly as compared to LFM radars. Based upon target and application specifics, it becomes sometimes necessary, despite the cost, to go in for such radars for high speed targets.

References

1. Pace, P. E., *Detecting and Classifying Low Probability of Intercept Radar*, Artech House, Norwood, MA, 2004.
2. Levanon, N., *Radar Principles*, John Wiley & Sons, New York, 1988.
3. Mahafza, B. R. and Elsherbeni, A. Z., *MATLAB Simulations for Radar Systems Design*, Chapman & Hall/CRC Press, Boca Raton, FL, 2004.
4. Barker, R. H., *Group Synchronizing of Binary Digital Systems in Communications Theory*, Butterworth, London, 1953, pp. 273–287.
5. Zhang, N. and Golomb, S. W., "Sixty-phase generalized Barker sequences," *IEEE Transactions on Information Theory*, Vol. 35, No. 4, April 1989, pp. 911–912.
6. Frank, R. L., "Polyphase codes with good nonperiodic correlation properties," *IEEE Transactions on Information Theory*, Vol. 9, 1963, pp. 43–45.
7. McClellan, J. H. and Purdy, R. J., "Applications of digital signal processing to radar," in Oppenheim, A. V. (ed.) *Applications of Digital Signal Processing*, Prentice-Hall, Englewood Cliffs, NJ, 1978, pp. 250–254.
8. Freedman, A. and Levanon, N., "Periodic ambiguity function of CW signals with perfect periodic autocorrelation," *IEEE Transactions on Aerospace and Electronic Systems*, Vol. 28, No. 2, April 1992.
9. Getz, B. and Levanon, N., "Weight effects on the periodic ambiguity function," *IEEE Transactions on Aerospace and Electronic Systems*, Vol. 31, No. 1, January 1995.
10. Kretschmer, F. F. and Lewis, B. I., "Doppler properties of polyphase coded pulse compression waveforms," *IEEE Transactions on Aerospace and Electronic Systems*, Vol. 19, July 1983, pp. 521–531.
11. Golomb, S. W., "Two-valued sequences with perfect periodic autocorrelation," *IEEE Transactions on Aerospace and Electronic Systems*, Vol. 28, April 1992, pp. 383–386.
12. Chu, D. C., "Polyphase codes with good periodic correlation properties," *IEEE Transactions on Information Theory*, Vol. 18, July 1972, pp. 531–532.
13. Frank, R. L., Comments on "Polyphase codes with good periodic correlation properties," *IEEE Transactions on Information Theory*, Vol. 19, March 1973, pp. 244.
14. Golomb, S. W. (ed.), *Digital Communications with Space Applications*, Prentice-Hall, Englewood Cliffs, NJ, 1964.
15. http://www.eng.tau.ac.il/~nadav/amb-func.html
16. Levanon, N., "CW alternatives to the coherent pulse train—signals and processors," *IEEE Transactions on Aerospace and Electronic Systems*, Vol. 29, No. 1, January 1993, pp. 250–254.
17. Painchaud, G. R., et al., "An experimental adaptive digital pulse compression subsystem for multi-function radar applications," in *IEEE Radar Conference*, 1990.
18. Lewis, B. L., Kretschmer, F. F., and Shelton, W. W., *Aspects of Radar Signal Processing*, Artech House, Norwood, MA, 1986.
19. Levanon, N. and Mozeson, E., *Radar Signals*, Wiley-Interscience, Hoboken, New Jersey, June 25, 2004.
20. Bowman, G. G., *Investigation of Doppler Effects on the Detection of Polyphase Coded Radar Waveforms*, Master's thesis, Report No. A773514, Air Force Institute of Technology, Wright-Patterson AFB, School of Engineering and Management, February 2003.

6

Frequency Hopped Waveform

6.1 INTRODUCTION

In this book we have reviewed two LPI radar signal waveforms, FMCW and phase-coded. The complexity of FMCW technology is minimal and it is very popular because of this reason. There are only two major obstacles that can be construed as a disadvantage in FMCW radars. These are the high time side lobes of the order of 13 dB down from the peak response and the nonlinearity in waveform generation for high bandwidths (and consequently high resolutions). The advantage of this waveform, however, lies in its Doppler tolerance, making it eminently suitable for use in aircraft target tracking radars. Phase-coded waveforms, on the other hand, are very easily adaptable to digital signal processing, being digital in nature, and polyphase codes produce relatively low time side lobes and as we have seen in some cases of polyphase codes, *no* time side lobes in the CW mode. The only problem is that these radars are relatively costly because of the complexity. The sub-pulse width defines the high range resolutions one can achieve using this waveform and unlike LFM waveforms, there are fewer constraints of nonlinearity in phase-coded waveform generation. The disadvantage with this waveform is that it has relatively poor Doppler tolerance against fast targets. We now study a third LPI waveform, which has found popularity in vehicular radars. This is the frequency hopped (FH) waveform. Generally, this class of signals transmits one frequency at each step. In doing this, it leaves the phase of the signal alone, that is, the frequency is transmitted with one continuous phase. The order of the frequencies transmitted at each step can be anything. This transmission can be one frequency at a time like in step-chirp in a rising or falling order (up-chirp or down-chirp) or a combination of frequency hops, the type of hop determining the type of ambiguity function. The most well known in this class of signals having hopped frequencies is the Costas signal for random hops and stepped frequency signal for systematic hops.

Frequency hopping radars are different from frequency agile radars, as the latter is a pulse radar using different frequencies on a pulse-to-pulse basis. The former, on the other hand, transmits a CW frequency hopped signal. FH waveforms are also called frequency shift keyed (FSK) waveforms. Both these terms will be used in this chapter. Similarly, the terms LFM and chirp will be used interchangeably.

6.2 FREQUENCY HOPPED SIGNALS AS LPI SIGNALS

LPI radars that use FH techniques hop or change the transmitting frequency in time over a wide bandwidth. The hop can be systematic or random. The random hop is based on a pseudo-random sequence and is intended to avoid interception and jamming. The frequency slots used are chosen from an FH sequence and it is this unknown sequence that gives the radar the advantage of LPI, because this sequence is not known to the intercept receiver, which consequently cannot follow the changes in frequency with a high probability. Furthermore, a large frequency hop set is available to choose from, making interception even more difficult. Therefore, this technique does not require controlling the power of emission that FMCW and phase-coded radars require, but rather relies on the randomness of its hopping sequence for LPI. Such radars are consequently easy to intercept but very difficult to exploit.

6.3 STEPPED FREQUENCY WAVEFORM

There are basically two types of FH waveforms, viz. systematic hopping and random hopping. A well-known systematic hopping waveform is the *stepped frequency waveform* (SFW). The well-known random hopping waveform is the *Costas code* [1]. We shall study this further down in this chapter. In Chapter 2 we had examined the SFW very briefly. We shall now study it in greater detail.

SFW may be used to produce a synthetic HRR target profile, by means of inverse discrete Fourier transform (IDFT). The target profile is synthetic because we use the IDFT to produce it. The IDFT is computed by means of frequency domain samples of the actual target range profile. Details are given in [2, 3]. This book is about CW radars. However, it is instructive to study the topic of step frequency as applied to ummodulated waveform and modulated waveform radars to give us a better understanding of this technique.

Briefly,

1. A series of N frequencies is transmitted. The frequency is stepped through a fixed frequency step Δf. Each group of N frequencies is called a burst.
2. The received signal is then sampled at a rate that coincides with the center of each frequency step (range-delayed sampling) to collect and digitize one pair of I and Q samples of the target's baseband response for each transmitted frequency.
3. The quadrature components for each burst are then collected and stored. It is assumed that the target does not change its aspect during the burst and the frequency step size is less than the reciprocal of the target range-delay extent.
4. We then resort to applying spectral weighting to the quadrature components to reduce the time side lobes. Corrections are applied for target velocity, phase and amplitude variations, quadrature sampling bias, and imbalance errors.
5. We then calculate the IDFT of the weighted quadrature components of each burst to synthesize the range profile for that burst. This process is repeated for N bursts to obtain consecutive synthetic HRR profile.

Figure 6–1 shows a typical SFW burst.

The target is assumed to be a point target with a velocity v_t towards the radar and is at an initial range R when time is zero. We analyze only one burst. The transmitted frequency is $s_n(t)$, while the received frequency is $s_n^{rx}(t)$. The echo delay of the moving target is $\tau(t)$.

One n-step burst of a stepped frequency transmitted waveform is [2],

$$s_n(t) = B_n \cos(2\pi f_n t + \phi_n), \quad nT_2 \le t \le nT_2 + T_1 \quad n = 0 \text{ to } N-1$$
$$= 0, \text{ otherwise} \tag{6.1}$$

where ϕ_n is the relative phase and B_n is the amplitude of the nth frequency step at frequency f_n.

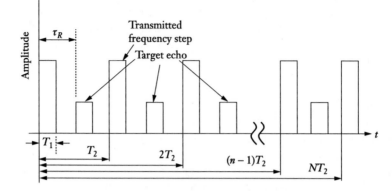

Figure 6–1 Stepped frequency waveform burst. (From [2], © Reprinted with permission)

The received signal is expressed as,

$$s_n^{rx}(t) = B_n^{rx} \cos\left\{2\pi f_n \left[t - \tau(t)\right] + \phi_n\right\}, \quad nT_2 + \tau(t) \le t \le nT_2 + T_1 + \tau(t), n = 0 \text{ to } N-1$$
$$= 0, \quad \text{otherwise} \tag{6.2}$$

where B_n^{rx} is the amplitude of the nth received signal at frequency step n. Range delay for the target with an initial range R at $t = 0$ is

$$\tau(t) = \frac{R - v_t t}{c/2} \tag{6.3}$$

The reference signal is expressed as

$$y_n(t) = B \cos\left(2\pi f_n t + \phi_n\right), \quad nT_2 \le t \le nT_2 + T_2 \quad n = 0 \text{ to } N-1 \tag{6.4}$$

where B is a constant. The received signal is down-converted to baseband in order to extract the quadrature components. This means that the received signal is mixed with the reference. After low pass filtering, the quadrature components are given by

$$\begin{pmatrix} x_I(t) \\ x_Q(t) \end{pmatrix} = \begin{pmatrix} A_n \cos\psi_n(t) \\ A_n \sin\psi_n(t) \end{pmatrix} \tag{6.5}$$

where A_n are constants and the phase of the mixer output is

$$\psi_n(t) = -2\pi f_n \tau(t) \tag{6.6}$$

From equation (6.3), the mixer output phase of equation (6.6) is expressed in terms of target range and velocity as

$$\psi_n(t) = -2\pi f_n \left(\frac{2R}{c} - \frac{2v_t t}{c}\right) \tag{6.7}$$

This is the total echo phase advance seen from transmission to reception for the nth frequency step.

For each waveform, the quadrature components are then sampled at

$$t_i = nT_2 + \tau_r + \frac{2R}{c} \tag{6.8}$$

where τ_R is the time delay associated with the range that corresponds to the *center* of the range profile. The phase of the sampled quadrature mixer output in equation (6.7) then becomes

$$\psi_n(t) = -2\pi f_n \left[\frac{2R}{c} - \frac{2v_t}{c}\left(nT_2 + \tau_R + \frac{2R}{c}\right) \right] \tag{6.9}$$

The sampled mixer outputs from both I and Q channels are expressed as

$$\begin{aligned} X_n &= A_n \left[\cos \psi_n + j \sin \psi_n \right] \\ &= A_n e^{j\psi_n} \end{aligned} \tag{6.10}$$

Equation (6.10) represents samples of the target reflectivity, due to a single burst in the frequency domain. This information can then be transformed into a series of range delay reflectivity (i.e., range profile) values by using the IDFT.

The IDFT is expressed as

$$H_p = \sum_{n=0}^{N-1} X_n e^{j(2\pi/N)np} \tag{6.11}$$

where N is the number of transmitted frequencies per burst and p is the slant-range position.

This form of the IDFT includes the gain N associated with coherent processing of the N stepped frequency waveforms. Therefore, we shall not divide the expression by N, that is, put $1/N$ outside the expression.

Substituting equation (6.10) into equation (6.11) we obtain

$$H_p = \sum_{n=0}^{N-1} A_n \exp\left(j\psi_n\right) \exp j\left(\frac{2\pi np}{N}\right) \tag{6.12}$$

The normalized synthetic response assuming $A_n = 1$ for all p, is expressed as,

$$H_p = \sum_{n=0}^{N-1} \exp\left(j\left(\frac{2\pi np}{N} + \psi_n\right) \right) \tag{6.13}$$

Equation (6.13) at range R and zero target velocity with ψ_n from equation (6.9) becomes

$$H_p = \sum_{n=0}^{N-1} \exp\left(j\left(\frac{2\pi np}{N} - 2\pi f_n \frac{2R}{c}\right) \right) \tag{6.14}$$

Now our derivation can take two turns, viz., stepped frequency LFM signal or stepped frequency signal which is unmodulated. Both are common. We shall at first investigate unmodulated stepped frequency.

This implies that for frequency step size Δf, $f_n = f_0 + n\Delta f$ where f_0 is the initial frequency. In that case

$$H_p = \exp\left(-j2\pi f_0 \frac{2R}{c}\right) \sum_{n=0}^{N-1} \exp\left(j\left[\frac{2\pi n}{N}\left(p - \frac{2NR\Delta f}{c}\right)\right]\right) \tag{6.15}$$

We use the following identity

$$\sum_{p=0}^{\beta-1} \exp(j\alpha p) = \frac{\sin\left(\dfrac{\beta\alpha}{2}\right)}{\sin\left(\dfrac{\alpha}{2}\right)} \exp\left(j\left((\beta-1)\frac{\alpha}{2}\right)\right) \tag{6.16}$$

For $\beta = N, p = n$ and $\alpha = 2\pi y/N$
where

$$y = p - \frac{2NR\Delta f}{c} \tag{6.17}$$

we obtain the synthetic response of equation (6.15) as

$$H_p = \exp\left(-j2\pi f_0 \frac{2R}{c}\right) \frac{\sin(\pi y)}{\sin\left(\dfrac{\pi y}{N}\right)} \exp\left(j\frac{N-1}{2}\frac{2\pi y}{N}\right) \tag{6.18}$$

Finally, the synthesized range profile is

$$|H_p| = \left| \frac{\sin(\pi y)}{\sin\left(\dfrac{\pi y}{K}\right)} \right| \tag{6.19}$$

In case of stepped LFM, at equation (6.15) we use $f_n = n\Delta f$, to finally obtain

$$H_p = \frac{\sin \pi y}{\sin\dfrac{\pi y}{N}} \exp\left(j\frac{N-1}{2}\frac{2\pi y}{N}\right) \tag{6.20}$$

which leads to the same result as in equation (6.19).

6.3.1 Range Resolution and Range Ambiguity

The range resolution is bound by the system bandwidth. Assuming SFW with n steps and step size Δf, then the corresponding range resolution is

$$\Delta R = \frac{c}{2N\Delta f} \tag{6.21}$$

Range ambiguity for an SFW system can be determined by examining the phase term corresponding to a point scatterer located at range R.

$$\psi_n(t) = 2\pi f_n \frac{2R}{c} \tag{6.22}$$

Therefore, differentiating with respect to frequency,

$$\frac{\Delta\psi}{\Delta f} = \frac{4\pi R}{c} \tag{6.23}$$

or

$$R = \frac{\Delta\psi}{\Delta f}\frac{c}{4\pi} \tag{6.24}$$

Inspection of equation (6.24) shows that range ambiguity exists with multiples of $2\pi m$ with $m \in \mathbb{Z}$ where $\mathbb{Z}$ designates the set $[-\infty, +\infty]$

Therefore,

$$R = \frac{\Delta\psi + 2m\pi}{\Delta f}\frac{c}{4\pi} = R + m\left(\frac{c}{2\Delta f}\right) \tag{6.25}$$

Hence, from equation (6.25), the unambiguous range window is

$$R_u = \frac{c}{2\Delta f} \tag{6.26}$$

If the target is located beyond R_u, then because of the IDFT, it will fold over. This means that scatterers located outside the unambiguous range will fold over and appear in the synthesized profile. In order to avoid this

$$\Delta f \leq \frac{c}{2D} \tag{6.27}$$

where D is the target extent in meters.

The frequency step size Δf must be less than the bandwidth of the waveform. If this condition is met, then the target will be completely profiled without any gaps. This will ensure that the clutter surrounding the target does not contaminate the synthesized target range profile.

$$\Delta f \leq B_{tx} \tag{6.28}$$

This is illustrated in Figure 6–2. In the figure, B_{tx} is the bandwidth of the waveform (chirp waveform sweep bandwidth, in this case), Δf is the frequency step size and B_t is the radar bandwidth. This is a case of $B_{tx} = \Delta f$. The aim is to reconstruct a wide portion of the target's reflectivity spectrum by piecing together several adjacent portions of the spectrum, each obtained by separate transmission and reception of pulses of bandwidth B_{tx}, but stepped appropriately in frequency by appropriate choice of the carrier frequency. If $B_{tx} < \Delta f$ then there will logically be gaps in the reconstruction. Hence, at the very minimum B_{tx} should be equal to Δf. It can be seen that the target is completely covered. Mathematically this required shift in the positive direction is computed as,

$$\delta f_i = \left(i + \frac{1-n}{2}\right)\Delta f$$

where n is a sequence of adjacent windows (indexed by $i = 0,\ldots,n-1$).

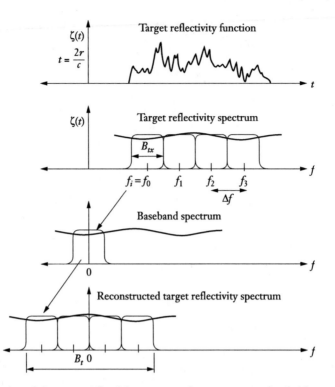

Figure 6–2 Reconstruction of the target reflectivity spectrum for $n = 4$ transmitted chirp waveform steps, each with carrier frequency f_i and bandwidth B_{tx}. (From [4], © IEEE 1998)

This is just illustrative of a four frequency step case. Before the IFFT, the target returns are arranged in the ascending order of frequency so as to obtain a complete spectral reconstruction. We take one complex sample of each waveform. This will then make a four sample set for four waveforms. We then carry out a 4-point IFFT. Do not forget that as we discussed earlier, resolution in FH radars depend upon the frequency difference across the band. This can happen only if we have a large number of steps. If the frequency steps are fine, then the target profiling is better as is readily apparent in Figure 6–2.

We illustrate this with two cases. We use the accompanying software entitled "*sfw_resolve.m*".

Design of Unmodulated SFW Radar: Table 6–1 deals with the first case of eight steps. This radar is an unmodulated waveform radar, that is, we are not transmitting chirp signals. The number of frequency steps determines the unambiguous range of the radar. Suppose we require an unambiguous range of, say, 2,400 m. We wish to cover this in eight steps. This means a signal length of

$$\frac{R_u}{8c} = \frac{2400}{8 \times 3 \times 10^8} = 1 \ \mu\text{sec}$$

Therefore, from equation (6.26)

$$\Delta f = \frac{c}{2R_u} = \frac{3 \times 10^8}{2 \times 2400} = 62.5 \ \text{KHz}$$

This satisfies equation (6.27) if we take 2,400 m (unambiguous range) as the target extent.

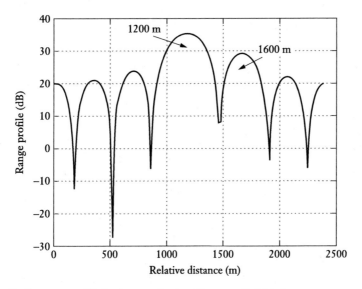

Figure 6–3 Synthetic range profile for two resolved scatterers and eight steps.

Table 6–1 Stepped Frequency Unmodulated Waveform Radar Parameters With Eight Steps

First center frequency	f_0	5.2625 GHz
Frequency step size	Δf	62.5 kHz
Number of steps	N	8
Total radar bandwidth	B_t	500 kHz
Signal length	T_1	1 μsec

This also satisfies equation (6.28) where

$$B_{tx} = \frac{1}{T_1} = \frac{1}{1 \times 10^{-6}} = 1 \text{ MHz}$$

where T_1 is the pulse width (see Figure 6–1)

The entire receiver window, in this example, has been taken as the unambiguous range. Hence, the initial range is zero. The target extent has, say, two scatterers located at [1200, 1600] meters with varying RCS of [100, 10], respectively. This scenario can be construed as two points on an aircraft carrier 500 m long. This can be the hull of the carrier and the island superstructure. The carrier is located 1,200 m from the radar.

The range resolution for these parameters is given by (6.21) as

$$\Delta R \geq \frac{c}{2N\Delta f} = \frac{3 \times 10^8}{2 \times 8 \times 62.5 \times 10^3} = 300 \text{ m}$$

Our scatterers are located 400 m apart. Hence, they should be resolved. This is confirmed in Figure 6–3. No window was used. The IFFT was a 128-point one with zero padding (8 + 120 zeros). This signal processing scheme assumes a central bandfilter whose bandwidth is small enough to remove the ambiguity associated with the transmitted pulse and the PRF. For example, if $T_1 = 1 \mu$sec and $T_2 = 2 \mu$sec (T_2 is the start of the next pulse, see Figure 6–1),

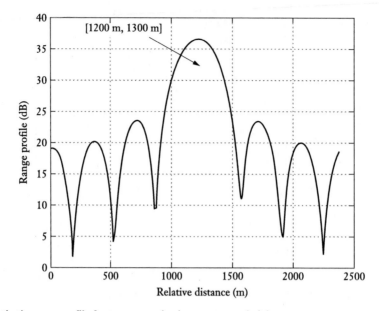

Figure 6–4 Synthetic range profile for two unresolved scatterers and eight steps.

the central bandfilter must have infinite cut-off at 500 KHz, which is unrealistic. Hence, due to these reasons, T_2 is usually 3 times T_1. This scheme has one drawback, in that because we are measuring fractions of a cycle per pulse, it is very susceptible to noise and very power inefficient. I am indebted to David Lynch[1] for this information.

If the scatterers are located at [1200, 1300] the targets cannot be resolved since they are 100 m apart, which is less than the range resolution for this radar. This is confirmed in Figure 6–4.

The targets are static. We now investigate the same targets with a 64-step radar. Due to the changed parameter, the waveform length has changed to 0.125 μsec, the signal bandwidth has changed to $B_{tx} = 8$ MHz, and the range resolution has improved to 37.5 m. Other parameters remained the same. Figures 6–5 through 6–7 show the drastic improvement in the results. Once again we did not use any weighting, but as before we used a 128-point IFFT with zero padding. Figure 6–7 shows us that with a 30 meter separation, the scatterers were unresolved as it is below the new range resolution of 37.5 m. Table 6–2 shows us the new parameters for the 64-step radar. Therefore, *higher steps give better range profiling*.

Design of LFM Waveform SFW Radar: We now examine the design procedure for the LFM (chirp) stepped frequency radar. The following are the salient steps.

Decide on the range resolution and the unambiguous range. Let us assume that we require an unambiguous range of 30 m with a resolution of 6 m.

1. We apply equation (6.26) to determine the frequency step Δf required to achieve this unambiguous range.

$$\Delta f = \frac{c}{2R_u} = \frac{3 \times 10^8}{2 \times 30} = 5 \text{ MHz}$$

1 Personal correspondence with the author.

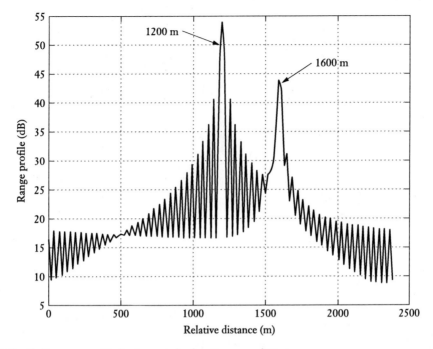

Figure 6–5 Synthetic range profile for two resolved scatterers and 64 steps.

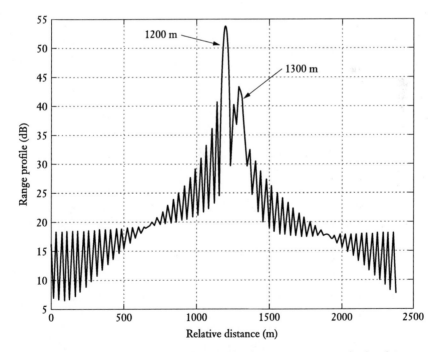

Figure 6–6 Synthetic range profile for two *formerly* unresolved scatterers, now resolved and 64 steps.

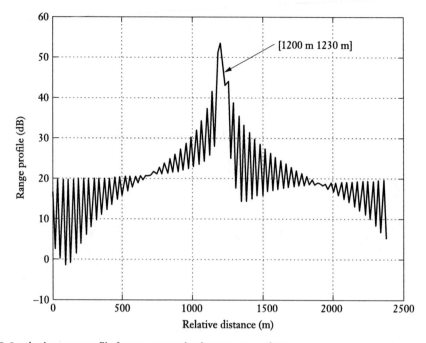

Figure 6–7 Synthetic range profile for two unresolved scatterers and 64 steps.

Table 6–2 Stepped Frequency Unmodulated Waveform Radar Parameters With 64 Steps

First center frequency	f_0	5.2625 GHz
Frequency step size	Δf	62.5 KHz
Number of steps	N	64
Total radar bandwidth	B_t	4 MHz
Signal length	T_1	0.125 μsec

2. We now determine the number of steps necessary to achieve our desired range resolution based on $\Delta f = 5$ MHz. Using equation (6.21), we obtain

$$N \geq \frac{c}{2\Delta R \Delta f} = \frac{3 \times 10^8}{2 \times 6 \times 5 \times 10^6} = 5 \text{ steps}$$

3. The number of steps need not be to the base two. We can always pad the IFFT to the desired number. In our case we use a 128-point IFFT, that is, 5 + 123 zeros. Such an IFFT will give us a smooth curve due to high IFFT resolution. Zero padding yields smooth curves.
4. Based on the available ADCs choose a chirp bandwidth that exceeds or at most is equal to the frequency step Δf, say, 5 MHz. Choose a convenient sweep time, say, 1 μsec.
5. The chosen PRF is the inverse of the sweep duration.

Therefore, our chirp signal has a bandwidth of 5 MHz and sweep duration of 1 μsec. This completes the design. Table 6–3 gives the final design results.

Note that in an FH system using stepped frequency, it is the bandwidth that affects resolution. This bandwidth is determined by the frequency step size and the number of steps, that is, $N\Delta f$. Therefore, we get better results if we increase the number of steps.

Table 6–3 Stepped Frequency Chirp Radar Parameters With 5 Steps

First center frequency	f_0	5.2625 GHz
Frequency step size	Δf	5 MHz
Number of steps	N	5
Total radar bandwidth	B_t	25 MHz
Sweep duration	T_s	1 μsec
Chirp bandwidth	B	5 MHz

Table 6–4 Stepped Frequency Chirp Radar Parameters With 64 Steps

First center frequency	f_0	5.2625 GHz
Frequency step size	Δf	10 MHz
Number of steps	N	64
Total radar bandwidth	B_t	640 MHz
Sweep duration	T_s	1 μsec
Chirp bandwidth	Hz	15 MHz

The chirp signal bandwidth may be equal to or *more* than the frequency step size, but not less than the frequency step size. In Table 6–3, the equal case is considered. The range resolutions for this type of radar are defined by equation (6.21) and *not* by the chirp signal bandwidth. Therefore, it will be advantageous to make $\Delta f = B$, the chirp bandwidth to get maximum benefit from the range resolution. The reader would have guessed by now that it is useful to use the chirp waveform in stepped frequency systems because it helps us in processing very fine resolutions, which would otherwise require very narrow unmodulated waveform widths, for example, 20 ns for 6 m range resolutions that are difficult to generate in an SFW processor.

In Section 6.5 below we discuss a chirp SFW system which uses *two* chirp waveforms in one frequency step, instead of one like in the systems we have studied so far. The idea here is to use the second chirp waveform with the first one as a reference in order to obtain exact range and Doppler readings without any of the range ambiguities due to range-Doppler coupling associated with chirp waveforms.

6.3.2 Effect of Target Velocity

Suppose we have the following imaginary chirp SFW radar (Table 6–4).

The defining factor in Table 6.4 is the frequency step size which is 10 MHz. We are, therefore, losing out on the benefit of the chirp bandwidth (not good!). Hence, the unambiguous range for this radar is 15 m, while the range resolution is 0.234 m. Like in earlier cases, we assume that we are using a central band filter. The initial range is 900 m and there are two scatterers at [908,910] meters with varying RCS of [100, 10] respectively. The target extent is, therefore, 10 m. It can be more than this, but it must satisfy (6.27) and should be less than the unambiguous range of 15 m. Otherwise, there will be folding of echoes beyond 15 m target extent. The two targets are clearly resolved as shown in Figure 6–8.

We now assume that the target has a velocity of 500 m/sec. This has broadened the waveforms of the scatterer returns (see Figure 6–9).

This is because of the second term in equation (6.9). It can be remedied if we decrease the PRI suitably. The reader can experiment with this as an exercise. Alternately, we can multiply the

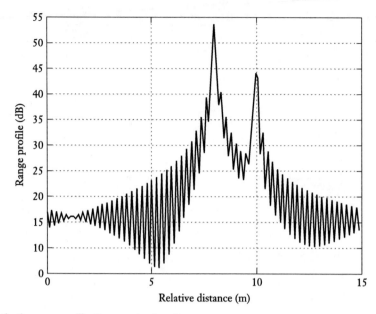

Figure 6–8 Synthetic range profile. Two resolved scatterers.

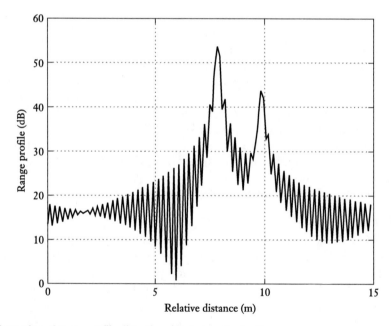

Figure 6–9 Illustration of range profile distortion due to target velocity.

complex received data at each waveform by the phase term (derived from equation (6.9))

$$\psi_n(t) = -2\pi f_n \left[\frac{2R}{c} - \frac{2v_t}{c} \left(nT_2 + \tau_R + \frac{2R}{c} \right) \right] \tag{6.29}$$

Implementing equation (6.29) is more difficult, because we require accurate values for target velocity and range.

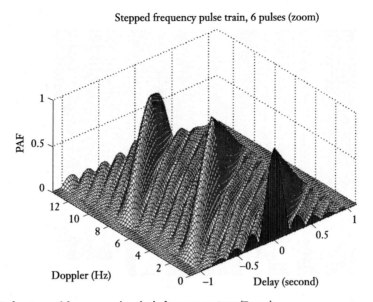

Stepped frequency pulse train, 6 pulses (zoom)

Figure 6-10 PAF of a stepped frequency signal, six frequency steps (Zoom).

Figures 6–3 through 6–9 have been obtained using the program "*sfw_resolve.m*" in the accompanying software. This same software is used for the unmodulated SFW case as well as for chirp SFW case, except that in the latter case, we need to follow the procedure of inputting parameters as outlined in this section.

A notable problem here is that in *unmodulated* waveform SFW radars, if the BT product $T\Delta f > 1$, the ACF of the signal exhibits grating lobes in addition to the main lobe (at $\tau = 0$) at

$$|\tau| = \frac{n}{\Delta f}, \quad n = 1, 2, \dots, n_{max} \text{ where } n_{max} = |T\Delta f| \tag{6.30}$$

This does not happen with LFM SFW. Details are given in [1] with methods to counteract it.

In Chapter 3 we had examined the ambiguity function of SFW signals. We noted that the signals have an ambiguity function like that of the CPT (not surprising, since this is also like a pulse train), but with each cardinal point having an ambiguity function that is skewed to the delay axis just like LFM. This is what imparts range-Doppler coupling to these signals. We now examine the PAF, ACF, and PACF. These figures have been obtained using the program "*ambfn7.m*" provided by Levanon [5]. These plots pertain to the zoom on one of the cardinal points.

We note from Figures 6–10 and 6–11 that SFW is not a perfect signal as the PACF has time side lobes just like LFM. One of the enduring problems with SFW is that it takes a long time to generate. In case of fast moving targets, it causes Doppler *smear* [6]. This is because SFW has to proceed step by step through a whole lot of steps, especially for large bandwidths. The transmission is serial, which takes time. This can now be avoided using the parallel concept discussed in Part III of this book.

6.4 RANDOM FREQUENCY HOPPED WAVEFORMS

We now investigate signals wherein the hopping is carried out in a random manner. Costas codes [1] belong to this class of signals. Costas signals (or codes) are similar to SFW, the only difference being that the frequencies for the steps are selected in a pseudo-random fashion, according to some

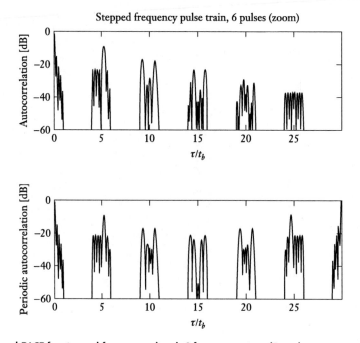

Figure 6–11 ACF and PACF for stepped frequency signal, 6 frequency steps (Zoom).

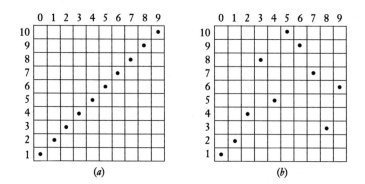

Figure 6–12 Frequency assignment for a burst of N subwaveforms (a) quantized LFM (b) Costas code of length $N_c = 10$.

predetermined logic and not in a rising fashion as has been done for SFW. The target returns are, however, processed by rearranging them in a rising fashion at the baseband just like SFW. The idea here is that Costas codes have a very sharp ambiguity function. This is what makes them very appealing.

Consider the $N \times N$ matrix in Figure 6–12. The rows are indexed from $i = N, N-1, \ldots, 2, 1$ and columns are indexed from $j = 1, 2, \ldots, (N-1)$. The rows are used to denote the steps and the columns are used to denote the frequency. A "dot" indicates the frequency value assigned to the associated step. Figure 6–12(a) shows the frequency assignment associated with SFW. In contrast, Figure 6–12(b) has pseudo-random frequency assignments. It will be appreciated that a matrix of size $N \times N$ has a total of $N!$ possible ways exist of assigning the "dots," that is, $N!$ possible codes.

The sequences of "dots" in Figure 6–12(b) are so chosen that the corresponding ambiguity function approaches an ideal or a "thumbtack" response. Indeed this can be easily verified. If we overlay a binary matrix representing the signal upon itself and then shift one relative to the other according to the delay (horizontal shifts) and the Doppler (vertical shifts) we note that at each combination of shifts, the number of coincidences between points of the fixed and the shifted matrix represents the relative height of the ambiguity function. This exercise for Figure 6–12(b) shows that except for the zero shifts case, when the number of coincidences is N, we cannot find any combination of shifts that will yield more than one coincidence. This is the criteria of Costas sequences, viz., those sequences of frequency hopping that will yield no more than one coincidence [1, 7].

The readers can verify that if this exercise is done for Figure 6–12(a), we note that for zero shift case the number of coincidences is $N = 10$. One shift to the right combined with one shift up, will result in $N - 1 = 9$ coincidences. Two shifts to the right and two up will result in $N - 2 = 8$ coincidences and so on. Thus the well-known diagonal ridge familiar from LFM signal is created [1].

Costas showed that the output of the matched filter is [7]. We use the same notation as in Chapter 3, with T for pulse duration and τ for delay.

$$\chi(\tau,\nu) = \frac{1}{N} \sum_{n=0}^{N-1} \exp(j2\pi n\nu T) \left\{ \Phi_{nn}(\tau,\nu) + \sum_{\substack{m=0 \\ m \neq n}}^{N-1} \Phi_{nm}\left[\tau - (n-m)T, \nu\right] \right\} \tag{6.31}$$

where

$$\Phi_{nm}(\tau,\nu) = \left(T - \frac{|\tau|}{T}\right)\frac{\sin\alpha}{\alpha}\exp(-j\beta - j2\pi f_m\tau), \quad |\tau| \leq T$$

$$0 \text{ elsewhere} \tag{6.32}$$

in which

$$\alpha = \pi\left(f_n - f_m - \nu\right)\left(T - |\tau|\right) \tag{6.33}$$

and

$$\beta = \pi\left(f_n - f_m - \nu\right)\left(T + \tau\right) \tag{6.34}$$

The ambiguity function for a length 7 Costas code (coding sequence 4, 7, 1, 6, 5, 2, 3) is shown in Figure 6–13. Note the sharp spike at the origin.

All side lobes have an amplitude less than or equal to $1/N$. The compression ratio of a Costas code is approximately N. This can be verified from the ACF plot in Figure 6–15(a). The pedestal is not smooth. Figure 6–14 shows the PAF for this code. It is not a perfect code. Figure 6–15 shows the ACF and PACF. Because of the spiky nature of this signal, Costas codes have little or no Doppler tolerance. In fact, this was originally developed as a noncoherent signal [7]. Construction algorithms for Costas sequences are given in [7, 8].

These figures have been obtained using the program "*ambfn7.m*" provided by Levanon [5].

6.5 TECHNOLOGY FOR SFCW

The preceding sections have dealt with the theory of this class of signals. Nothing has been said about the signal processing hardware and design details. These aspects have been extensively investigated in Part III of this book on Pandora radar. In fact, Chapter 10 especially deals with SFCW radar design. The Pandora radar is a multi-channel radar. The commonly used hardware design for SFCW is just any one single channel of this radar.

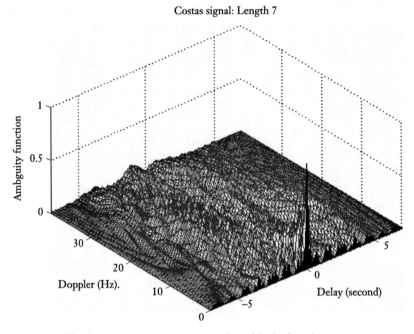

Figure 6–13 Costas signal (coding sequence 4, 7, 1, 6, 5, 2, 3): Ambiguity function.

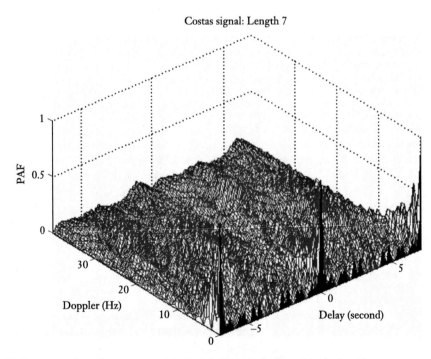

Figure 6–14 Costas signal (coding sequence 4, 7, 1, 6, 5, 2, 3): Periodic ambiguity function.

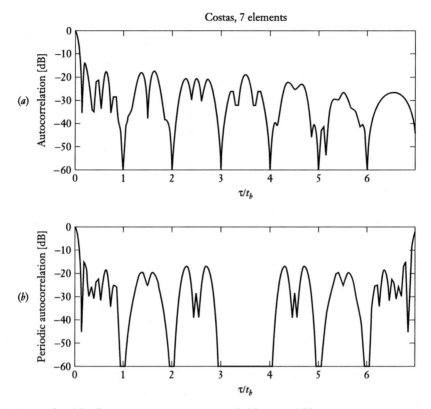

Figure 6–15 Costas signal (coding sequence 4, 7, 1, 6, 5, 2, 3): (*a*) ACF and (*b*) PACF.

6.6 HYBRID FSK/LFM TECHNIQUE

This book is about LPI radar design. We have seen that both LFM as well as FSK signals lend themselves to LPI usage. These are pure signals. However, if we generate a hybrid signal, then, depending upon the specific application, one can expect a more enhanced performance, than would otherwise have been given by these individual constituent signals. There are many such variations of hybrid signals. Pace ([9], p. 177) discusses hybrid PSK/FSK systems in his book on LPI radars so as to achieve a high time–bandwidth product, enhancing the LPI features of the radar. In using hybrid signals, the ambiguity properties of the constituent signals are retained. Admittedly, in this example of automobile radars, we are not worried about the stealth aspect. It is interesting to see as to how one can use both the FSK and LFM signaling methods to advantage in a multitarget environment.

In automobile radars there is a need for quick and accurate readouts of target range and Doppler in a dense target environment. The resolutions in range are typically 1 m. The technology behind this is usually pulse radars using ultra short pulse length of around 10 nsec (chirp pulse radars) or SFCW radars both umodulated and LFM based. Consider the LFM case for chirp SFCW radars. If we require a range resolution of 1 m, we would require a bandwidth of 150 MHz. Now if we require the true Doppler value of the target, we would require a retransmission of the chirp waveform with a different slope with a view to comparing the phases of the two target returns. Sometimes the transmission of a triangular waveform is preferred as discussed in Chapter 4. In multiple target environments we would require multiple chirp signals with different slopes. Such retransmissions are time consuming, typically 50–100 msec. Furthermore, the requirement for correction of range-Doppler coupling persists. We cannot gloss over it by widening the range gates so as to accommodate this coupling due to Doppler. Many of these automobile radars are linked to other systems like Collision Avoidance (CA) wherein a passenger's life may depend upon accurate knowledge of the range and Doppler of the automobile ahead. The clearances are often a matter of a few feet! Retransmission and complex signal processing of the chirp waveform takes time, which in most cases is never available. In view of the above, a lot of research has been done with a view to improving the quality of automobile radars. In this section we shall discuss one such noteworthy effort on the part of the Hamburg–Harburg Technical Institute in Germany. This and subsequent sections have been reproduced here with permission from IEEE.

In automotive radar systems, the maximum range for automotive radars is usually 200 m, the range resolution is 1 m and the velocity resolution is 2.5 km/hr [10]. In order to meet these requirements specific waveforms need to be designed. FMCW radar systems are generally preferred because of their low measurement time and low computational complexity. The preferred waveforms for automobile applications are usually, LFM or FH waveforms. The following discussion is reproduced with permission from IEEE based on the paper by Rohling and Meinecke [10 © IEEE 2001]. The radar under consideration operates at 77 GHz.

The authors considered the use of pure FH and pure LFM for this application and made the following observations. The pure FH approach is shown in Figure 6–16. This uses two frequencies f_A and f_B, the so-called two frequency measurement [11], in the transmit signal. Each frequency is transmitted inside a coherent processing interval (CPI) of length $T_{CPI} = 5$ msec. Using a homodyne, the echo signal is down-converted to baseband and sampled N times. The frequency step $\Delta f_{step} = f_B - f_A$ is small and due to this a single target will be detected at the same Doppler frequency position in the adjacent CPIs but with different phase information on the two spectral peaks. The phase difference $\Delta\varphi = \varphi_B - \varphi_A$ in the complex spectra is the basis for target range estimation R. The relation between target distance and phase difference is given by [10]

$$R = \frac{c\Delta\varphi}{4\pi\Delta f_{step}} \qquad (6.35)$$

The maximum unambiguous range for such a radar is required to be 150 m with a frequency step $\Delta f_{step} = 1$ MHz. The target resolution depends upon CPI length T_{CPI}. The technically simple VCO modulation is an additional advantage of this waveform. But a two frequency step signal with $\Delta f_{step} = 1$ MHz has very poor range resolution given by $\Delta R = c/2n\Delta f_{step} = 3 \times 10^8 /(2 \times 2 \times 1 \times 10^6) = 75$ m which is clearly unacceptable. Hence, multiple automobiles cannot be resolved.

Radars which apply pure LFM technique, modulate the transmit frequency with a triangular waveform. The sweep bandwidth Δf_{sweep} is typically 150 MHz, yielding a range resolution of 1 m.

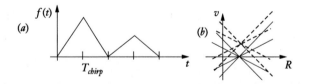

Figure 6–16 (a) Waveform for use in multitarget situations and (b) corresponding example $R - v$ diagram for a two target situation and the related intersection points. (From [10], © IEEE 2001)

However, a single sweep of the chirp waveform gives an ambiguous measurement in range R and velocity v. The down-converted receive signal is sampled and Fourier transformed inside a single CPI. If a spectral peak is detected in the Fourier spectrum at index n (normalized integer frequency) the ambiguities in target range and velocity can be described in an $R - v$ diagram by the following equation [10]

$$n = \frac{v}{\Delta v} - \frac{R}{\Delta R} \Leftrightarrow \frac{v}{\Delta v} = \frac{R}{\Delta R} + n \qquad (6.36)$$

where ΔR is the range resolution and Δv is the velocity resolution given by

$$\Delta v = \frac{\lambda}{2T_{Chirp}} = 0.8 \text{ m/sec}$$

λ is the wavelength of 4 mm at 77 GHz radar frequency and $T_{Chirp} = 2.5$ msec.

The chirp time of 2.5 msec will be explained below. Due to reasons of range-velocity ambiguities further measurements are necessary with different chirp gradients in order to achieve unambiguous range-velocity measurement even in a multitarget situation. The well-known up/down chirp principle depicted in Figure 4–10 is described in detail in [12]. LFM waveforms can be used in multitarget environments, but the extended measurement time is an important drawback of this technique. In multitarget situations a waveform as shown in Figure 6–16(a) is used which consists of four different chirp signals. In general, the frequency modulation will be different in each of the four chirps. In each chirp signal all targets are detected which still fulfill equation (6.36). The detected spectral lines from all four chirp signals can be drawn in a single $R - v$ diagram, as shown in Figure 6–16(b) where the gradient of a single line is dependent on the chirp sweep rate. In multiple target situations many intersections between lines of different and adjacent chirps appear as in the example in Figure 6–16(b). If such an intersection point occurs which has no physical representation of a reflection object, it is called a ghost target. A real target is represented by an intersection point between all considered four lines.

Concept of Combined FH and LFM Waveforms: In view of the above, the Technical Institute of Hamburg–Harburg came up with a new proposal of combining FH and LFM signals [10]. This proposal offers the possibility of an unambiguous target range and velocity measurement simultaneously. The transmit waveform consists of two LFM up-chirp signals (the intertwined signal sequences are called A and B). The two chirp signals are transmitted in an intertwined sequence (ABABAB…), where the stepwise frequency modulated sequence A is used as a reference signal while the second up-chirp signal is shifted in frequency with f_{Shift}. The received signal is down-converted into baseband and directly sampled at the end of each frequency step. The combined and intertwined waveform concept is illustrated in Figure 6–17.

It can be seen that essentially there are two LFM SFW waveforms A and B intertwined. These waveforms are transmitted by turns. This means that first, A is transmitted (only one sweep) and then B (only one sweep). This is again followed by one sweep of A and so on. This is different from the usual LFM SFW sweep discussed earlier in this chapter wherein there was one chirp waveform per frequency step. In this case we have *two* chirp waveforms per frequency step, first A then B. Each signal sequence A or B is processed separately by using the Fourier transform and CFAR target detection techniques. A single target with a specific range and velocity will be detected in both sequences at the same integer index $n = n_A = n_B$ in the FFT output signal of the two processed spectra. In each sequence A or B the same target range and velocity ambiguities occur as described in equation (6.36). But the measured phases φ_A and φ_B of the two complex

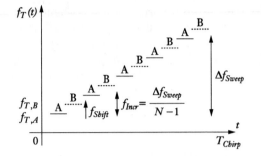

Figure 6–17 Combined FH-LFM CW waveform principle. (From [10], © IEEE 2001)

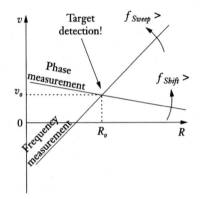

Figure 6–18 Graphical resolution principle of ambiguous frequency and phase measurements. (From [10], © IEEE 2001)

spectral peaks are different and include the fine target range and velocity information, which can be used for ambiguity resolution. The phase difference $\Delta\varphi$ can be evaluated for target range and velocity estimation as $\Delta\varphi = \varphi_B - \varphi_A$. It is given by [10]

$$\Delta\varphi = \frac{\pi}{N-1} \cdot \frac{v}{\Delta v} - 4\pi R \frac{f_{Shift}}{c} \tag{6.37}$$

where N is the number of frequency steps in each transmit signal sequence A and B. This is the nomenclature convention that we have been following all along in this book. The ambiguity of $\Delta\varphi$ can be resolved graphically as shown in Figure 6–18.

The analysis leads to an unambiguous target range R_0 and relative velocity v_0 [10]:

$$R_0 = \frac{c\Delta R}{\pi} \cdot \frac{(N-1)\Delta\varphi - \pi n}{c - 4(N-1)f_{Shift}\Delta R} \tag{6.38}$$

$$v_0 = \frac{(N-1)\Delta v}{\pi} \cdot \frac{c\Delta\varphi - 4\pi f_{Shift}\Delta R n}{c - 4(N-1)f_{Shift}\Delta R} \tag{6.39}$$

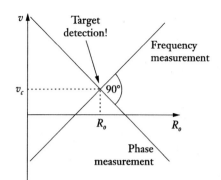

Figure 6–19 R – v diagram for the combined waveform with optimized frequency shift. (From [10], © IEEE 2001)

This new intertwined waveform shows that unambiguous target range and velocity measurements are possible even in a multi-target environment. An important advantage is the short measurement and processing time.

Example

Suppose the signal bandwidth is $\Delta f_{sweep} = 150$ MHz for a range resolution of 1 m. The stepwise frequency modulation is split into $N = 256$ separate bursts of $f_{Incr} = (150 \text{ MHz}/255) = 588$ kHz each. The increment is 588 kHz, meaning that A and B will have sweep bandwidths of 294 kHz each. The measurement time inside a single burst A or B is assumed to be 5 μsec resulting in a chirp duration of the intertwined signal of $T_{Chirp} = 5 \times 10^{-6} \times 2 \times 256 = 2.56$ msec. The factor 2 is necessary to account for the fact that A and B *together* comprise a total of 512 steps. This value results in a velocity resolution of

$$\Delta v = \frac{\lambda}{2T_{Chirp}} = \frac{0.004}{2 \times 2.56 \times 10^{-3}} = 2.81 \text{ km/h} \tag{6.40}$$

The equation at (6.40) is derived from the basic Doppler equation ($V = 2f_D/\lambda$) discussed in earlier chapters and where $f_D = 1/T_{Chirp}$. It was found that the parameter f_{Shift} needed to be optimized on the basis of high range and velocity accuracy. The highest accuracy occurs if the intersection point in the $R - v$ diagram results from two orthogonal lines as illustrated in Figure 6–19. Hence, the frequency shift between the signal sequences A and B is

$$f_{Shift} = -\frac{1}{2} f_{Incr} = -294 \text{ kHz} \tag{6.41}$$

The waveform pertaining to equation (6.41) is shown in Figure 6–20.

In view of this modified f_{Shift} equations (6.38) and (6.39) take the form [10],

$$R_0 = \frac{c\Delta R}{\pi} \cdot \frac{(N-1)\Delta\varphi - \pi n}{c - 4(N-1)f_{Shift}\Delta R} \quad \text{(from 6.38)}$$

We know from (6.41)

$$f_{shift} = -\frac{1}{2} f_{incr}$$

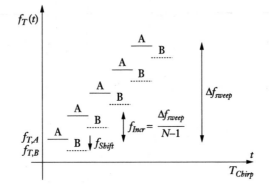

Figure 6–20 Combined FH-LFM waveform with optimized frequency shift. (From [10], © IEEE 2001)

and from Figure 6–20,

$$f_{incr} = \frac{f_{sweep}}{N-1}$$

By definition,

$$\Delta R = \frac{c}{2f_{sweep}}$$

Substituting in (6.38) we obtain,

$$R_0 = \frac{\Delta R}{\pi}\left(\frac{(N-1)\Delta\varphi - \pi n}{2}\right), \text{ or}$$

$$\frac{R_0}{\Delta R} = \frac{1}{\pi}\left(\frac{(N-1)\Delta\varphi}{2} - \frac{\pi n}{2}\right)$$

$$= \frac{(N-1)}{2\pi}\Delta\varphi - \frac{n}{2}$$

$$\therefore \quad \frac{R_0}{\Delta R} = \frac{N-1}{2\pi}\cdot\Delta\varphi - \frac{n}{2} \tag{6.42}$$

Similarly,

$$\frac{v_0}{\Delta v} = \frac{N-1}{2\pi}\cdot\Delta\varphi + \frac{n}{2} \tag{6.43}$$

This new waveform was tested by the institute by fitting it out on an experimental car in realistic street situations. Figure 6–21 shows the test car equipped with a 77 GHz FMCW radar sensor with separate transmit and receive antennas, a smart brake buster and throttle control system for automatic driving. The radar sensor detects all targets inside the observation area and measures target range and velocity simultaneously. The relevant target is selected by signal processing and the car control system is activated by this information. In this case the car followed the detected relevant object with controlled distance [10]. The institute claims that the car has done more than 40,000 km on public streets [10]. The target azimuth angle is measured by the receive signal in three adjacent and overlapping beams. All this takes place within a short measurement time.

Figure 6–21 Experimental car of the Technical University of Hamburg–Harburg equipped with a 77 GHz far range radar sensor. (From [10], © IEEE 2001)

One of the major problems in such systems is multiple echoes and the requirement for a large apertures and scanning in angle.

References

1. Levanon, N., Radar Principles, John Wiley & Sons, New York, 1988.
2. Wehner, D. R., High Resolution Radar, Artech House, Norwood, MA, 1987.
3. Mahafza, B. R. and Elsherbeni, A. Z., *MATLAB Simulations for Radar Systems Design*, Chapman & Hall/CRC Press, Boca Raton, FL, 2004.
4. Wilkinson, A. J., Lord, R. T., and Inggs, M. R., "Stepped-frequency processing by reconstruction of target reflectivity spectrum," in *Proceeding of the 1998 IEEE South African Symposium on Communications and Signal Processing (COMSIG '98)*, ISBN 0–7803-5054–5, September 7–8, 1998, pp. 101–104.
5. http://www.eng.tau.ac.il/~nadav/amb-func.html
6. Nathanson, F. E., *Radar Design Principles*, 2nd edn., McGraw-Hill, New York, 1991.
7. Costas, J. P., "A study of a class of detection waveforms having nearly ideal range-Doppler ambiguity properties," *Proceedings of IEEE*, Vol. 72, 1984, pp. 996–1009.
8. Golomb, S. W. and Taylor, H., "Constructions and properties of Costas arrays," *Proceedings of IEEE*, Vol. 72, 1984, pp. 1143–1163.
9. Pace, P. E., *Detecting and Classifying Low Probability of Intercept Radar*, Artech House, Norwood, MA, 2004.
10. Rohling, H. and Meinecke, M.-M., "Waveform design principles for automotive radar systems," in *2001 CIE International Conference on Radar*, Proceedings, October 15–18, 2001, pp. 1–4.
11. Artis, Jean-Paul and Henrio, J.-F., "Automotive radar development methodology," in *International Conference on Radar Systems*, Brest, France, 1999.
12. Rohling, H., Meinecke, M.-M., Klotz, M., and Mende, R., "Experiences with an experimental car controlled by a 77 GHz radar sensor," in *International Radar Symposium*, IRS-98, Munich, 1998.

PART II

Theory and Design of Calypso FMCW Radar

7

Calypso FMCW Radar[*]

7.1 INTRODUCTION

This chapter pertains to an FMCW navigation radar named Calypso. The name is the author's pseudonym for an actual FMCW navigational radar [1, 2] developed for navigation jointly by Philips Research Laboratory (PRL), Redhills, Surrey, England and Hollandse Signaal Apparaten B.V. (now Thales Netherlands B.V.), Hengelo, The Netherlands. This radar was the precursor of today's PILOT and Scout navigational radars. The contents of this chapter have been reproduced from [1] and [2] with permission. The details from [1] pertain to the radar after its first trials, while [2] pertains to its configuration and performance at the end of the research phase. In doing so, care has been taken to ensure that the language of the original documents has been retained as well as the drawings therein. Where necessary, additional explanatory notes have been added, without destroying the flavor of the original document. The reader was initially introduced to this radar in Chapter 1. In this chapter, we shall examine in detail certain remaining parameters of this radar. We shall then examine issues like, power budget, noise figures, noise cancellation, ADCs, calibration and verification, MTIs, and so forth. We shall also investigate issues like single antenna operation and reflected power cancellers (RPCs) as well as removal of range ambiguities.

This is a dual antenna system, that is, it employs one antenna for transmission and one for reception. Unlike pulse radars, this is *often* the preferred mode in CW radars, since the transmission is continuous. In pulse radars, the radar gets into the listening mode after transmission. Hence, there is no chance of the transmitting signal swamping the receiver. However, there will be transmission leakage into the receiver. These issues will be further examined in this chapter. But in CW radars, the transmission and reception are continuous. This makes the isolation of the transmitter and receiver chains extremely critical. Typically, this isolation is better than 60 dB. We can use a single antenna system in CW radars, but the problem is a complex one and will also be discussed in this chapter. The frequency of transmission is in the X-band. The signal source for this radar is a Yitrium–Iron–Garnet (YIG)-tuned oscillator [2]. Current techniques also use DDS as discussed in earlier chapters, for the sake of better linearity control in the sweep.

The power output is 3.5 W average. The system has measured values of AM/FM noise figures, 166 kHz from the carrier. This will become clearer as we progress with this chapter. If we use equation (1.13) and calculate the energetic ranges for this radar for two different targets, 1 m² (for aircraft—the smallest expected target for such a radar) and 100 m² (for a ship target), we obtain 7.2 kms and 22.6 kms respectively. Hence, it is clear from Table 7–1, that the maximum analyzable range of 36 kms is the instrumented range. The IF bandwidth for this radar extends from 300 Hz

[*] This chapter has been jointly written with Andy G. Stove, Research Manager, Technical Directorate, Thales UK, Aerospace Division, Manor Royal, Crawley, W. Sussex RH 10 9PZ, UK.

Table 7–1 Calypso Radar Parameters

Antenna gain	28 dB
Antenna 3 dB	1.65° azimuth, 32° elevation
Rotation speed	24 rpm
Isolation between antennas	>60 dB
Frequency	8.5–9.1 GHz
Transmitter (YIG tuned)	X-band
Power output (CW)	3.5 W
AM noise (166 kHz from carrier)	-160 dBc Hz^{-1}
FM noise	-110 dBc Hz^{-1}
Active sweep bandwidth	4.3, 8.5, 17.0, or 34.0 MHz
Sweep repetition interval (SRF)	6.144 msec
Center frequency	8.5, 8.7, 8.9, or 9.1 GHz
Maximum analyzable range	36 km
Minimum analyzable range	9 m
Receiver noise figure	3 dB
IF bandwidth	320 Hz–166 kHz
	(+12 dB/Octave swept gain to 15 kHz)
FFT processor	2,048 points in < 6 msec
FFT weighting loss	1.5 dB
RF losses	4.9 dB
System losses	4.3 dB
Signal-to-noise (single sweep)	12.8 dB
	1 m^2 at 7.2 km
	$P_d = 50\%, P_{fa} = 10^{-6}$
	(Swerling case 1)

From [1]: © Reprinted with permission.

to 166 kHz. The interesting point to note is that there is a swept gain in this radar up to 15 kHz. This is to compensate for near range returns, which will otherwise saturate the amplifiers. The losses shown are very typical in such radars. We shall examine each of these aspects in greater detail.

7.2 CALYPSO DESIGN PARAMETERS

This radar was designed on the lines discussed in the earlier chapters. The final numbers on this radar are shown in Table 7–1. The numbers shown are typical of what one can expect to encounter.

The schematic for this radar is shown in Figure 7–1.

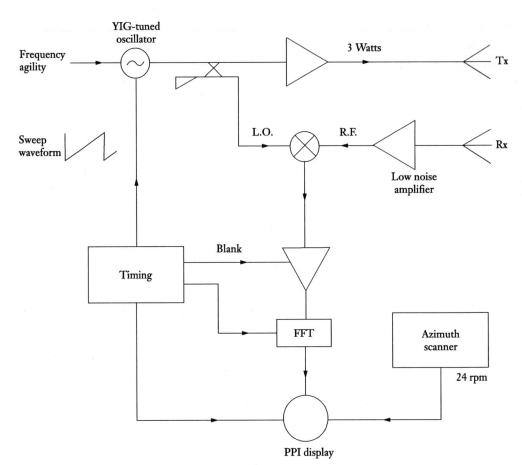

Figure 7–1 Calypso FMCW radar schematic. (From [1], © Reprinted with permission)

7.3 DOPPLER TOLERANCE

The Doppler frequency in radians/sec is given by

$$\omega_d = \frac{4\pi f v}{c} \tag{7.1}$$

where f is the radar frequency and v is the target radial velocity.

If we assume $f = 9$ GHz and $v = 20$ m/sec (approximately 40 Knots, i.e., a fast patrol boat), we obtain a Doppler frequency of 1.2 kHz.

This gives a proportional range error (assuming no Doppler correction is made), as given by

$$r_e = \frac{\omega_d}{\omega_b} = \frac{fv}{r(\Delta f / \Delta t)} = \frac{fv}{r\mu} \tag{7.2}$$

where ω_b is the beat frequency and $\mu = \Delta f / \Delta t$ is the sweep rate.

If we assume a range of 5 km and a frequency sweep of 20 MHz in 6.75 msec, this would produce a range error of 1.2%. Such an error should present no problems to a marine radar, where most targets of interest are slow moving ships. If fast moving targets are to be tracked, then this Doppler range error must be corrected by either increasing the frequency sweep or by correcting for Doppler error in the processing stage or by confining the Doppler error to one range bin.

The sweep pattern and center frequency of the transmitter are both selectable via a digital control link. The sweep interval is calculated by taking a frame of 2,048 data samples at a rate of 3 μsec per sample. This number of 3 μsec was defined by the capability of the available FFT at that time. This then gives the maximum IF signal frequency, which is half the sampling frequency of 332 kHz.

The FFT processor consists of two identical 2,048-point FFT processors with appropriate buffer memory and interface cards. The output is in log modulus format and forms the "A" scope display of the radar [3]. This then feeds the "Z" modulation of the PPI display [3]. The azimuthal antenna position is sent to the PPI by a 12-bit shaft encoder [3] connected to the rotating transceiver. The transceiver is located at the antenna pedestal in order to reduce the system noise figure due to extraneous noise "injection."

The FFT loss and the system losses are shown in Table 7–1. The RF losses consist of 0.8 dB for waveguide losses to and from the transmit and receive antennas. A further 1.1 dB is introduced by an RF power limiter before the receiver amplifier. Finally, a 3 dB loss is included for the double sideband noise into the RF preamplifier. This could be eliminated by using an image reject mixer as will be discussed further down in this chapter. This gives the total RF loss of 4.9 dB.

7.4 BEAM PATTERNS/COVERAGE DIAGRAM

The antennas are slotted waveguides, providing a fan-beam pattern with a horizontal beamwidth of 1.65° and a vertical beamwidth of 32°. The antenna electrical boresight is tilted by 15 degrees. The pattern has low side lobes (values are classified).

We now examine what are called coverage diagrams. This is the locus of the energetic range of targets of various RCS ($P_D = 50\%, SNR = 12.8$ dB). The target RCS values are 0.1, 0.5, 1, 5, 10, and 50 m². Such a contour diagram or coverage diagram for this radar is shown in Figure 7–2.

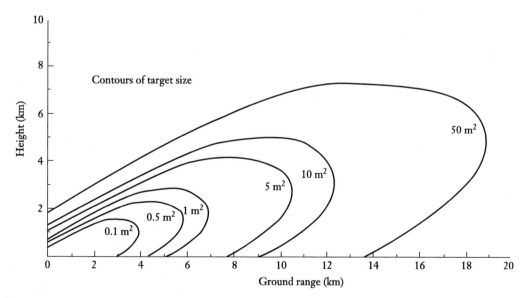

Figure 7–2 Calypso coverage diagram. (From [1], © Reprinted with permission)

The diagram (it does not take into account multipath and clutter limited detection) clearly shows that the beam shape is not ideal for detecting aircraft over land for two reasons. Firstly, the height at which the aircraft (of say 10 m²) can be detected is low, a maximum of 4.3 Kms (14,000 ft). Secondly, and more importantly, the beam shape illuminates a large area of ground clutter. This second consideration, especially without Moving Target Indication (MTI) implemented could limit the performance of the system. Ideally for air-surveillance, a "cosecant-squared" beam is required [4]. This gives very good height coverage with a very sharp cut-off at ground level in order to reduce ground clutter interference. However, this measure is not necessary in this radar, because this is a marine navigation radar.

7.5 FMCW DESCRIPTION

The power supplies for this radar are housed in the base of the mast, while the microwave transceiver, signal processing, and control units are mounted on the rotating platform. The antenna (see Figure 7–3) is a slotted array [5] configured in two rows, the top row being the transmit antenna and the bottom row being the receive antenna with better than 60 dB isolation between them. This figure shows a typical antenna of this type, but the actual antenna is classified. The power unit and the transceiver unit are connected by slip rings, so that signals/supplies freely pass through while the antenna rotates. The transceiver is electrically screened and all communication with the transceiver is digital to improve noise immunity. Connection to the FFT processor and display is by two fiber optic links again to minimize interference.

The transmit antenna is mounted above the receive antenna and the required isolation is achieved by placing a serrated plate between the antennas. The separation between the two antennas, center line to center line is 20 cm. One popular trick to determine whether the isolation is sufficient is to point the radar vertically upwards toward free space. We then insert a 40 dB attenuator (some large value) into the receiver input. There should be no change in the receiver

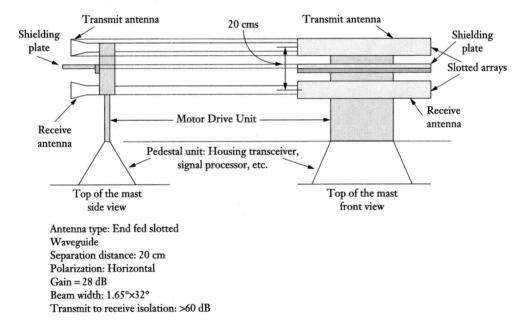

Antenna type: End fed slotted
Waveguide
Separation distance: 20 cm
Polarization: Horizontal
Gain = 28 dB
Beam width: 1.65°×32°
Transmit to receive isolation: >60 dB

Figure 7–3 Typical FMCW radar slotted array antenna. This configuration pertains to separate transmitter and receiver antennas. Note the shielding plate used to enhance the isolation between the antennas.

noise level. If the noise figure of the system had been degraded by leakage of transmitter noise sidebands directly into the receiver, then adding attenuation into the leakage path would have reduced it and altered the noise level. A null reading, therefore, indicates that the transmitter leakage is negligible. Leakage within the transceiver unit or by sidebands on the mixer LO signal, can be detected by "Y" factor measurements [6]. Finally, it should be noted that slotted arrays generally have a "squint" [5]. Typically, this is 1/100 MHz. The total should not exceed a nominal figure across the entire bandwidth. The actual value of squint for this radar is classified.

The majority of the microwave system is contained within the case mounted directly beneath the dual antenna assembly (see Figure 7–3). The transmitter is frequency agile in two respects:

1. The center frequency can be chosen from one of four frequencies.
2. The amount of frequency sweep on the transmitter can be chosen from one of four sweeps (Δf).

Both of these parameters are controlled by digital signals (8 bit word).

The transmitter oscillator is a YIG-tuned oscillator [2]. The amount of frequency sweep applied to it governs the range resolution and the maximum analyzable range (instrumented range). The IF bandwidth was 166 kHz and therefore, the maximum range on any range setting must correspond to a beat frequency of 166 kHz. We have seen in earlier chapters that the range resolution is given by

$$\Delta R = \frac{c}{2\Delta f} = \frac{c}{2\left(\dfrac{\Delta f}{\Delta t}\right)\Delta t} \tag{7.3}$$

Using equation (7.3) we can adjust Δt value to 63 μsec corresponding to a range of 9.45 km. We need to obtain at a range of 36 kms, the maximum beat frequency of 166 kHz. In such an event, a target at 9.45 kms should yield a beat frequency of 43.575 kHz. We create this delay of 63 μsec using a glass delay line as an artificial target. The figure of 63 μsec is used because such delay lines are cheap and readily available in the TV industry. We then adjust the sweep till we obtain a beat frequency of 43.575 kHz. This means that at 36 km we can obtain a beat frequency of 166 kHz. The three remaining sweeps, being selected by a binary sequence, give ranges of 18, 9, and 4.5 km.

We now need to measure the linearity of the YIG-tuned oscillator by examining the spectral width of the beat frequency from the glass delay line. This should be 0.1% in order to yield 1,024 resolvable range bins from the FFT processor. The sufficiency of this linearity is determined according to methodology given in Chapter 4. The four selectable frequencies and sweeps are shown in Table 7–2. Also, included is a measurement of the oscillator FM noise at each center frequency. This was measured at 166 kHz from the carrier frequency (because 166 kHz is the maximum beat frequency). If FM noise at this frequency offset is acceptable, we will then know that it will be tolerable at all other lesser beat frequency values, since FM noise competes with the targets at shorter range, which will return more power for the same target size. Note also that the best resolutions are available at the shortest ranges, which is as per theory. Consequently, the power is also consequently, the minimum at the shortest range scale. An 8-bit control word is accepted by the radar transceiver to switch ranges and sweeps.

The power amplifier has a gain of 36 dB from 8.5 to 9.5 GHz. It is a class A amplifier with a measured 1 dB compression output of 36.7 dBm (4.7 W) at 9.0 GHz. This amplifier will actually limit at 38.4 dBm (6.9 W) if driven hard. However, the radar operates in the linear region to reduce AM-to-FM noise conversion (explained later in this chapter).

A PIN attenuator is fitted before the power amplifier. It is used to reduce the power output of the transmitter at short range. This is an electronic counter measure (ECM) consideration, in that

Table 7–2 Sweep/Frequency Values

Center Frequency (GHz)		FM Noise (dBc Hz^{-1} at 166 kHz)
9.096		−117
8.860		−113
8.705		−114
8.492		−110
Active Frequency Sweep (MHz)	Range Resolution (km)	Maximum Range (km)
34.0	4.5	4.5
17.0	9	9
8.5	18	18
4.3	36	36

From [1], © Reprinted with permission.

Table 7–3 Output Power vs. Range Setting

Active Frequency Sweep (MHz)	Maximum Range (km)	Output Power (dBm)
4.3	36	35.2
8.5	18	35.2
17.0	9	33.2
34	4.5	25.9

From [1], © Reprinted with permission.

a quiet radar should never transmit more power than is required, to avoid detection by hostile ESM systems. This variation with range setting is shown in Table 7–3. An interesting observation here is that in fact for real LPI radars, power should be reduced by nearly 36 dB and not less than 10 dB as shown. However, LPI capability is not a priority for this Calypso radar under discussion.

These measurements were taken at 8.9 GHz. The variation with transmitter frequency is ±0.1 dB. Incidentally, the latest versions of PILOT/Scout radars have manual control of output power, independent of the range scale setting.

The total transmission loss between the output of the power amplifier and the input of the transmit antenna, including the waveguide feeds is 0.4 dB. This loss is also repeated in the feed from the receive antenna flange to the input of the limiter.

The amplifier has a gain of 14 dB and a 1 dB compression output power (see Part III) of +20 dBm. This is used to drive a delay line (loss of 6 dB) and provides a local oscillator drive level of +8 dBm to the mixer. The delay line is used for path length correlation of FM noise between the transmitter and the receiver. This is discussed in the next section.

The receiver is formed by the limiter, low noise amplifier, and mixer in that order. The IF amplifiers are housed in the signal acquisition rack on the mast. The X-band limiter has an insertion loss of 1.1 dB and a damage level of approximately 1 W CW. The limiter begins to operate at approximately +20 dBm (100 MW) input power. The low noise amplifier has a gain of +20 dB and a noise figure that varies between 2.65 dB at 8.5 GHz and 2.98 dB at 9.1 GHz. The mixer conversion loss (see Appendix K) is 6 dB (this is typical of most mixers) at an LO drive level of +8 dBm. The 1 dB compression point for the RF input of the mixer is +4 dBm, so care must be taken not to saturate the input of the receiver.

7.6 RECEIVER NOISE FIGURE

The reader is advised to study Appendix L on noise figure calculations before proceeding with this section.

The noise figure of the receiver chain was measured from the waveguide flange of the transceiver case to the output of the first amplifier in the IF amplifier chain. This was measured using the *Y factor method* [5]. This consists of placing an accurately calibrated noise source at the input to the receiver chain and comparing the noise at the chain output with the source ON (HOT) and the source OFF (COLD). The "Y" factor is defined as

$$Y = \frac{\text{Noise Output (HOT)}}{\text{Noise Output (COLD)}} \tag{7.4}$$

The noise figure of the system is then

$$\text{NOISE FIGURE (dB)} = \text{ENR}_{dB} - 10\,\text{Log}_{10}\,(Y - 1) \tag{7.5}$$

where ENR is the excess noise ratio of the precision noise source at the RF frequency of interest, here extending from 8.5 to 9.1 GHz.

The ENR of the noise source is 14.1 dB over this band. The output noise was measured at an IF frequency of 100 kHz and gave a *Y* factor of 10 dB.

Note: The receiver noise is double sideband in nature. Therefore, not only will noise in the IF bandwidth of 0 to +166 kHz be seen, but the image noise sideband of 0 to −166 kHz will be folded into the IF bandwidth. This gives a 3 dB degradation in the receiver noise floor. It can be reduced if we use image reject mixers. We will discuss this aspect later on in this chapter.

The IF amplifier behind the receiver mixer, also contributes to the receiver noise figure. The above *Y* factor measurement was made at the first stage output of this amplifier at an IF frequency of 100 kHz. This amplifier has a $1/R^4$ frequency response, up to 15 kHz, equivalent to swept gain in pulse radars. This is shown in Figure 7–4.

The reduced gain at very low IF frequencies will degrade the receiver noise figure at low beat frequencies, that is, short range. By removing the RF preamplifier and observing the noise level at the end of the IF amplifier chain, the effect of the IF noise could be determined for lower beat frequencies. The measured results are plotted in Figure 7–5 and shown in Table 7–4. The system noise figure is constant over about 75% of the total IF bandwidth. This is important, because if the noise figure rises with decreasing beat frequency, a weak target near the maximum indicated range may be lost by switching to a longer range scale, where the corresponding target beat frequency becomes less. The flat noise figure over 75% of the IF bandwidth ensures that this is not a problem.

7.7 AM NOISE CANCELLATION

The problem of transmit–receive noise leakage is regarded as one of the most severe problems facing the FMCW radar designer. The radar requires maximum sensitivity at the higher end of the sweep corresponding to maximum range. This corresponds to a maximum IF beat frequency. We need, therefore, to control the noise level at this beat frequency. There are basically two

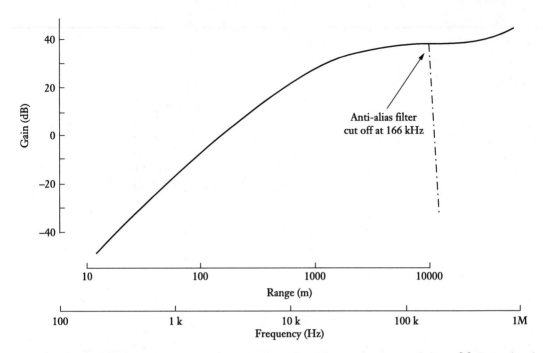

Figure 7–4 IF amplifier frequency response (Range scale set for 10 km maximum range). (From [1], © Reprinted with permission.)

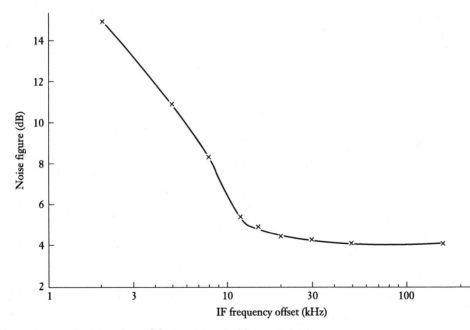

Figure 7–5 IF noise figure. (From [1], © Reprinted with permission.)

Table 7–4 IF Noise Figure at Different Frequencies

IF Frequency (kHz) Offset from Bandcenter	Noise Level Degradation (dB)			Overall Noise Figure (dB)
	Overall Noise Level	IF Only	IF Contribution	
160	0	−17	0.1	4.1
100	0	−17	0.1	4.1
50	0	−18	0.1	4.1
30	−5	−20	0.3	4.3
20	−9	−20	0.4	4.4
15	−12	−20	0.8	4.9
12	−14	−20	1.3	5.4
8	−18	−20	4.3	8.3
5	−19	−20	6.9	10.9
2	−20	−20	>10	>14

From [1]: © Reprinted with permission.

types of noise that we have to contend with, viz., AM noise and FM noise. AM noise exists in the transmitted waveform and therefore appears throughout the IF bandwidth after detection and is approximately the same at all IF frequencies. The FM sidebands give rise to noise also, but FM noise is not constant throughout the IF bandwidth. It decreases with rising IF frequency (beat frequency). These noises need to be reduced/cancelled. There is also a third category of noise, and that is FM-to-AM noise. This arises as a result of imperfections in the frequency responses of the various components in the RF path. This imposes an upper limit to the degree of cancellation, which can be achieved. It can be proved that though FM noise is the stronger in the receiver, it is highly correlated to the noise in the transmitter. Hence, at the output of a mixer, wherein we take the difference frequency, it is actually much less than the AM noise one encounters in typical radar systems [7]. Therefore, AM, rather than FM noise cancellation becomes extremely critical in FMCW radars (even if we use balanced mixers to control AM noise), though FM noise is the stronger. Finally, it is pointed out that FM noise power decreases with rising beat frequency, while the degree of its cancellation also reduces with rising beat frequency. These counteract each other to get more or less constant FM noise power as a function of beat frequency. This is providing we have FM phase noise correlation between the transmitter and receiver. While normally such a correlation exists for any radar, (since the transmitted replica is correlated with the received echo) there also exists a problem of leakage path length from the transmitter to the receiver input. This leakage path phase noise needs to be correlated with the phase noise in the replica for effective cancellation. We therefore, artificially delay the leakage signal separating the transmit and receive antennas by a minimum precalculated distance [8]. The precalculated separation will ensure correlation between the leakage signal phase noise and the phase noise in the replica, by making their path lengths identical. It is clarified that the path length of the phase noise in the replica is from the local oscillator in the transmitter to the mixer in the receiver. The path length of the leakage path is from the transmit antenna to the receiver antenna, that is, much shorter. Hence, we need to suitably delay (increase) this leakage path length. This will effectively cancel FM phase noise and in fact make it much less than the AM phase noise, though it started out as the stronger noise contributor. However, the AM phase noise still needs to be reduced. This is

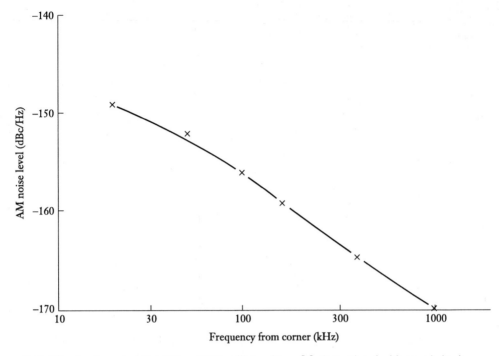

Figure 7–6 AM noise characteristic of X-band YIG oscillator. (From [1], © Reprinted with permission.)

achieved by reducing the AM noise in the leakage signal. If we were to increase the path length, the attenuation of the leakage signal will be more and hence, AM noise will reduce. But this measure will spoil the correlation effect for FM noise cancellation. To get around this problem we use separators as we have done in this radar. Separators will reduce the strength of the leakage signal to such levels that AM phase noise will also cease to be an issue. Once we control noise in this fashion, there will still remain noise due to:

1. FM-to-AM noise.
2. Radar returns from targets will always be uncorrelated in terms of phase noise with the phase noise in the replica since the round trip propagation time is much greater than the leakage path separation distance.
3. Noise from multiple leakage paths, due to poor engineering.

These above-mentioned noise sources ultimately contribute to the noise in an FMCW radar receiver. Normally in radars, the typical noise levels are −160 dBc/Hz for AM noise and −115 dBc/Hz for FM noise [9]. This is considered acceptable for radar operation. It can, of course, be even better. Further details are given in Appendix E.

The measured AM noise of the YIG oscillator is shown in Figure 7–6. Farther away from the carrier the noise floor is about −170 dBc/Hz, but closer to carrier the AM noise becomes worse due to the $1/f$ noise in the GaAs FET oscillator transistor. It was decided that −160 dBc/Hz at 160 kHz would be the acceptable AM noise level. This would imply a 60 dB transmit/receive isolation (−25 dBm leakage into the receiver). This means that the detected AM noise level would be −180 dBm/Hz s.s.b. (single side band) or 10 dB below the receiver noise floor of −170 dBm/Hz s.s.b. with 4 dB noise figure (the actual measured noise figure for this radar was, in fact, 3 dB).

7.8 FM NOISE CANCELLATION

To prevent the receiver noise figure being significantly degraded by oscillator FM noise leakage from the transmitter, the YIG oscillator must have an FM noise level given by (see Appendix E)

$$P_t + N_{FM} + R + C = -174 + N_R - 9 \tag{7.6}$$

where P_t is the transmit power (dBm), N_{FM} is the FM noise figure (dB), N_R is the receiver noise figure (dB), R is the transmit/receive isolation (dB), and C is the cancellation of the FM noise achieved by equalizing the path lengths of the leakage signal and of the local oscillator signal.

We know from Appendix E that

$$C = 4 \sin^2 \pi f_{max} \tau \tag{7.7}$$

where f_{max} is the maximum beat frequency (166 kHz) and τ is the permissible time delay between the local oscillator and the FM leakage into the mixer.

For

$$P_t = 35 \text{ dBm (3.5 W)}$$
$$N_{FM} = -113 \text{ dBc/Hz}$$
$$N_R = 7 \text{ dB (this includes limiter loss and image noise)}$$
$$R \le 60 \text{ dB}$$

we obtain $C = -38$ dB. Substituting in equation (7.7), we obtain $\tau = 12.1$ nsec. This corresponds to a path difference of 3.62 m. In practice, the path length was matched to 0.5 m and inserted as a delay line of length 4.2 m between the local oscillator and the mixer of the receiver.

This is shown in Figure 7–7.

Readers should not get confused that the delay line length in the mixer path is 4.2 m while the antenna separation is 20 cm. The overall electrical path lengths of both the leakage and the mixer need to be equal. This requires trials and testing in the field. During noise measurements

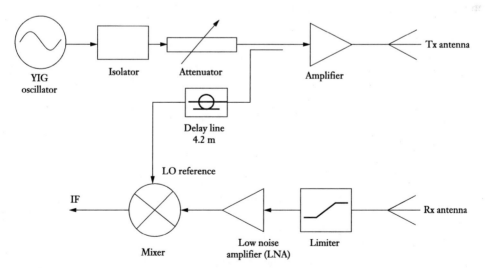

Figure 7–7 Microwave transceiver. (From [1], © Reprinted with permission.)

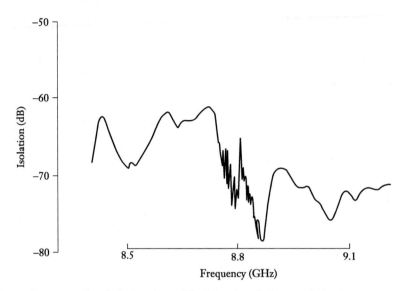

Figure 7–8 Transmitter to receiver isolation. (From [1], © Reprinted with permission.)

there should not be any large reflectors in the near field of the antenna. These reflectors would otherwise reflect the transmitter noise sidebands back into the receiver and upset the calculations. It has been found that relatively small cross sections will reflect large amounts of FM noise into the receiver. In the near field [3], the FM noise progressively becomes more and more correlated with the target returns as we move closer to the radar. For this radar, this occurs at distances <470 m. The dynamic range of an FMCW radar, when limited by the FM noise is the ratio between the power in the largest received signal, normally a low beat frequency and the noise power due to the sidebands from that target at the maximum IF frequency of the radar. In our case, the FM noise of the oscillator is typically, about −113 dBc/Hz at 166 kHz and the effective noise bandwidth is 162 Hz × 1.4 spreading due to FFT weighting, or 229 Hz (23.6 dB Hz).

If the beat frequency of the large target is assumed to be small then the noise limited dynamic range of the radar can be assumed to be 113 − 23.6 = 89.4 dB, assuming no correlation of the noise from the target. If the target comes too close to the radar, the radar enters the near field of the target (<20 m) and path loss is proportional to $1/r$. This imposes a serious limit on the size of close-in targets, which can be tolerated by the radar. To improve on this, it is necessary to use a quieter oscillator with less FM noise, that could improve the FM noise limited dynamic range of the radar and hence, its ability to handle large close-in targets. Another way is to use as high a beat frequency as possible, since FM noise levels reduce at high beat frequencies. This will, however, require faster FFTs.

The isolation achieved in this radar is shown in Figure 7–8. It is interesting to see that there are essentially two stages in the isolation readings. The first stage extends up to 8.9 GHz and the second stage beyond 8.9 GHz. In each stage we note that with rising frequency, the degree of cancellation decreases, that is, the mean is a rising ramp. This is as per the theory discussed in Chapter 4. At 8.9 GHz, the cancellation was optimized, yielding a value around −79 dB. This could be due to the fact that the path delays were optimal at this frequency, since in all radar calculations in this chapter 8.9 GHz has been used everywhere as the sample carrier frequency.

7.9 IF AMPLIFIER

The schematic for the IF amplifier is shown in Figure 7–9. The amplifier consists of five stages. The first stage acts as a preamplifier, the second and fifth stages as blanking amplifiers, while the third and fourth stages act as automatic gain control (AGC) amplifiers. Two parameters of concern here are harmonic distortion and noise figure.

Very low harmonic distortion is required to prevent the generation of false targets at double the range of actual targets. This is because a strong return signal will generate harmonics by driving the amplifiers into the nonlinear region of operation. The specification for harmonic distortion is −60 dBc (<0.1%) over the IF frequency band and is specified for a 1 V p-t-p output from the whole chain. Since the harmonic content is increased by reducing the gain in each stage, the worst case distortion occurs when the AGC system is set to the minimum gain of 0 dB. These distortions decrease with increased feedback from the AGC circuit and therefore, increased gain. The noise figure of the first stage was measured as 3 dB at 100 kHz. A quiet amplifier is required here in order not to corrupt the overall noise figure, including the RF chain.

The AGC system detects the output signal level (D1) and then compares it against two thresholds (upper (*TH1*) and lower (*TH2*)) in a comparator. In practice the AGC is fast enough to cope with any target at three sweeps per beamwidth rotation rate. Recall that at a rotation speed of 24 RPM and an azimuth beamwidth of 1.65°, two sweeps are the usual number per dwell time. Therefore, the AGC is adequate. The AGC gains range from 37.4 dB down to 6.8 dB in eight steps. Manual AGC facility also exists at preset values. The timing for the AGC is governed by the blanking timing. Recall from previous chapters that in order to reduce the nonlinearities, we need to blank the sweep at the ends. At the commencement of each sweep the blanking pulses are removed and the AGC starts operating. The blanking ratio is −41 dB.

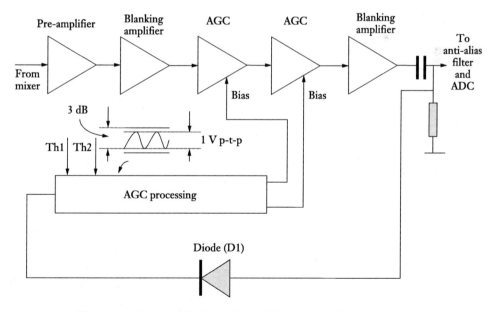

Figure 7–9 IF amplifier schematic (From [1]: © Reprinted with permission.)

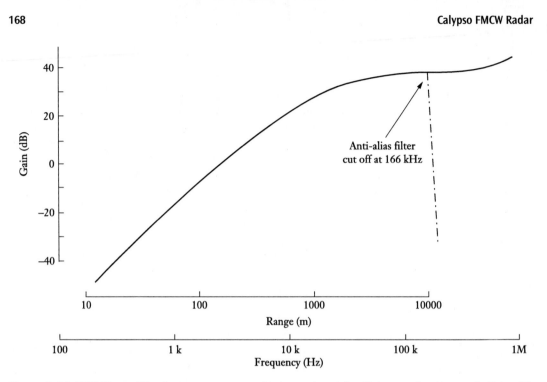

Figure 7–10 S.S.B IF amplifier frequency response (Range scale set for 10 km maximum range). (From [1], © Reprinted with permission.)

The amplifier frequency response is shown in Figure 7–10. This is the total gain across the five stages, with the two AGC stages set to minimum gain. The gain rises with frequency at a rate of 40 dB per decade up to about 15 kHz. This is to give an R^4 swept gain characteristic in order to equalize returns from near and distant targets. The overall response is limited to 166 kHz by the anti-alias filter.

7.10 ANTI-ALIAS FILTER AND ADC

The anti-alias filter is required to tailor the IF bandwidth to the Nyquist frequency to prevent the aliasing back of targets and noise outside the unambiguous bandwidth of the FFT. The filter is passive, so it gives extremely low harmonic distortion and does not suffer from limiting. The filter insertion loss is 6 dB.

The filter output is then amplified and then given to an ADC. The ADC satisfies the Nyquist criterion and has a sampling rate of better than 330 kHz and a dynamic range of 70 dB. It is a 12-bit ADC.

7.11 CONTROL CIRCUITRY

The control functions in this radar have been implemented based on a Texas Instrument signal processor and the control functions are split into three main areas:

1. The control of the sweep interval, blanking interval, and ADC sampling rate (3 μsec sampling rate).
2. The reception and decoding of the serial eight bit word sent by the FFT processor.
3. The parallel to serial conversion of the 12-bit IF digital data stream.

7.12 CALIBRATION AND VERIFICATION

The aim of FMCW radar calibration is to ensure that the target returns across the IF bandwidth, that is, at all beat frequencies (ranges) measure up to expected values.

The principle difficulty in measuring the sensitivity of a medium power radar system is in finding a suitable reference target. The reference target must be substantially larger than any of the clutter around it. It must also be at a relatively short range for good alignment. In this case, the large radar cross-section (RCS) and the short range mean that the expected SNR of the target will approach the instantaneous dynamic range of the system at the IF frequency corresponding to the target range.

To calibrate this radar, a square trihedral corner reflector of size 50 cm with a theoretical RCS of 33.2 dBm^2 at 8.9 GHz was used. During calibration, the radar was tilted up by 16°; so that the target was seen at the one-way half power point of the antenna (i.e., 3 dB below maximum). The target was set up on a tower 600 m from the radar.

The theoretical SNR of a 1 m^2 target at 7.2 km is 12.8 dB (see the example in Chapter 1). This includes 1.6 dB beamshape loss. The radar was in the nonscanning mode during the measurements, that is, looking steadily at the target. The expected SNR can be thus calculated as:

1 m^2 at 7.2 km scanning	12.8 dB
No scanning loss	+1.6 dB (since the radar is not scanning, there will be no beamshape loss [2])
33.2 dBm^2 (RCS of target)	+33.2 dB
600 m range	+33.4 dB
At half-power point of elevation coverage	−6 dB
Expected SNR	75 dB

This well exceeds the dynamic range of the radar. We need to select such a suitable sweep from the list of available sweeps, that target beat frequency should be clearly seen and should be above the noise floor so that we can estimate the SNR of the target. The longest sweep of 34 MHz was selected and the target beat frequency at 21.5 kHz was clearly visible on the spectrum analyzer. The power at this frequency (see Table 7–3) is 25.9 dBm. We then insert a 40 dB attenuator in the receive waveguide. This makes the expected SNR to be around 35 dB. It was measured as 35 dB ± 0.5 dB. The radar, therefore, has the required sensitivity at the lower power level.

An indication of whether it worked with higher transmitter powers could first be obtained by switching to the 10 km range scale. On this scale, the transmitter power is increased by 7.3 dB but the target frequency is reduced to 11 kHz, degrading the noise figure by a further 1.4 dB. The SNR should, therefore, improve by about 6 dB on switching to the longer scale. An improvement of about 8 dB was noted. The discrepancy is partly due to inaccuracies in calculating the noise degradation and inaccuracies in estimating relative noise levels. The agreement is, however, considered reasonable. The radar can now be said to be operating correctly at full power if the SNR improves by 40 dB when he attenuator is removed.

On the short-range setting (25.9 dBm transmitter power) the SNR increased by 37 dB for a 40 dB increase in receiver sensitivity. On the long-range setting (35.2 dBm transmitted power), the SNR increased by 30 dB for a 40 dB increase in receiver sensitivity. The sensitivity on full power is, therefore, degraded presumably by noise leakage from transmitter to receiver, by the effects of FM noise sidebands on the signals received from strong targets at short range. It is pointed out that the target range is only 600 m. At such a range the FM noise reflected from it is very strong and uncorrelated. This increases the noise floor. Hence, it is recommended that the target be kept as far away from the radar as is possible without compromising on the quality of measurement. Similarly, when setting up the radar, the beam shape should be elevated, so as to minimize the clutter feedback of FM noise. There should be no reflecting surfaces nearby, for similar reasons.

7.13 IMAGE REJECT MIXERS

The radar, as built, suffers from a 3 dB loss due to the noise at the RF image frequency being added to the RF noise when both are down-converted to zero IF. The effect cannot easily be removed by RF filtering because of the closeness of the RF and image frequencies, but can be simply removed with an image reject mixer. Image reject mixers are discussed in [3].

7.14 MOVING TARGET INDICATION

FMCW radars possess stable CW transmitters and the relatively low IF means that digital MTI can easily be implemented. The basis of the application of MTI to an FMCW radar is that the output of each cell of the FFT (each range cell of the radar) consists of two signals, the I and Q components that are exactly analogous to the I and Q components of a coherent video of a pulse radar. If the target is stationary, the phase and amplitudes of the return (amplitudes and ratio between I and Q channels) will be the same from one sweep to the next and subtracting will then lead to a cancellation as for a pulse radar. MTI on FMCW radars relies entirely on changes in the range of the target from sweep to sweep of the order of a fraction of a wavelength, which show themselves up as changes in the phase of the return from sweep to sweep in a manner exactly analogous to a pulse radar.

There are two cases we need consider:

1. If the target is stationary during the sweep but jumps in range on the flyback from one sweep to the next, then the radar will not see the jump but will see a target which is not moving during the sweep but which moves from sweep to sweep. This movement will be seen as a change in phase of the return from one sweep to the next and the target will pass through the MTI filter exactly as in a pulse radar. We are talking of a coherent pulse-to-pulse MTI, relying on movements of the order of 1 cm/msec rather than an area scan-to-scan MTI which requires a movement from one range cell to another from scan to scan.

2. If the target moves during the sweep. In such a case we need to consider the effects of range–Doppler cross coupling. We encounter the phenomenon of the blind speed of the MTI, wherein this is a speed at which the Doppler shifts the apparent range by an *integral* number of range cells. This means that a blind speed is one at which the target phase moves through an integral number of cycles from one sweep to the next.

We now briefly analyze the behavior of FMCW radars in the presence of target Doppler. This analysis is performed for a single target. The effect of multiple targets can be easily extrapolated because the system is linear with respect to the received signals. End effects due to the flyback will be neglected [7]. We use the following symbols:

t = time since start of the sweep

f_t = instantaneous transmitter frequency

f_0 = transmitter frequency at time $t = 0$

μ = chirp rate

τ = time of flight of the signal from the transmitter to the target and back

r_1 = range of the target at time $t = 0$

r_0 = mean range of the target during sweep

v = radial velocity of target

T = sweep repetition interval of the radar

c = velocity of light

The radial velocity of the target is assumed to be constant. Now, $f_t = f_0 + \mu t$. The standing phase of the transmitted signal is

$$\phi = 2\pi \int_0^t f_t \, dt$$
$$= 2\pi[f_0 t + (1/2)\mu t^2] \tag{7.8}$$

assuming $\phi = 0$ at time $t = 0$.

We note that in the following equations a_0, b_0, and c_0 are constants being the amplitudes of the signals. The instantaneous amplitude of the transmitted signal is

$$a(t) = a_0 \sin 2\pi[f_0 t + (1/2)\mu t^2] \tag{7.9}$$

The received signal from the target is delayed and attenuated

$$b(t) = b_0 \sin 2\pi[f_0(t - \tau) + (1/2)\mu(t - \tau)^2] \tag{7.10}$$

The IF signal is

$$c(t) = c_0 \cos 2\pi[f_0 \tau + \mu t \tau - (1/2)\mu \tau^2] \tag{7.11}$$

If the target is moving

$$r(t) = r_1 + vt \tag{7.12}$$

and

$$\tau = 2r(t)/c \tag{7.13}$$

After some manipulation

$$c(t) = c_0 \cos 2\pi[2\mu r_1 t (1 - 2v/c)/c + 2f_0 vt/c + 2\mu vt^2(1 - v/c)/c + 2(f_0 - \mu r_1/c)r_1/c] \tag{7.14}$$

The expression contains "frequency" terms which are time varying and "phase" terms which are not. The first frequency term $2\mu r_1(1 - 2v/c)/c$ is the range beat, which is proportional to the range of the target. It is normally assumed to be equal to $2\mu r_0/c$. The second frequency term $2f_0 vt/c$ is the Doppler shift. The third frequency term $2\mu vt^2(1 - v/c)/c$ is a cross-term, which may either be interpreted as chirp on the range beat due to the changing range or as chirp on the Doppler due to the changing transmitter frequency. The final term $2(f_0 - \mu r_1/c)r_1/c$ represents a constant phase term.

From the point of view of MTI, the IF signal for two successive sweeps can be written as

$$c_1(t) = c_0 \cos 2\pi \left[f_0 \tau_1 + \mu \tau_1 t - (1/2) \mu \tau_1^2 \right] \qquad (7.15)$$

and

$$c_2(t) = c_0 \cos 2\pi \left[f_0 \tau_2 + \mu \tau_2 t - (1/2) \mu \tau_2^2 \right] \qquad (7.16)$$

The simplest form of MTI processing is a simple canceller, which subtracts the returns from the two successive sweeps

$$d(t) = c_2(t) - c_1(t) \qquad (7.17)$$

Substituting equation (7.15) and equation (7.16) into equation (7.17), we obtain

$$D(t) = 2C_0 \cos 2\pi \left[f_0 \tau_0 + \mu \tau_0 t - (1/2) \mu \tau_0^2 - (1/2) \mu \delta \tau^2 \right]$$
$$\times \sin 2\pi \left[f_0 \delta \tau + \mu \delta \tau t - (1/2) \mu \tau_0 \delta \tau \right] \qquad (7.18)$$

where $\tau_0 = (1/2)(\tau_1 + \tau_2)$ is the mean time of flight of the signals for the two sweeps and $\delta \tau = (1/2)(\tau_1 - \tau_2)$ is half the change in the time of flight between two sweeps.
Now

$$\tau_1 = 2(r_1 + vt)/c, \tau_2 = (r_1 + vt + vT)/c$$

and

$$\delta \tau = Tv/c$$

The velocity is assumed constant, so $\delta \tau$ is not a function of time and $-(1/2)\delta \tau^2$ can be written as a constant phase ϕ, so [7]

$$D(t) = C_0 \sin 2\pi \left[f_0 \tau_0 + \mu \tau_0 t - (1/2) \mu \tau_0^2 - \phi \right]$$
$$\times 2 \sin 2\pi \left[f_0 \delta \tau + \mu \delta \tau t - (1/2) \mu \tau_0 \delta \tau \right] \qquad (7.19)$$

Except for the phase term, the first half of equation (7.19) is the same as the right-hand side of equation (7.11), which was the expression for the signal received from a single sweep. The second term is the effect of the MTI filter.

In a conventional pulse radar, the MTI term is $2 \sin 2\pi f_0 \delta \tau$. The FMCW case contains two additional terms. One, $+\mu \delta \tau t$ is time varying and represents the change in transmitter frequency during the sweep. The other, $-(1/2)\mu \tau_0 \delta \tau$ represents the fact that the range beat frequency is slightly different between one sweep to the next because the range has changed. Since both these extra terms are functions of $\delta \tau$, they have no effect on the static cancellation, for which $\delta \tau = 0$.

The terms $\mu \delta \tau t$ and $(1/2)\mu \tau_0 \delta \tau$ are negligible for most practical purposes [7]. It can, therefore, be seen that a simple MTI canceller behaves the same for an FMCW radar as it does for a pulse radar [7].

Figure 7–11 shows the operation of the MTI canceller on an experimental S-band FMCW radar built at Philips Research Laboratories [7]. The upper trace shows the video "A-scope"

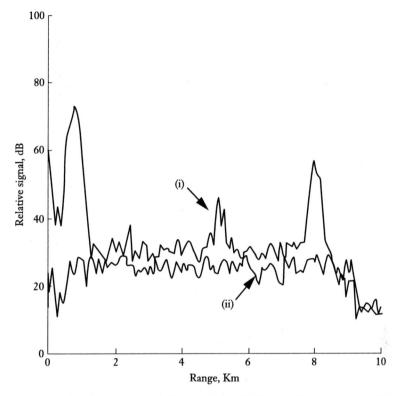

Figure 7–11 Performance of a three-sweep canceller on S-band FMCW radar (i) Uncancelled, (ii) Cancelled. (From [7], © Reprinted with permission.)

picture from one sweep of the radar. The lower trace shows the signal at the output of a digital three-pulse MTI canceller. It can be seen that the static targets are indeed cancelled and that a static cancellation of better than 40 dB has been achieved. This analysis can be extended to cover multiple cancellers and staggered sweep repetition frequencies. It is pointed out that during stagger, the sweep bandwidth will necessarily be shorter than otherwise. This will decrease the range resolution.

7.15 SINGLE ANTENNA OPERATION

A fundamental difficulty in all FMCW radars is the requirement of simultaneous transmission and reception. The transmitter and receiver must be isolated from one another to prevent transmitter power and noise leakage from desensitizing the receiver. Toward this end, a bistatic configuration is preferred. However, sometimes it might be necessary to mount an antenna on a rotating platform. In this radar, though, we still have a bistatic system even though we have mounted it on a rotating platform.

For a single antenna system, a circulator would be used to feed the antenna. Transmit-to-receive isolation now corresponds to antenna return loss and circulator leakage [3]. A typical return loss may be 15 dB, so to achieve the required 60 dB return loss a further 45 dB is required. This could be provided by an RPC scheme. Alternatively, the transmitter and receiver could be switched on and off in anti-phase. This is termed Interrupted FMCW or FMICW and was discussed in Chapter 2.

7.15.1 Reflected Power Canceller

The RPC discussed in this section is based on PIN diode technology and was specifically developed for the PILOT radar (discussed in Chapter 1). The canceller works by sampling the transmit signal and feeding it into the receiver with the same amplitude, but in anti-phase to the reflection from the antenna. The device used for this is called a vector modulator. Destructive interference then occurs and a null in the return loss of the receiver is generated. The PILOT radar parameters impinging on the RPC design are [10]:

Transmitter power	1 W continuous
Sweep rate	1 kHz
Transmit bandwidth	50 MHz maximum
Center frequency	9.375 ± 30 MHz
Antenna	1.2:1 VSWR (22 dB return loss)

Based on these parameters, given the noise sideband levels of the transmitter, it was calculated that 50 dB of transmitter leakage power cancellation was required. The antenna provides 22 dB of this leaving the RPC to improve this by approximately 30 dB. It is important that the RPC does not inject any excess noise into the receiver. This rules out the use of amplifiers within the canceling path, which would inject their own thermal noise into the receiver. In addition to this, the loop must be able to cancel the reflected power throughout the 50 MHz sweep, that is, it must track the variations in the antenna return (wide loop bandwidth). The RPC should be maintenance free and should adapt to the aging of the antenna or to a replacement antenna being fitted.

This canceller is shown schematically in Figure 7–12. There are two separate channels, I and Q, which operate independently. The residue of the I (or Q) component of the uncancelled signal is detected by the mixer. It is then amplified and filtered to provide the I (or Q) drive for the vector modulator. Each loop, I and Q, is therefore a zero-order control loop. Lack of orthogonality between the two channels will cause one channel to inject a disturbance into the other. This is not desirable, but it does not impinge upon the operation of the loop because the orthogonality requirements on the loop components are relatively modest. The function of an I and Q vector modulator is to simultaneously control the phase and amplitude characteristics in the processing of a microwave signal. This device will convert a signal to a desired vector location via a digital command. The theory of operation is to divide the input signal into two equal signals 90° apart I (inphase) and Q (quadrature). This allows the magnitude of each signal to be relocated along its vectors' axis. The two signals are then combined. Using the Pythagorean theorem the sum of the vectors produces the resultant output signal. These vector modulators are available commercially as an IC (integrated circuit) package.

The vector modulator approach is preferred because [10]:

- The noise characteristics of the input signal must not be altered, that is, the signal passing through it must not saturate or distort.
- The modulator requires quadrature inputs.
- The modulator has a low insertion loss.
- It must be capable of being driven at the maximum response frequency needed for the loop. For PILOT radar, this means a flat response up to at least 10 kHz.

We now examine the mixer and the low frequency circuits.

Mixer: The mixer is a double-balanced quadrature mixer. The important features are that it has a good orthogonality between the I and Q channels and a constant low DC offset across the RF bandwidth. To maintain a good orthogonality in the mixer, careful control of the relative path lengths and hence, relative phases of the canceling and local oscillator signals is necessary. The

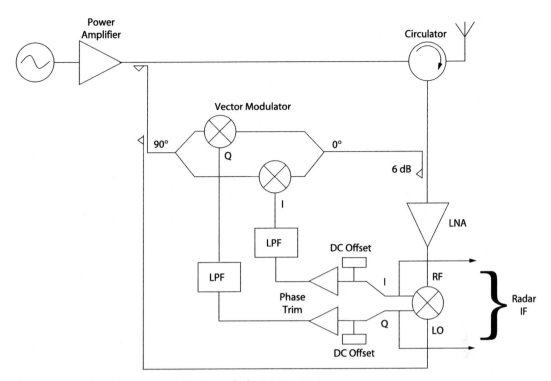

Figure 7–12 Reflected power canceller. (From [10], © IEEE 1990.)

loop is stable if the total phase error around the two quadrature channels is less than 90°. The paths need to be matched across the RF bandwidth. For a cancellation bandwidth of 2 GHz, the LO path to the mixer has to be matched to the RF path to the mixer (via the modulator) within about 3.5 cm (in air) [10]. In practice, the path lengths were matched to give a bandwidth much greater than the power amplifier bandwidth.

Low Frequency Circuit: The first IF amplifier is a low noise DC-coupled amplifier with 200 kHz bandwidth. The DC offset potentiometer serves to set the 0.6 V offset needed for the modulator drive voltage. The phase trimming stage synthesizes the correct drive for each channel of the modulator from the two signals available from the mixer. This is done in order to minimize the phase shift around each channel. The IF filters (not shown) remove high frequency noise from the loop and prevent spurious modulation of the canceling signal by such noise. The unity gain frequency response of the loop is set at about 10 kHz. This is high enough to track the variations in return loss in a 1 kHz sweep but not too high as to cancel out close-in targets [10].

7.15.1.1 Performance of the RPC The RPC was tested with a 7 feet end-fed nonresonant slotted waveguide navigation antenna complete with rotating joint. The return loss was measured on a network analyzer and is shown in Figure 7–13.

The level of the return loss was mainly due to the rotating joint (which is usually the case). The antenna is designed to operate at 9.375 GHz and the return loss is better than 20 dB over about 250 MHz. The RPC was tested over 1 GHz to examine the effects of the poor return loss at the edges of the band. The result is shown in Figure 7–14 for a sweep rate of 50 MHz/msec, that is, the maximum sweep rate used by the PILOT radar. The RPC cancels the power by more than 33 dB over 400 MHz, that is, well within the specified 30 dB over 50 MHz. The transmitter

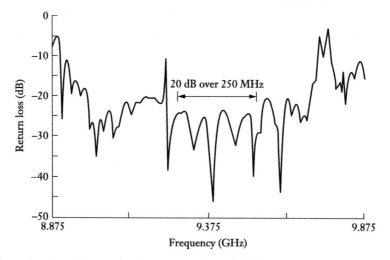

Figure 7–13 Return loss from 7 feet navigation antenna. (From [10], © IEEE 1990.)

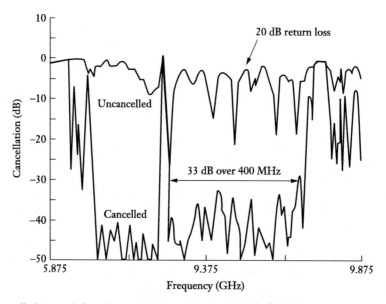

Figure 7–14 Cancellation result for 7 feet navigation antenna. (From [10], © IEEE 1990.)

noise sidebands were also cancelled and the residue had only a negligible effect on the receiver noise figure.

The final effect of the RPC can be seen in this trial of the PILOT radar with a single antenna and RPC evaluated on board a Royal Swedish Navy Fast Attack Craft. Figure 7–15 shows the PPI recorded in the strait between Sweden's mainland and the island of Oland with a bridge approximately 50 meter high at 15 nanometer. The range scale is 24 nanometer with range rings every 4 nanometer.

In the section we briefly examined the design of the RPC for the PILOT radar. We noted that the RPC improves the transmit/receive isolation from about 20 dB (without the RPC) to about 50 dB, which is comparable with the isolation obtained from a dual antenna system.

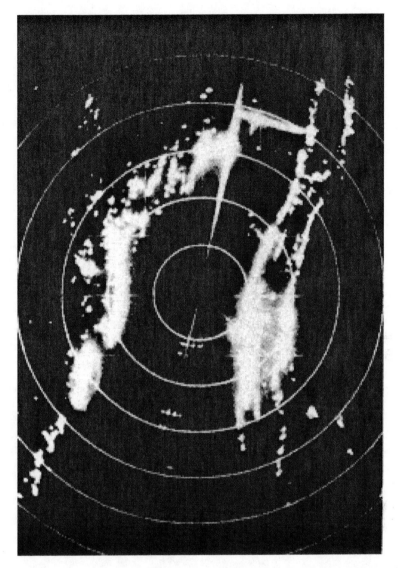

Figure 7–15 PPI recorded in Sweden with a single antenna PILOT radar. (From [10], © IEEE 1990.)

7.16 AMBIGUOUS RETURNS

In order to avoid this problem altogether, a prudent radar designer should always ensure that under all conditions, the radar energetic range must always be much less than the instrumented range. In such a situation, second time around echoes will rarely occur. This will eliminate the ambiguity problem altogether. If the ranges are comparable, then we will need to resolve range ambiguities as and when they arise. Ambiguities arise in CW radars, when part of the return from one period of the modulation, returns during the next modulation period if the target is at any range other than zero. This section briefly discusses the ambiguity problem and surveys the popular approaches toward resolving it. The Calypso radar, however, did not require resolving

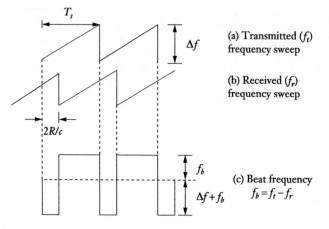

Figure 7–16 Classical FMCW.

ambiguities, because the instrumented ranges in this radar (provided the range scale is suitably selected by the user) are much larger than the energetic ranges.

The FMCW signal processing approach in generating a beat frequency proportional to range is only efficient when the time of flight to the target is short compared with the sweep time. This is illustrated in Figure 7–16.

Here the correct beat frequency $f_b = 2(\Delta f / \Delta t)R/c$ is generated for a time from $t_d = 2R/c$, the time of flight of the signal, to time T_s, the length of the radar sweep period, where $(\Delta f / \Delta t)$ is the sweep rate, R is the target range, and c is the velocity of the signal. For a time $t = 0$ to $2R/c$, a "second time around" return is received with a beat frequency $-\Delta f + f_b$, where Δf is the total frequency excursion of the sweep. It can easily be seen that as $2R/c \rightarrow T_s$, the amount of the expected beat signal tends to zero and the second time around beat becomes lower in frequency.

Summarizing, the beat frequency and the corresponding range to the target are given by

$$f_b = f_t \pm f_d \tag{7.20}$$

where f_t is the transmitted frequency and f_d is the Doppler shift.

Also,

$$R = \frac{c(f_t \pm f_d)T_s}{2\Delta f} = \frac{cf_t T_s}{2\Delta f} \pm \frac{cf_d T_s}{2\Delta f} \tag{7.21}$$

where T_s is the sweep time and the positive and negative signs are for negative and positive Doppler shifts, respectively.

Equation (7.21) shows that the measured range represents the true range (first term) and an error due to Doppler shift (second term).

$$R_{measured} = R_{true} \pm R_{Doppler} \tag{7.22}$$

The range error due to Doppler shift can be rewritten in terms of range bins as

$$\frac{R_{Doppler}}{\delta R} = \pm \frac{f_d}{WRF} \tag{7.23}$$

where δR is the range error and *WRF* is the waveform repetition frequency and is equal to $1/T_s$ Hz. This shows that the measured range is decreased (or increased) by one bin as the Doppler shift is increased (or decreased) by a frequency equal to the waveform repetition frequency. For example, if the target has a Doppler shift of +20 Hz using a WRF of 5 Hz, then the measured range should be corrected by subtracting four range bins.

Therefore, it can be seen that unlike in pulsed radars, in FMCW radars, an ambiguous Doppler frequency is not folded within the same time delay resolution cell, but is shifted to a nearby time delay cell. This is referred to as the range/Doppler ambiguity of this waveform. The waveform repetition period T_s limits the unambiguous Doppler shift range to $\pm 1/2T_s$ Hz. If the expected target Doppler shift is less than $\pm 1/2T_s$ Hz, the range and Doppler information can be estimated using a double FFT. The first FFT over each sweep yields the range information and the second FFT for each range bin over a number of sweeps provides the Doppler frequency. The duration over which the second FFT is carried out determines the Doppler resolution. Extending the Doppler shift range by reducing the waveform repetition period reduces the radar range capabilities and consequently creates range ambiguities. Alternatively, employing a waveform with adequate duration to detect the furthest echo resolves the range ambiguity and introduces Doppler ambiguity due to fast moving targets. In pulsed radars, this problem is often overcome by using staggered PRFs [3]. The problem is more complex for FMCW systems. A number of solutions have been proposed. These include LFM ranging [11], triangular FMCW as was discussed in Chapter 4 and the three-cell structure proposed by Poole [12].

The LFM ranging system [11] consists of an LFM signal section whose duration is chosen to avoid range ambiguity. This is followed by a CW transmission that provides an independent measure of Doppler shift. When we measure the instantaneous difference between the frequency of the received echoes and the frequency of the transmitter during the modulation part, an algebraic sum of the target Doppler and range is obtained. The target's range is subsequently determined from the difference frequency between the modulated section and the CW section.

The three-cell structure proposed by Poole [12] transmits three sweeps, each corresponding to a cell. The first two sweeps have identical start and stop frequencies, while the start frequency of the third sweep is offset by Δf whose value is chosen to cover the highest expected Doppler frequency. In this technique the Doppler frequency and the range are determined from the phase change between the cells. The Doppler resolution for this system is limited by the error in the measurement of the phase, which depends upon the SNR. Hence, higher transmitter powers are needed to improve the system performance.

An alternative approach has been suggested in [13], wherein it is proposed to transmit three blocks of regular FMCW signals with different WRFs during a single target illumination. The duration of each WRF and the number of sweeps during each WRF are chosen so as to resolve adequately the range ambiguities and to achieve the required Doppler resolution, respectively. The Doppler shifts of targets are estimated sequentially during each block. Since the WRFs of the three waveforms are different, aliasing can cause the estimated Doppler shifts to be different during each WRF. The Chinese remainder theorem can be then utilized to calculate the true Doppler shift where the maximum unambiguous Doppler range is limited by the least common multiple (LCM) of the selected WRFs. The three WRFs are so chosen that they have an LCM frequency greater than the expected target Doppler shift. This guarantees that the estimated Doppler shifts during each WRF are unique.

All the above methods are time intensive and are not suitable for multitarget environment. A quicker approach involving hybrid frequency hopped LFM signals has already been discussed in Chapter 6.

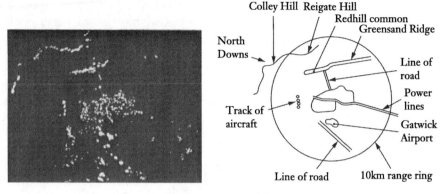

Figure 7–17   Left: PPI picture; Right: Key to PPI picture. (From [2], © IEE 1987.)

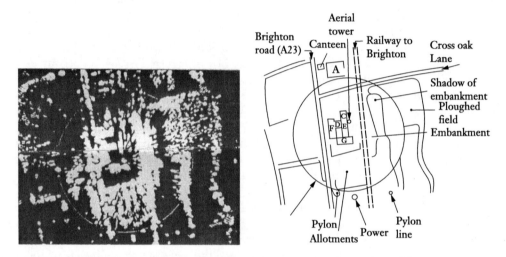

Figure 7–18   Left: Close range PPI picture; Right: Key to close range PPI picture. (From [2], © IEE 1987.)

7.17 PERFORMANCE OF CALYPSO FMCW RADAR

The performance of this radar can readily be seen in Figures 7–17 and 7–18. Figure 7–18 shows the highest resolution mode. These pictures were taken from the roof of the laboratories at Redhill, Surrey, England.

7.18 SUMMARY

We have now come to the end of our trail in studying FMCW radars for single frequency (single channel). This type of radar is most common and was developed for its high quality of range discrimination and LPI capability. As an example we picked up the trials report of an existing radar and traced its fortunes from the design stage up until field trials. *The reader is cautioned, however, that this radar is an old radar and constitutes the initial efforts in this area.* Radars are a very competitive field and it is more than likely that many of the ideas discussed here are obsolete and have since been replaced by bigger and better ideas. The reader can gain clues on these latest techniques by

studying the latest papers on this subject. However, from our point of view this radar presents many opportunities in that it is a real radar and not a theoretical one. Hence, the numbers shown here are real ones. The radar design directly applies many of the ideas that we had acquired in the earlier chapters and is, therefore, from this point of view very instructive to study. During this process, we were introduced to the multifarious problems usually encountered by the designer in the designing of such radars and we also observed based on the measured results, how close the ground reality followed the theory studied in earlier chapters. We investigated the importance of AM/FM noise in such radars and the problems encountered in mitigating them. The problems encountered in the design of IF amplifiers and anti-alias filters and the choice of a suitable ADC were also examined. We have seen how the sampling time, receiver frequency resolution (studied in Chapter 4), speed of the ADC, and the choice of sweep time were all interrelated. This analysis then lead to the choice of the bandwidth of the IF amplifier. We next examined issues like radar calibration and verification. We saw that issues like near field clutter and nearby reflecting surfaces play a crucial role in determining the FM noise floor in such radars. The problem of targets with Doppler was next investigated and we examined the technology behind the MTI capability of such radars. We learned that it came at the expense of range resolution caused due to the short sweeps involved in MTI stagger. Finally, we studied single antenna operation and the complexities in the design of RPCs. We concluded our study with an analysis of range ambiguity problems in such radars and how they can be solved. We are now in a position to better appreciate the next problem which is the design of multifrequency LPI radars in Part III of this book.

References

1. Barrett, M., Beasley, P. D. L., and Stove, A. G., *An Advanced FMCW Radar*, Philips Research Laboratories (PRL), Tech. Report 3344, 1986.
2. Barrett, M., Reits, B. J., and Stove, A. G., "An X-band FMCW navigation radar," in *Proceedings of the IEE International Radar Conference on Radar '87*, IEE Conf. Publ. 281, 1987, pp. 448–452.
3. Skolnik, M. I., *Introduction to Radar Systems*, 3rd ed., McGraw-Hill, Boston, MA, 2001.
4. Kraus, J. D. and Marhefka, R. J., *Antennas*, 3 ed., McGraw-Hill Science/Engineering/Math, New York, November 12, 2001.
5. Balanis, C. A., *Antenna Theory Analysis and Design*, Harper and Row, New York, 1982.
6. Bowick, C., *RF Circuit Design*, Newnes (now Elsevier Inc., Burlington, MA, USA), March 1997.
7. Stove A. G., "Linear FMCW radar techniques," *IEE Proceedings-F*, Vol. 139, No. 5, October 1992.
8. Stove A. G. and Vincent, R. P., *Noise Reduction in CW Radar Systems*, European Patent 0 138 253 B1.
9. Scheer, J. A., et al., *Coherent Radar Performance Estimation*, Artech House, Inc., Norwood, MA, 1993.
10. Beasley, P. D. L., Stove, A. G., Reits, B. J., and As, B. O., "Solving the problems of a single antenna frequency modulated CW radar," *Radar Conference, 1990. Record of the IEEE 1990 International*, May 1990, pp. 391–395.
11. Schleher, D. C., *MTI and Pulsed Doppler Radar*, Artech House, Inc., Norwood, MA, 1991.
12. Poole, A. W. V., "Advanced sounding 1. The FMCW alternative," *Radio Science*, Vol. 20, No. 6, 1985, pp. 1609–1616.
13. Musa, M. and Salous, S., "Ambiguity elimination in HF FMCW radar systems," *IEE Proceedings on Radar, Sonar & Navigation*, Vol. 147, No. 4, August 2000, pp. 182–188.

PART III

Theory and Design of Pandora Multifrequency Radar

8

Design Approach to
Pandora Radar

8.1 INTRODUCTION

High-resolution radar (HRR) design has long been a quest for radar engineers. There is no one single definition of HRR. One definition of high-resolution radar is a range resolution better than 1 meter. There are several ways to achieve HRR, such as FMCW, short pulse, linear frequency modulated (LFM) pulse compression, and stepped frequency waveform (SFWF) (pulse-to-pulse discrete frequency coding) pulse compression. An enormous amount of literature exists on HRRs using stepped frequency waveforms. In the present state of technology, such large *continuous* sweep bandwidths are difficult to achieve without a broadband linearizer. Normally, HRR is required in range profiling cases. In such an event, nonlinearity in the sweep waveform plays an important role, as it can deteriorate the quality of time side lobes. In range profiling cases, we require very low time side lobes.

The one source for information on HRR is the book by Wehner [1]. This is an old publication and very out of date. The inch resolutions discussed in the book are routinely achieved these days. In this chapter, we carry out our analysis based on LFM waveforms. However, SFWF can also readily be applied without any changes. The radar, in fact, can be made switchable between these two waveforms.

8.2 QUALITATIVE REQUIREMENTS

8.2.1 Problem Definition

HRR requires that we resort to large sweep bandwidths in FMCW radars. But we still require ultra-pure signal sources with very low nonlinearities so as to yield low range side lobes. This aspect is also extremely difficult to attain and involves a lot of research activity and expenditure. Hence, the need arose to find a way of building a radar from commercial components (Commercial Off The Shelf—COTS) so as to keep the cost minimal and yet not lose out on high resolution, even if it is a synthetic one (since we process the signal using an IFFT, as discussed in Chapter 6). It was keeping this aspect in mind, that this radar was conceived.

The multifrequency FMCW radar under discussion in this book has been called the PANDORA. This is an acronym for Parallel Array for Numerous Different Operational Research Activities. Essentially, we are trying to achieve wideband capability using multiple narrow band FMCW radars (in our specific case eight radars corresponding to eight channels) *operating simultaneously*. This allows us to study wideband processing techniques without the need for wide instantaneous bandwidth. This automatically implies that we have the capability to study coherent, noncoherent,

and mixed processing concepts in a wideband environment using *narrow band FMCW radars*. All this has been achieved without the attendant engineering problems of wideband radars. Finally, we can alter the phase, amplitude, and frequency of the individual channels to vary the transmitted power spectra for matched illumination [2]. This is based on the original work done by Gjessing [3]. In this book, we shall examine the overall concept of Pandora, its working principle, and some of the key technologies that go into its design. We shall then discuss the experimental results achieved during their implementation. It is pointed out that in Chapter 10, we discuss the Pandora radar that utilizes a bandwidth of 4 GHz (400–4,800 MHz, yielding a resolution of around 3.4 cm) in the X-band. By definition ([4], p. 2), this makes it an ultra-wideband (UWB) radar. We have achieved this without the attendant problems in UWB radar design and using the COTS approach.

8.2.2 Program Requirements

The design of the system is constrained within certain basic requirements. These revolve around three aspects:

1. Requirement based on the environment, that is, the overall system
2. Requirement based on radar functioning
3. Requirement based on cost

We now examine each category.

8.2.3 Overall System Requirements

Table 8–1 shows the basic system constraints.

Since this is an experimental radar, the maximum instrumented range is kept at 1 km. This is due to the gain of the transmitting antenna being kept at 1 dB, that is, omni transmission (see the assumed parameters, below). The receiver antenna has a gain of 30 dB.

8.2.4 Specific Radar Requirements

In designing the radar to constraints listed above, certain other constraints have been imposed on the radar design. These are listed in Table 8–2.

Additionally, we have assumed certain parameters for the purposes of calculations.

8.2.5 Assumed Parameters

In order to reduce hardware complexity, the Doppler option will be switchable between channels. In view of target dwell time limitations, Doppler cannot be implemented even at the specified integration time of 10 msec as it is insufficient to measure target Doppler. Doppler option can only be exercised if the integration time is increased to 200 msec or higher (see Appendix B) for

Table 8–1 Basic System Constraints

Target RCS	2 m²
Target doppler	300 m/sec
Maximum instrumented	
Range	1 km
Polarization	Vertical

Courtesy IRCTR: Reprinted with permission.

Table 8–2 Specific Radar Requirements

Number of channels	8 channels
Switchable Doppler option	Between channels
Probability of detection	0.50
Probability of false alarm	10^{-6}
Integration time	10 msec
SNR (single sweep)	12.8 dB
Type of target	Swerling 0/1
Average power/channel	5 W
Wavelength	0.032 m
Desired coarse range resolution	3.125 m
Fine range resolution	<0.4 m

Courtesy IRCTR: Reprinted with permission.

Table 8–3 Assumed Parameters

Antenna bandwidth	1 GHz
System losses	10 dB
Rx noise figure	3 dB
Transmitter antenna gain	1 dB
Receiver antenna gain	30 dB
System noise temperature	400 K

Courtesy IRCTR: Reprinted with permission.

the measurement of *target* velocity of 300 m/sec (Coarse mode of operation (para 1, Appendix B)). The coarse range resolution of 3.125 m is the *nominal* resolution of each of the eight channels prior to the high-resolution FFT. The resolution decreases with range.

8.2.6 Cost Requirements

The cost of the demonstration model, that is, the single-channel radar, should be kept as low as possible. It is expected that unless this cost is controlled, the complete multi-channel radar will become extremely costly. Hence, the COTS approach is used for its implementation.

8.3 THE PANDORA RADAR DESCRIPTION

The Pandora project is a flexible multi-channel multifrequency FMCW radar in the X-band. It comprises the following essential blocks:

1. FMCW waveform generator.
2. Power combiner block.
3. Wideband low noise amplifier.
4. Power resolver block.
5. Stretch processing for each FMCW channel.

6. Noncoherent processor.

7. High-resolution FFT or channel FFT.

The concept involves generating FMCW signals in narrow bandwidths for each individual channel, with the proviso that the center frequency of each channel differs from the adjacent channels with the frequency deviation of the basic signal plus a guard band. During reception, the received signal is correlated with the basic signal (stretch processing [5, 6]) to yield a beat signal that is one per range gate per channel. The different beat signals are separated by bandpass filters, such that the whole RF frequency band can be filled, for example, 9.378–10.154 GHz, in steps of 48 MHz sweeps in eight channels (48 × 8 = 384 MHz, plus the guard band of 49 MHz each between channels, totaling 776 MHz). The outputs from these eight channels may be synthesized to cover a band of 776 MHz. This synthesis is achieved by using a high-resolution FFT, which effectively divides the range FFT resolution (of 3.12 meters for a 48 MHz sweep) by eight. Putting it another way, for our particular example, the 48 MHz sweep yields a resolution of 0.19 m. If we correct this for Hamming weighting, we multiply by a factor of 1.8 to yield 0.34 meters. These high-resolution target returns are then sent to a range-bearing display where we display a *synthetic* 2D image of the target, *but in real time*. It can be seen in Figure 8–1 that the FMCW source is split into eight different channels each having a separate LFM sweep. The LFM sweeps have a constant bandwidth of 48 MHz but have different start and end frequencies. These are then combined in a power combiner. There are three techniques investigated to combine these frequencies, an active combiner, a passive Wilkinson combiner, and a passive waveguide combiner.

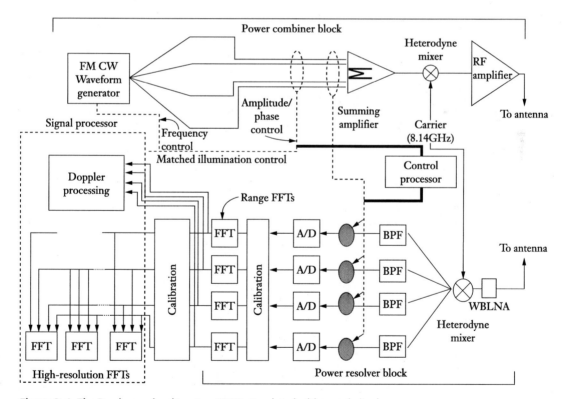

Figure 8–1 The Pandora radar. (Courtesy IRCTR: Reprinted with permission.)

We shall investigate the performances of each of these later on in this book. The combined signal is then heterodyned to X-band (I-band), amplified and then transmitted. The receiver is exactly the reverse. The incoming radar return is initially amplified in a wideband LNA and then fed to the power resolver block. The power resolver is built around a bank of Chebyshev filters centered at the respective sweep center frequencies. The rejection ratio of the nearest adjacent channel is better than −50 dB compared to the input signal level. This ensures that the sweep frequencies are correctly received. The sweeps are then subjected to stretch processing [5]. The resulting beat frequencies, corresponding to the various ranges, are then fed to a range FFT. The range FFT is basically a spectrum analyzer. We ensure a high sampling rate, typically 4 MHz, and carry out a 4 K FFT. This will ensure a very high degree of range resolution, since the FFT bin resolution will be high for every beat frequency and the beat frequencies correspond to range. However, the achievable resolution is limited by the sweep bandwidth. We call this range resolution as "coarse" range. For a particular range cell, each coarse range is available at eight different beat frequencies, corresponding to the eight channels. Each of these coarse range cells is switchable to a Doppler FFT channel for target Doppler resolution. Collectively, these cells are subjected to a high-resolution FFT. In this mode, the I and Q outputs of the respective range FFTs are fed to yet another FFT.

This is an eight-channel FFT and it effectively divides the range cell by eight, yielding a high-resolution synthetic target profile. Also, other processing strategies may be adopted. In accordance with its name, this radar among other applications is intended for developing matched illumination algorithms on the lines suggested by Gjessing [3]. The concept of using low-resolution radar signals at different frequencies for synthesizing a high-resolution result is not new. In his paper, Gjessing [2] describes some experiments with a multifrequency CW radar system. He illustrates the potential of this radar concept in regard to detection, coarse ranging, and identification of low-flying aircraft against sea-clutter background. Our approach is similar, with the difference being that we transmit FMCW instead of CW only.

A word on the overall system: The radar operates on two antennas, one for transmission and one for reception. This will have no impact on the performance of the combiner/resolver as these are buffered by amplifiers, the former by the RF amplifier and the latter by a wideband LNA. We ensure 60 dB of isolation between the antennas in order to ensure a reduction of AM phase noise. Range–Doppler coupling effects are controlled by ensuring that the range and Doppler are unambiguous. This will ensure that the effects of range–Doppler coupling are confined to within one range bin. This will be the same for every channel. The range is in any case unambiguous in FMCW radars if we ensure that the instrumented range is in excess of the energetic range. The ambiguity of Doppler is controlled by ensuring that the change of Doppler across the sweep for a particular target is less than the Doppler resolution. Hence, it is confined to within one range bin. In Figure 8–1, we have shown two calibration blocks. In the first block, just after the A/D converter, we have predetermined delay lines which correct for the various phase discrepancies between the channels, so that for a particular point target, all the target returns fall within the same range bin. In the second calibration block, it is ensured using a look-up table, that all the I and Q outputs from the coarse channels have the same phase. This is a necessary prerequisite for the channel FFT. Finally, we have shown a *control processor* that handles the "Household" functions, that is, the start and stop of the sweeps, the matched illumination parameter settings, the commencement of the range FFTs, the channel FFT, and so forth. Furthermore, since this is a target profiling radar, we cannot add to our complications due to clutter returns. We will see at Appendix A, that unless we have a good MTI facility, clutter will drastically reduce the detection ranges from as much as three km to 300 m though a more detailed analysis would have to take into account imperfections in MTI performance.

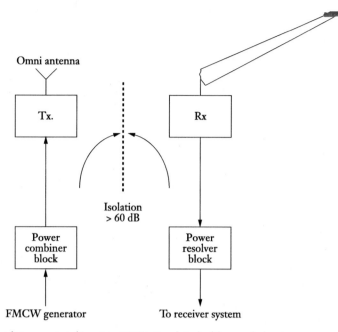

Figure 8–2 The Pandora concept. (Courtesy IRCTR: Reprinted with permission.)

8.4 BASIC EQUATIONS OF FMCW RADARS

We had already discussed the FMCW radar theory in Chapter 4. These concepts are included at Appendix A under the heading *CW Radar Range Equations*. We shall apply the above equations using a program called "*cw.cpp*". This software calculates the energetic ranges for both FMCW and phase-coded radars. Thereafter, in the section on power resolver blocks, we shall further discuss the signal processing of the Pandora with special emphasis on the range resolution aspects and problems associated with the design of the baseband filters. We shall finally develop a fresh set of equations bearing in mind the signal processing of the Pandora and develop a method for calculating ranges for the Pandora, which will have two channels, a coarse channel and a fine channel. Finally, a program called "*pandora.cpp*" is included at Appendix B with explanatory notes. This program designs the Pandora parameters on the lines discussed.

8.5 DESIGN APPROACH

The ideal range resolution is given by

$$\Delta R = \frac{c}{2\Delta f} \tag{8.1}$$

where c is the velocity of light and Δf is the sweep bandwidth.

In practice this is not realized owing to the reduction of the transmitted bandwidth caused due to signal transit time and sweep recovery time problems. These aspects and their implications were examined in earlier chapters.

The choice of radar waveform will affect radar performance in terms of range, Doppler, and angle measurements as well as the radar system's detection performance. In the earlier chapters we had analyzed the performance of various types of signals through their ambiguity and periodic ambiguity functions. In this section, we will apply that knowledge to determine the optimum signal waveform suiting our needs. In doing so, we will necessarily cover old ground, but this exercise is being undertaken so as to acquaint the reader with the *modus operandi* of waveform selection, viz., the reasoning one needs to take and the parameters to look for in waveform selection.

8.5.1 Waveform Selection

There are essentially two main types of coding techniques that we shall take into consideration for Pandora in this chapter, viz., Linear Frequency Modulation (LFM) and Phase Shift Keying (PSK). FSK modulation has been considered for Pandora in a different application in Chapter 10 (specifically SFCW). However, within these, there are different variants, which are selected based upon the user requirement. In the light of our radar, we need to carry out a similar exercise to determine as to what type of waveform is most suitable for our use. This aspect depends upon the following factors:

- What is the target environment that the radar must contend with in terms of the number of simultaneous targets, their range, range rate, and RCS?
- Which parameters need to be measured and with what accuracy?
- What is the range and range rate resolution?
- How does range–Doppler coupling affect our system?
- How much is the integration time available?

We investigate these aspects through the radar ambiguity function [7]. The radar ambiguity function, as we have seen in earlier chapters, expresses the magnitude of the system output as a function of range and Doppler shift of the target. It is represented as a 3D diagram, where the magnitude of the radar system output $|\chi|$ is epresented as amplitude above the range (t)—Doppler (ω_d) plane. This is graphically illustrated in Figure 8–3.

In Figure 8–3, any point on the 3D surface will exhibit an ambiguity (strictly speaking, this depends upon the nature of the ambiguity function, e.g., even functions do not exhibit any ambiguity but odd functions do, as we shall see below) other than along the cardinal axes i.e.,

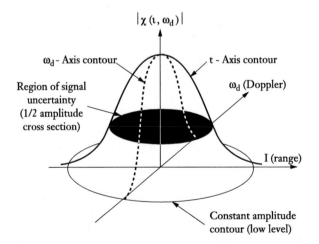

Figure 8–3 The approximation to the ideal ambiguity function. (Courtesy IRCTR: Reprinted with permission.)

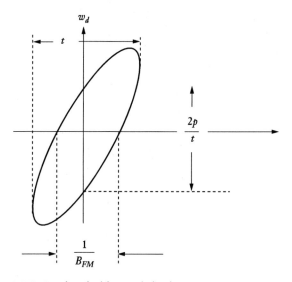

Figure 8–4 LFM. (Courtesy IRCTR: Reprinted with permission.)

along range and Doppler axes. This means that a change in Doppler will yield a change in range and vice-versa. This is the range-Doppler coupling. It is only along the cardinal axes that there will be no coupling i.e., a case when the target Doppler is zero or the range is zero. The business of waveform selection, therefore, involves selecting such an ambiguity function so that during the normal operation of the radar, either there is no range-Doppler coupling (this means that the ambiguity function does not exist or has an extremely low value in the range-Doppler of interest, and hence, no coupling) or the range-Doppler coupling is controlled to acceptable limits as a signal processing error, which is the usual case. In this book there is no room to go into analysis of various types of waveforms with regard to their ambiguity functions. However, there are two broad classes of signals viz. frequency modulated such as LFM, VFM (this is basically up and down chirp juxtaposed next to each other, yielding a V shape giving rise to it's name. Such a waveform is an even function and yields accurate estimates of target range and Doppler *on a one frequency step basis*), Quadratic FM, etc., and discrete coded signals using various types of phase coding like Barker codes, etc. If we see the half-amplitude cross sections (the black cross-section in Figure 8–3) of LFM and Quadratic FM (even-function) we note that Quadratic FM ambiguity function in Figure 8–5 exhibits a sharp spike as compared to LFM in Figure 8–4.

This means that we can use such waveform for non-dense objects because it will yield accurate estimates of target range and Doppler *on a one frequency step basis*. However, the autocorrelation function for the QFM waveforms is more strongly spread over the correlation interval as compared to LFM. This is apparent in Figure 8–6. The QFM waveform has higher amplitude over the 2T domain as compared to LFM. The single tone waveform is the highest, but it does not interest us as we are investigating waveforms suitable for CW transmissions and not pulse transmissions. Hence, QFM will cause "self-clutter" against point targets. It is, therefore, not suitable for a dense target environment. The same argument works for VFM and other types of even-function FM waveforms.

Though LFM works well in a dense target environment, it has a strong range–Doppler coupling as compared to QFM. The inclination of the ambiguity function to the range axis reveals this. But the range interval of this ambiguity is restricted as can be seen in Figure 8–4. The ambiguity interval is given by $1/B_{FM}$ which is the range resolution, and B_{FM} is the sweep bandwidth. This

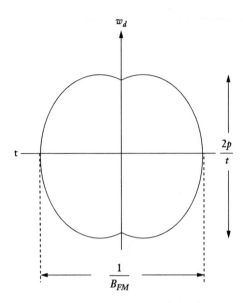

Figure 8–5 Quadratic LFM. (Courtesy IRCTR: Reprinted with permission.)

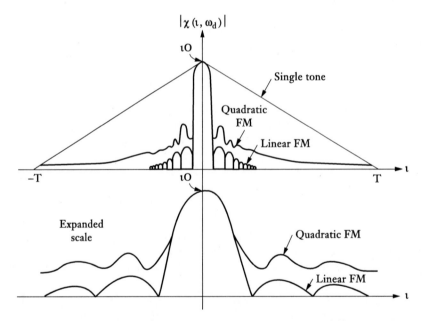

Figure 8–6 Autocorrelation of LFM, QFM, and single tone waveforms. (Courtesy IRCTR: Reprinted with permission.)

means that within the range resolution of an LFM signal we can get range errors due to target Doppler. This is an important observation and from the point of view of the Pandora, we can confine this ambiguity to within one range bin. These issues have already been studied in previous chapters. In view off the structure of the ambiguity function, LFM waveforms are resilient to Doppler. Due to range–Doppler coupling, as the Doppler increases, the response output peak will

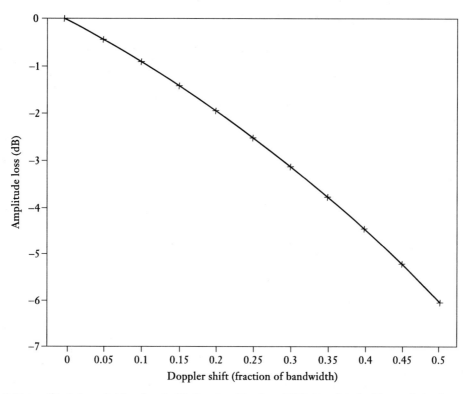

Figure 8–7 Amplitude loss of chirp signal with Doppler. (Courtesy IRCTR: Reprinted with permission.)

change in delay and will be maximum along the line defined by $f_d = -\mu t$, the inclined line where μ is the chirp rate. The result of this effect is that a chirp signal can tolerate considerable Doppler shift before it completely decorrelates, although there will be a range–delay shift corresponding to the Doppler. This is clear from the graph in Figure 8–7. We can see from the figure, that a frequency shift of 0.5 B is required for a 6 dB amplitude loss. In Figure 8–4 it can also be seen that the ambiguous range–Doppler cell is defined by $(1/B_{FM} \times 2\pi/\tau)$ where τ is the chirp signal duration. Such resilience comes in very useful for signal processing. Hence, in the final analysis, it is now clear why LFM is popular in radars. The ambiguity function in Figure 8–5 belongs to the major class of "thumb-tack" ambiguity functions, while the one in Figure 8–4 belongs to the class of "knife-edge" ambiguity functions for obvious reasons. Even-function FM waveforms like QFM and VFM do not exhibit any range–Doppler ambiguity, that is, their range–Doppler coupling is zero. This is, because of their "thumb-tack" nature. Hence, one can get uncorrelated range and Doppler estimates on a one pulse basis. "Knife-edge" waveforms like LFM on the other hand, exhibit range–Doppler coupling. However, it has the best Doppler tolerance of any waveform, which is a very good thing in the radar scenario where targets are often moving fast.

There is another class of frequency modulated waveforms called nonlinear FM. For many applications in which the expected range of Doppler shifts is quite small, the signal autocorrelation waveform is of paramount importance. It is desired that this function exhibit very low range side lobes in order to reduce the masking effect of large signals on smaller signals not at the same range. If we use LFM, then we need to weight the received signal in order to reduce the side lobes. However, if we use nonlinear FM like quadratic (with odd symmetry), quartic or cubic (again with odd symmetry), we will not require any weighting at the receiver. However, nonlinear FM is not resilient to Doppler as an inspection of its ambiguity function at Figure 8–5 will show due to its spiky nature. Though it exhibits the same type of ridge like structure as LFM, the addition of curvature to the FM function causes the peak of the ridge to decrease more rapidly as a function of Doppler shift than for the LFM case. However, one does not suffer mismatch loss due to frequency weighting at the receiver. It is because of its weakness to Doppler that nonlinear FM waveforms have not been popular. If the target Doppler is excessive, the side lobes reappear. In the Pandora radar, there is one more reason why we need to reject it. The equation for nonlinear, nonsymmetric FM waveform for 40 dB Taylor time side lobe pattern is given by [8]

$$f(t) = \omega \left(\frac{t}{T} + \sum_{n=1}^{7} K_n \left| \sin \right| \frac{2\pi nt}{T} \right) \tag{8.2}$$

where K is a constant whose value depends upon the value of n and ω is the sweep bandwidth.

Now that we have examined the LFM waveform as the best candidate from the class of frequency modulated waveforms, let us compare it to phase-coded waveforms.

A signal that contains a discrete phase code is usually characterized by an ordered set of predetermined phases. The conventional notation is $\{X_i\}$, $i = 1, 2, 3, \ldots, N$. For each of the N positions in this sequence, a phase is selected from a finite "alphabet" of phases. The most common phase code alphabet is the binary set of 0° and 180°. Much of the effort in the study of phase-coded signals has been directed at deriving sequences that have good autocorrelation properties. One such is the Barker code. This code has the property that the side lobe level never exceeds ±1. However, Barker codes in excess of length 13 are not known. The ambiguity function for Barker code of length 13 is shown in Figure 8–8.

The length limitation of Barker code is a severe restriction in that it limits the detection range of targets due to low code compression gain. This gave rise to polyphase codes like Frank codes.

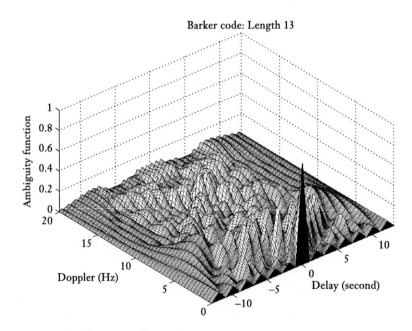

Figure 8–8 Ambiguity function for Barker code: Length 13.

These codes were already discussed in earlier chapters. There is one common aspect with regard to phase-coded signals. They are unsuitable for high resolution radars, because the resolution of the phase-coded signal is the inverse of the segment width. Too small segment widths require antennas with large instant bandwidths. In terms of technology, this is difficult to achieve. However, recently there has been an impressive innovation in this area of HRR using phase-coded signals. This approach is based on OFDM modulation and uses multifrequency phase-coded signals [9]. This approach does yield high resolutions, but yet again there is a limitation. The limitation stems from the poor Doppler tolerance of phase-coded signals. A phase-coded waveform like the Barker code, for example, will result in a sharp ambiguity function near the origin. The zero range–Doppler response will fall off as a sinc function, where the first null is $1/T_s$ and T_s ($T_s = 13t_b$, where t_b is the bit duration) is the time length of the entire sequence. The range resolution is given by the time length of one bit or "chip." This causes phase-coded waveforms to have poor Doppler resilience. In fact, if the phase shift due to target Doppler exceeds 90°, the waveform decorrelates. On the other hand, in phase codes like P3 and P4, the ridge remains high up to much higher Dopplers. For example in a P4 of length 25, the ridge drops to a value of 0.5 (compared to a peak of 1 at the origin) at Doppler that is approximately ten times the inverse of the pulse duration. Nevertheless, due to the delay–Doppler coupling, the point under discussion occurs at a delay of about $10t_b$ (out of $T_s = 25t_b$). Frank code is different from the P3 and P4 family because its ambiguity function exhibits two more ridges parallel to the main ridge (N. Levanon and M. Jankiraman, personal communication, August 2005). Doppler tolerance is a subject of concern, and depending upon the chosen type of code, significant losses may occur at high Dopplers. Hence, in the final analysis if we are talking about fast targets like aircraft, LFM appears the optimum choice for the Pandora radar.

8.6 DISCUSSION ON SUB-SYSTEMS

It was shown in Figure 8–1, that this radar comprises various sub-systems. These sub-systems incorporate specialized requirements, which need to be catered to, in order to ensure proper system integration. We shall examine these in Sections 8.6.1 through 8.6.8.

8.6.1 FMCW Waveform Generator

This item is very crucial to the development of the Pandora radar. It can be seen with respect to the discussion at Appendix D, that we need to achieve extremely low nonlinearity errors. Typically, for a resolution of 40 cm, we need a source having a nonlinearity error of better than 0.003%. If we run the program "*pandora.cpp*" for a range resolution of 20 cm, we find that we need a signal source having a nonlinearity error of better than 0.0006%. All these figures demand special techniques for FMCW generation. It is difficult to achieve adequate linearity over a wide bandwidth. Nevertheless, since we are synthetically generating high-resolution images, we need only concern ourselves with bandwidths not exceeding 48 MHz. This is sufficient for our purposes as in the Pandora type of signal processing, taking into consideration a total bandwidth of 776 MHz and Hamming weighting, it will yield a resolution of typically 34 cm.

LFM waveforms may be generated by several means. Typical sources are:

1. SAW oscillators
2. YIG oscillators
3. Gunn oscillators

and many more. All these sources have different phase noise levels, but it is extremely difficult to achieve our desired signal purity levels. No doubt, the job becomes easier if the sweep bandwidths

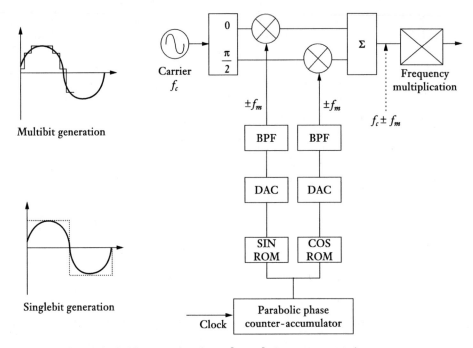

Figure 8–9 Multi-bit and single-bit generation. (From [10, 11], © IEE 1984, 1987.)

are limited. There is, however, another technique, viz., direct digital synthesis (DDS). Here, the sequence of waveform samples is precomputed, stored, and clocked out of memory. Such methods have important advantages.

1. It is waveform bandwidth, rather than time–bandwidth product that is technology limited, so very long chirp signals can be generated.
2. The digital circuitry is less susceptible to temperature changes.
3. The waveform bandwidth can be altered by simply altering the digital clock rate. In SAW oscillators, for example, separate delay lines are needed for this purpose.

There are two principal approaches to digital generation, single-bit generation [10] and multi-bit generation [11]. In both techniques (see Figure 8–9), since the waveform bandwidths required cannot be generated by digital means directly, frequency multiplication becomes necessary. This will inevitably introduce some phase and amplitude distortion, which may need to be minimized. Eber and Soule [12] have shown that although the multi-bit hardware is complex, the bandwidth obtainable for a given clock rate is far superior.

Eber and Soule have built a DDS generator for a bandwidth of 35 MHz and BT product of 1,000 and have achieved a time side lobe of −33 dB. This is exactly suited to our requirements and is a demonstrated technology. Since, the Pandora system is based on the quality of the DDS, we examine it in greater detail in Appendix G.

8.6.2 Power Combiner Block

In this block, we need to *simultaneously* generate FMCW signals on eight channels from a common signal source in the interests of phase coherence. The frequency deviation will be calculated so as to cover the entire bandwidth of interest. For example, we can have 50 MHz sweeps in eight channels covering the entire bandwidth from 9.378 to 10.154 GHz allowing for guard band between the channels. The center frequencies of these sweeps will be staggered accordingly. This implies that we need to additively combine these, prior to transmission. This combining cannot be done at microwave frequencies in the normal manner, since these signals are not mutually coherent being different frequencies. It will be appreciated that at microwave frequencies, signal combining using for example, Wilkinson combiners or Magic T's can only be carried one frequency at a time, and the signals should be coherent. Since we are talking about using eight different frequencies, this condition is obviously not satisfied. If we still intend to combine at microwave frequencies, we need to use filters. This means high insertion losses. Hence, the problem of power combining for such a radar is a complex one.

It can be seen in Figure 8–1, that we have initially used one generator followed by an eight-way splitter. This generates eight basic sweeps. We then proceed to impart bandwidth coverage to these sweeps by using mixers, one for each bandwidth. We then give these eight sweeps to a power combiner to generate an additively mixed signal for the RF transmitter power amplifier. We can implement the combiner in three ways:

1. Active combiner.
2. Passive waveguide combiner.
3. Passive Wilkinson combiner

Active Combiner: In this approach, we add the low power low-frequency signals in active components (op-amps) and mix the combined signal to RF, where it is power amplified. The unity-gain bandwidth of the op-amps used should be twice the sum of the sweep bandwidths of the individual channels plus the guard bands. In our case, it should be around 1,600 MHz. The Burr-Brown OPA 641 satisfies this requirement, as it has a unity-gain bandwidth of 1.6 GHz. The circuit configuration is shown in Figure 8–10. The gain does not exceed two at any stage.

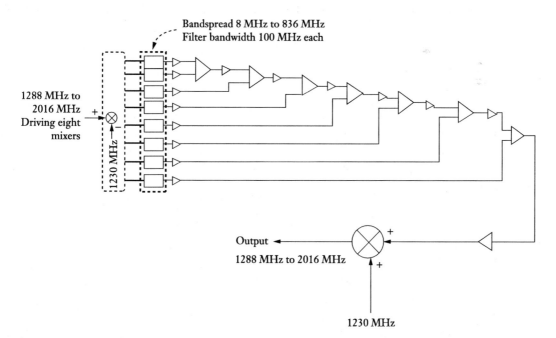

Figure 8–10 Op-amp-based power combiner. (Courtesy IRCTR: Reprinted with permission.)

Table 8–4 Active Power Combiner (LO frequency: 1,230 MHz)

Filter	Center Frequency (MHz)	Output Bandspread (MHz)	Spurious Harmonics in Passband
1	2,016	736–836	4RF–6LO
2	1,912	632–732	6RF–10LO
3	1,808	528–628	3RF–4LO, 5RF–7LO
4	1,704	424–524	4RF–6LO
5	1,600	320–420	4RF–5LO, 6RF–8LO
6	1,496	216–316	4RF–5LO, 5RF–6LO, 6RF–7LO
7	1,392	112–212	6RF–7LO, 8RF–9LO
8	1,288	8–108	2RF–2LO, 3RF–3LO

Courtesy IRCTR: Reprinted with permission.

The buffer amplifiers used are OPA 640 having unity-gain bandwidth of 1.3 GHz. The important thing here is that the op-amps should be stable when they are operating at the specified gains, that is, two for OPA 641 and one for OPA 640 (buffers). The advantage of such a circuit is that the amplitude distortion across the eight individual sweeps is extremely low making for high signal purity. The phase shifts across the op-amps are not critical, as they get eventually corrected when we correct the group delays across the filters of the transmitter and receiver tract. The primary disadvantage in this approach is that we need to use nine extra mixers, since the input frequency spread from the FMCW generator ranges from 1,288 to 2,016 MHz [13], in the L Band, well above the op-amps capability of handling. We cannot directly generate frequencies ranging from 8 to 836 MHz from the FMCW generators, because of filtering problems requiring very steep filters. The second disadvantage is that op-amps generate second and higher harmonic distortions which are typically 30 dB down as compared to the basic signal level of 0 dBm and lie in the pass band and consequently we cannot eliminate it. The frequencies need to be chosen judiciously so as to ensure a low intermodulation (IM) distortion due to the mixers. Table 8–4 illustrates this case. The worst case is that a fourth-order IM product exists in the 8th channel.

Passive Waveguide Combiner: A way of combining powers of different frequencies is by using manifold multiplexers. This concept requires filters to isolate the various frequency channels. There are three types of multiplexers: (1) Multiplexers with circulators [14], (2) Multiplexers with hybrids [14], and (3) Manifold multiplexers [15, 16]. In this chapter, we shall discuss the work done in Delft University on Manifold multiplexers [17].

A manifold multiplexer is shown in Figure 8–11. When a signal from source 1 enters the manifold, the signal does not travel to the other sources, because the other filters are reflecting. Thus, the signal is delivered to the load. The problem here is that the load is not matched to the source. The load of one filter is a combination of all the other filters and the load Z_L. When the impedances of the equivalent circuits of all the source–filter combinations and the dimensions and the other characteristics of the manifold are known, one can calculate the load of filter 1. This load is, in general, not equal to the load Z_L of the multiplexer. Matching this load can be done by inserting a matching network between the filter and the manifold. The multiplexers are dependent on the filters that are used. A waveguide filter is built because this type of filter has fewer losses than most other types [18]. The filter has a Chebycheff characteristic. The maximum ripple of the loss in the

On combining....

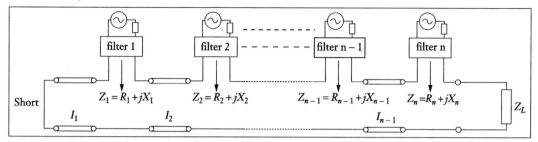

Figure 8–11 Manifold multiplexer. (Courtesy IRCTR: Reprinted with permission.)

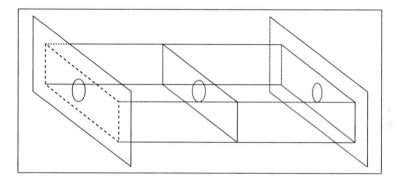

Figure 8–12 Resonator. (Courtesy IRCTR: Reprinted with permission.)

pass band is 0.01 dB. Though the Pandora will require more resonators, the filter was tested using two resonators with iris couplings between the resonators. The resonator is a half wavelength of transmission line. The iris coupling is a metal plate with a circular iris in its center (Figure 8–12).

Figure 8–12 is the second-order shunt inductance iris coupled waveguide filter.

Figure 8–13 is the simulated (solid line) and measured (dashed line) transfer of the filter of Figure 8–12. The top part of the figure is the power transfer function and the bottom part is the phase transfer function.

Two properties of this filter were calculated and measured (see Figure 8–13):

1. The power and phase transfer functions.
2. The internal impedance of the equivalent circuit of the source and filter.

The calculation and measurement of the transfer functions agree within 6%, that is, the center frequency deviates approximately 0.1% and the loss in the previous and next channel deviates less than 6%. The filter was simulated without taking into account the thickness of the irises. When this is allowed for, the resonance frequency will be too high. This, however, can be tuned down by tuning screws.

Passive Wilkinson Combiner: The eight-way combiner for the Pandora is shown in Figure 8–14.

In this approach, we utilize a passive Wilkinson combiner. The combiner has eight inputs and one output which is the sum signal. The combiner used for this radar has been marketed by Pulsar Microwave Corp., USA. The frequency spread for this combiner is ranging from 1,227 to

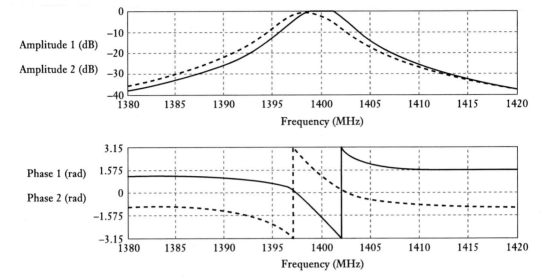

Figure 8–13 Filter insertion loss. (Courtesy IRCTR: Reprinted with permission.)

Figure 8–14 Passive Wilkinson combiner.

2,077 MHz and the variation of insertion loss across this bandwidth does not exceed 0.3 dB. The isolation is better than 20 dB. We now check the combining capability of the combiner. We mix two CW signals at 1,288 and 2,100 MHz, that is, across the bandwidth of interest. We can see the output from the combiner in Figure 8–15.

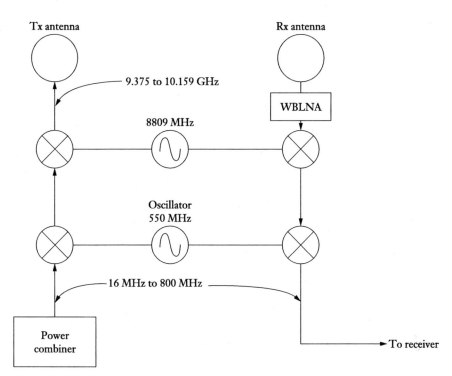

Figure 8–15 Heterodyne receiver. (Courtesy IRCTR: Reprinted with permission.)

Both signals were calibrated for equal input power (–11 dBm). A total of 1,288 MHz was given to pin 1 and 2,100 MHz was given to pin 8. The lower frequency was found to have a power level of –20.7 dBm, while the upper was at –22.7 dBm. But this is across the entire bandwidth of interest. It was found that across the sweep bandwidth of 48 MHz, the difference in power levels was not exceeding 0.4 dBm.

Hence, the Wilkinson combiner appears in principle as the most accurate and versatile combiner as it can also function as a power splitter at the transmitter and receiver inputs (see Figure 8–1). It also has the minimum parts count as compared to the active combiner. However, there is a power loss of –9 dB for eight channels (configuration loss). A detailed analysis of this combiner is given in Appendix J.

8.6.3 Wideband Low Noise Amplifier

This amplifier should have a wide dynamic range typically 70 dB or better. The noise figure for this amplifier should be around 2.5–3 dB. The design for this is discussed in Appendix F.

8.6.4 Power Resolver Block

The receiver antenna should have ideally 360° coverage. Such an antenna is not envisaged in the short term. At this stage of the project, we resort to a static nonrotating planar array of gain 1 dB centered at a nominal frequency of 9.78 GHz having a bandwidth of 8% (around 800 MHz). The justification for this figure is discussed later on in this book (see Appendix B). This array will have a limited number of preformed beams. The output of each preformed beam is connected to one receiver channel. Each receiver channel commences with a wide band low noise amplifier. All receiver channels are identical and so we shall discuss only one channel.

Depending upon the target bearing, one or two adjacent beams will receive the radar return. Assume, for the purposes of argument that only one beam has received it, that is, the target lies along the maximum of one beam. This return signal comprises a mixture of eight different frequencies all individually frequency modulated. After amplification in the WB LNA (see Figure 8–1) the signal is then fed to a heterodyne mixer. The incoming signal is mixed with the transmitted waveform and the output basic FMCW signal frequencies are split into different radar channels by bandpass filters. Henceforth, each radar channel is processed individually by correlating them with their respective basic FMCW signals. This is carried out in a mixer ("Stretch" processor). The output of this mixer is the difference frequency (beat frequency). This is then sampled via a low pass filter (anti-aliasing) and fed to an A/D converter via an IF amplifier. The IF amplifier has a rising frequency response of between 6 dB/octave to 12 dB/octave in order to give more gain to weak, long range, high frequency returns like an STC in pulsed radars. The output of this A/D converter is then fed to an FFT processor for carrying out a multipoint FFT. This is called a range FFT as the output is displayed as per the beat frequency values that correspond to range. This type of processing is called stretch processing and yields very high compression ratios and has already been discussed in previous chapters. The filtering effect of the FFT reduces the noise in the range bin and thereby yields a compression gain (in dB) approximately $\log_{10}$(FFT length). Waveform compression is achieved by putting the beat frequency signal in a range bin whose width in frequency domain is defined by the size of the FFT.

Before we feed the digitized signal to the range FFTs, we need to ensure that the channels are calibrated, so that the low-resolution cells of all channels represent the same range. This can be achieved by having delay lines between the output of the RF amplifier of the combiner and the output (input to WB LNA) of each preformed beam. These delay lines will simulate an echo. We can then use this input to calibrate the channels so that they read the same range cell. Alternately, we need to position a corner reflector on a specific range bin. It should be ensured that the same range bins in all the FFTs should have the same phase values prior to implementing the high-resolution FFTs. If not, we need to incorporate a look-up table for holding the correction values prior to carrying out a second FFT for high resolution. The outputs of the range FFTs (low-resolution outputs) are then routed for Doppler extraction using Doppler FFTs and for further different types of noncoherent processing, for example, we can utilize the multifrequency facility for target recognition using techniques proposed by Gjessing [3]. Simultaneously, the outputs from the low-resolution channels are routed to the high-resolution FFT, henceforth called Channel FFT for visual target recognition using high resolution (34 cm) range profiling or 2D synthetic image. This is carried out in the high-resolution FFT or channel FFT unit.

8.6.5 Heterodyne Receiver and the Range Resolution Problem

The range resolution in a heterodyne receiver gets degraded relative to the ideal range, because of the reduced effective processed transmitted bandwidth that results from using the undelayed transmitted signal as the local oscillator signal. We can avoid this problem by using a separate sweeping LO that delays the start of the demodulation sweep by the propagation time. Of course, a separate highly linear LO adds to the complexity, but phase noise correlation between transmitted and received signal will not be there. This is the negative aspect. The FM noise power decreases with increasing beat frequency. These offset each other to get more or less constant noise power as a function of beat frequency. These details were studied in Chapter 4. This balance will be upset if there is no phase noise correlation [6, 19]. The choice before us is whether to choose heterodyne or homodyne mixing for our radar. We examine this issue with respect to the frequencies specified in Figure 8–15, by way of illustration. *The frequencies shown in the figure are hypothetical and do not pertain to the actual frequencies in the radar. These are given merely by way of illustration.*

Mixers: We have available suitable mixers in the industry. For example, Model DM1-18A marketed by ANZAC is a double-balanced mixer with an RF, LO port range of 1–18 GHz, and an IF port range of DC to 500 MHz. However, there is an image frequency problem.

In our discussion, we will see that prior to giving the signal to the heterodyne mixer, the bandwidth extends from 16 to 798 MHz. When such a signal is demodulated, we have:

$f_{imag} = f_{sig} + 2f_{IF}$—this applies to the case when the LO frequency is higher than the RF.

$f_{imag} = f_{sig} - 2f_{IF}$—this applies to the case when the LO frequency is lower than the RF.

Suppose we require a transmitted bandwidth extending from 9.375 to 10.159 GHz. If we allow for an RF filter roll-off of 300 MHz, this becomes 9.075 to 10.459 GHz.

Now we require that at both these extreme frequencies, there should be no image frequencies within the pass band. We now determine the minimum possible IF.

At a frequency of 10.159 GHz, the LO is lower than the signal.

$$\therefore f_{imag} \leq 9.075 \text{ GHz}$$

$$f_{IF} = \frac{f_{sig} - f_{imag}}{2} = \frac{10.159 \times 10^9 - 9.375 \times 10^9}{2} = 550 \text{ MHz}$$

At a frequency of 9.375 GHz, LO is higher than the signal.

$$\therefore f_{imag} \geq 10.459 \text{ GHz}$$

$$f_{IF} = \frac{f_{imag} - f_{sig}}{2} = \frac{10.459 \times 10^9 - 9.375 \times 10^9}{2} = 550 \text{ MHz}$$

We now take the LO frequency as 8.809 GHz.

We examine the image frequency problem for the two extreme cases of IF viz., 16 and 800 MHz, that is, across the bandwidth.

For IF = 16 MHz,

$$f_{16} = 16 \times 10^6 + 550 \times 10^6 + 8.809 \times 10^9 = 9.375 \text{ GHz}$$

$$\therefore f_{imag} = f_{sig} + 2f_{IF} = 9.375 \times 10^9 + 2 \times (16 + 550) \times 10^6 = 10.507 \text{ GHz}$$

For IF = 800 MHz,

$$f_{800} = 800 \times 10^6 + 550 \times 10^6 + 8.809 \times 10^9 = 10.159 \text{ GHz}$$

$$\therefore f_{imag} = f_{sig} + 2f_{IF} = 10.159 \times 10^9 + 2 \times (800 + 550) \times 10^6 = 12.859 \text{ GHz}$$

We note that both the extreme image frequencies lie outside the pass band of 9.075–10.459 GHz (allowing for a 300 MHz roll-off).

This implies that we need to shift the IF spectrum, that is, adopt heterodyning. If we go in for a baseline IF of 550 MHz, it should prove to be adequate.

This means that the Pandora radar requires heterodyning instead of homodyning. The latter will only work for narrow bandwidth signals.

The final received signal will comprise 48 MHz sweeps extending from a carrier frequency of 16 MHz to around 800 MHz. This will now be split into their respective channels by the passive baseband filter bank.

Range Resolution: We had already examined the range resolution problem in FMCW radars in Chapter 4. We noted that the range resolution is a function of the sweep bandwidth. We needed to make the range bin resolution equal to or better than the range resolution. We came to the conclusion that we will initially increase the beat frequency to the extent possible (based on the speed of the ADCs in the market) and then subsequently improve the range bin resolution by reducing the nonlinearities as this is more difficult.

The program *"pandora.cpp"* in Appendix B does this for you and if you decide not to increase the beat frequency (sampling rate) further, it will tell you the desirable nonlinearity levels for the required resolution, by solving this equation transcendentally provided nonlinearities are responsible for the deterioration of resolution. Otherwise, it will tell you to anyhow increase the sampling rate.

8.6.6 Noncoherent Processor

This is actually a misnomer. In fact, the processing is coherent throughout, but is so named to differentiate it from the channel FFT. It is the processing adopted for the coarse mode. In the coarse mode, we carry out Doppler processing after the range FFT. The output of the Doppler FFT is then noncoherently integrated. The signal is then fed to the detector and then finally to the display. The coarse mode is switch able between channels as otherwise it will become hardware intensive (see Figure 8–16).

In the diagram, the portion to the right of the dotted line is the switchable part of the coarse channel. There will be main lobe and side lobe clutter, but these will occupy the first two filters (in case our platform is moving at 25 knots, see Appendix B). These will need to be censored out by the censor unit. The ambiguity resolver resolves the Doppler folding at the output of the Doppler filters. We have eight channels operating at different frequencies. The range is in any case unambiguous. To resolve the Doppler ambiguity, we need to stagger the sweep rates in each channel in the ratio of prime numbers. Typically, we will need four channels operating at different stagger rates. *The constraint here is that the target Doppler should not change bins due to sweep frequency change across these four staggered channels.* This is ensured if we take the channels

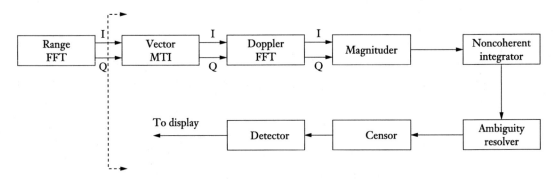

Figure 8–16 Coarse mode signal processing. (Courtesy IRCTR: Reprinted with permission.)

adjacent to each other. A simple calculation will show that the overall frequency change across these four channels should not exceed 500 MHz. The fifth channel will cause such a change in the value of the target Doppler that it will appear in a different bin. Hence, the fifth channel is no use to us. This means that the PRF rate for the Doppler filter banks that follow will be different. This will cause the real target returns to appear in the same bin in all these four channels, but the folded frequencies will appear in different bins as the PRF rate is different. Hence, if we switch our filter bank between the channels, the same targets will appear in different bins if they are folded. We then use Chinese remainder theorem to resolve the ambiguities. We need not use MTI as we have suitable AD converters in excess of 12 bits. Hence, we can design Doppler filter banks with extremely low side lobes. This will ensure that far off filters do not pick up main lobe clutter via their side lobes. This makes notching out main lobe clutter unnecessary. Since this is an eight-channel radar, we can achieve eight noncoherent sweeps in one sweep as the extreme channels will not be correlated. This means that we will take that much less time to achieve a specified noncoherent gain.

8.6.7 High-Resolution FFT or Channel FFT

In the event we employ an eight-channel receiver, each range cell from each of the eight range FFTs, will be connected to one eight-point channel FFT. If there are 32 range cells in each of the eight range FFTs, we will consequently have 32-channel cell FFTs, one per each low-resolution range cell. Alternately, to save on hardware, we can switch the channel FFT between the low-resolution range cells. The output of the channel FFT will in effect be a high-resolution profile of the target returns from a target located in a particular range cell. We then procure a target image with a resolution of 34 cm (see Figure 8–17). We shall discuss the mathematics of this later (see Appendix B).

It should be noted that we are not carrying out any Doppler processing, as it is not needed as we are only carrying out range profiling. The section to the right of the dotted line is switched between range gates.

It is advantageous to carry out Doppler processing with a view to eliminating clutter. In that case, every range FFT will output directly to a Doppler FFT. Since, the integration time is less, typically 10–40 msec, the aim of our Doppler processing will be to isolate clutter. However, the target Dopplers will get folded. This needs to be resolved as discussed earlier and the specified Doppler cell outputs can then be given to the channel FFT. This can be done very fast in the Pandora architecture, as we are having eight channels in one sweep. Hence, Chinese remainder theorem can be implemented on line. Otherwise we can do it off line if we cannot find fast enough processors.

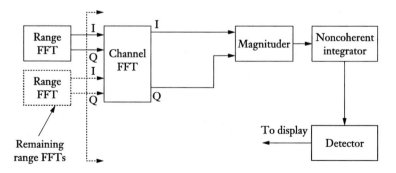

Figure 8–17 Fine mode signal processing. (Courtesy IRCTR: Reprinted with permission.)

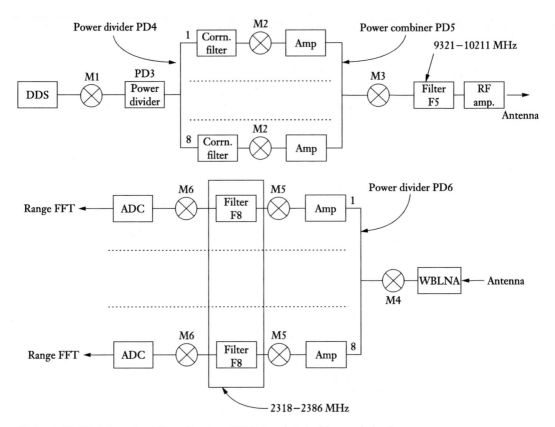

Figure 8–18 Effect of baseband filters. (Courtesy IRCTR: Reprinted with permission.)

8.6.8 Baseband Filters

This is a passive filter bank. We cannot make it an active bank as op-amps do not perform well at frequencies of more than 800 MHz. The bandwidths of these filters have been so chosen, so as to be uniform throughout the bandwidth. The basic approach to the filter problem is shown in Figure 8–18.

The basic signal scheme is shown above in simplified form. We generate a DDS sweep of 200–248 MHz and step up the signal to 2,328–2,376 MHz using mixer M1. The output of the mixer is then given to a two-way power divider PD3. One arm of the divider output carries the reference to mixer M6 in the receiver and the other arm constitutes the transmission signal. The signal is then split eight ways using a power divider PD4 and then given to mixer M2 of each channel via a correction filter. This correction filter is used to compensate for filter group delays and is extensively discussed further down in this chapter and Appendix I. The mixer M2 of each channel then down mixes the basic signal to the proper channel frequency. These frequencies are shown in Table 8–5. Each mixer M2 is fed with local oscillator frequency specified in column three to generate frequency spreads centered at frequencies specified in column two. The channel bandwidths are identical at 100 MHz. These frequencies need to be chosen judiciously so as not to generate spurious frequencies in the pass band of order lower than four. It is obvious that we have been successful in this effort, from column four.

Table 8–5 Spurious Harmonics

Filter	Center Frequency (MHz)	Local Oscillator (MHz)	Spurious Harmonics in Passband
a	2,016	4,368	Nil
b	1,912	4,264	8RF–3LO, 9RF–5LO
c	1,808	4,160	8RF–4LO
d	1,704	4,056	Nil
e	1,600	3,952	4RF–1LO, 6RF–3LO, 9RF–3LO
f	1,496	3,848	Nil
g	1,392	3,744	7RF–2LO, 9RF–4LO
h	1,288	3,640	Nil

Courtesy IRCTR: Reprinted with permission.

The methodology is detailed in [13]. These frequencies are then combined using combiner PD5 and then translated to X-band. The receiver is exactly the reverse and built around the divider PD6. The frequencies are once again upconverted to the 2,318–2,386 MHz bandwidth at filters F8. The individual sweeps in each channel are centered at 2,352 MHz with a bandwidth of 68 MHz. The idea behind this strategy is to take advantage of the steep skirts of the F8 baseband filters that work optimally at 2,352 MHz. These are Type 1 Chebyshev cavity filters with a pass band ripple of 0.01 dB. The nearest spurious frequency from the adjacent channel is manifested as 2,272 MHz, which is rejected by these filters at −38 dBc. To achieve better performance, we use two F8 filters. In such a case, the rejection is −63 dBc. This is detailed in Appendix I. The combiners and dividers are identical and have been already discussed. The accompanying software "*mixer.m*" analyses the IM products at the output of mixers.

8.6.9 Stretch Processing

FMCW radars achieve high resolutions easily, thanks to correlation detection employed in stretch processing. For example, if the sweep bandwidth of LFM signal is 50 MHz and the sweep time is 1 msec, then the compression ratio (time–bandwidth product) required will be 50,000. In stretch processing, the only requirement is that the target range and range rate should be approximately known ([6], p. 592). Putting it another way the target range should be unambiguous and the Doppler ambiguity should be confined to within one range bin. If the Doppler ambiguity is not so confined, the range–Doppler coupling will cause errors in range. In such a case, waveform compression may be accomplished synchronously via correlation. This is in fact, what is happening in FMCW radars as they are operating to within the resolution of the uncompressed waveform. Therefore, for FMCW radars, the target Doppler need not be known. It is sufficient if the FMCW radar ambiguity in Doppler is less than one range bin, because, then the range–Doppler coupling gives range errors of less than one range cell [20]. As long as the range–Doppler coupling error is confined to within one range bin accuracy, range–Doppler coupling errors will have no effect. The range is unambiguous due to the nature of FMCW radars, which operate at limited ranges, being defined by the sweep time, sweep bandwidth and maximum beat frequency. Hence, we are assured that the range is always unambiguous. The Doppler remaining unambiguous, however, is a problem. This is not automatic, but has to be deliberately ensured by selecting the waveform parameters carefully. We will take an example to illustrate the problem. Suppose we have a sweep bandwidth of 50 MHz and a sweep time of 20 msec. Suppose also, we are seeking to detect a fighter at Mach 1 (about 300 m/sec). Suppose further that this sweep bandwidth is varying from 9.3 to 9.35 GHz. In that case, the target Doppler $\left(f_d = 2V/\lambda \right)$ for a constant velocity of 300 m/sec varies through 18,600–18,700 Hz, that is, 100 Hz across the sweep bandwidth. If our Doppler resolution is 50 Hz, that is, nearly half of the sweep bandwidth, then the Doppler is no longer unambiguous, as there is a 100 Hz change in Doppler across the sweep. We want a Doppler resolution in excess of 100 Hz for unambiguity. If instead we have a Doppler resolution of 200 Hz, then the 100 Hz change across the sweep bandwidth will be unnoticed and the Doppler for this sweep bandwidth will be *effectively* unambiguous. If now we carry out stretch processing, it will be effective and accurate, to the nearest range bin, the range–Doppler coupling being confined to *within* one range bin.

Furthermore, in our multifrequency radar, the Doppler shift for the same constant target velocity is different for each frequency channel, but if we follow the above rules, it will be confined to the same range bin in all the channels. During calibration of the radar, we need to adjust delay lines in the signal path of each channel, so that for a point target, they all read the same range bin.

Hence sweep bandwidths should not be very large for a given Doppler resolution, as the Doppler might change over the sweep for a constant velocity target. Since the signal processing is coherent, we can extract the target Doppler by carrying out Doppler FFT on every range cell of the range FFT [20].

Stretch processing has already been discussed extensively in Chapter 4.

8.7 THE SAMPLING QUESTION

Until now we have basically concentrated on the energetic range of FMCW radars. We have determined the average power levels required for different types of targets at various ranges. However, the design is not yet complete. Having determined the energetic range, we now need to choose the signal processing parameters *within* this energetic range, for example, the sampling frequency, sweep bandwidth, and so forth.

The following equations define the various relationships for the Pandora type of radars:

$$R_{\max} = \frac{Nc}{4\Delta f}$$

where N is the number of samples/sweep (range gates $= N/2$), c is the velocity of light, Δf is the sweep bandwidth in Hz, and $R_{\max}$ is the maximum instrumented range. This should be more than the calculated energetic range for the available power levels. The software in Appendix B, gives the minimum sampling frequency for the chosen $R_{\max}$.

$$\Delta R = \frac{c}{2\Delta f}$$

This gives the coarse range resolution ΔR from the point of view of the Pandora.

We now need to correct this for range resolution adequacy, as discussed earlier in Chapter 4. Having determined upon the sampling rate and nonlinearity, we now need to divide by Φ, where Φ is the number of parallel channels, to obtain the fine range resolution:

$$\therefore \Delta R = \frac{c}{\Delta f \Phi}$$

However, this is not the final ΔR. It will be recalled that the overall sweep across eight channels is 776 MHz. This means that we are in actuality sampling this sweep at eight points. The range resolution for this sweep is 0.19 m and allowing for Hamming weighting, this works out to 0.34 m. This figure is the best one can get. The ΔR as defined by the above equation is the worst one can get *at zero range*. The final figure for range resolution will lie somewhere in between these values.

The antenna bandwidth 'B' is calculated as discussed in "Fine Mode," Appendix B. In choosing these parameters, there are three points we need to consider:

1. The range R_{max} should be maximum.
2. The range resolution should be high, that is, $\Delta R \rightarrow$ minimum.
3. The antenna bandwidth B should be minimum.

Considering the above three requirements, the most important requirement should be that R_{max} should be maximum. There is no advantage in having high resolutions for point blank ranges. Having determined upon R_{max}, we should next try our best to accommodate the other two parameters, that is, ΔR and B within this limitation.

In practice, even 1.5 meters resolution gives a reasonable target range profiling. Ideally, we would like to go below the high range resolution barrier of 1 meter, but we cannot do that *easily* due to antenna bandwidth limitations. This is the penalty we pay for Hamming weighting, because we need to control the range side lobes in such range profiling radars. There is no point in using nonlinear FM waveform, as such waveforms transmit an extra sweep bandwidth to compensate for PACF broadening caused due to weighting, even though weighting is not actually carried out in the receiver. This aspect has already been discussed earlier. In reality, if we can control our nonlinearities in the waveform, we can achieve better resolutions. But the deterioration of receiver frequency resolution due to weighting cannot be avoided.

8.8 NEED FOR A DELAYED SWEEP OSCILLATOR

During discussions in Chapter 4 regarding range resolution, we determined that it will be advantageous to have large sweep times to reduce the loss in range resolution due to round trip delays. Even if this does not satisfy us, we will be compelled to adopt a delayed sweep local oscillator, which would delay the transmitted sweep to the SSB demodulators by an amount equal to the round trip delay. The advantage of this approach is that we achieve the highest possible range resolution depending upon the sweep bandwidth at *all* ranges. Normally, in FMCW radars, the range resolution is best at zero range and progressively decreases with range. This "delayed sweep" concept is hardware intensive. This is discussed in detail at Appendix C. However, this concept has not been implemented in the prototype discussed in Chapter 10, as there was no need for it in that kind of application.

8.9 FMCW WAVEFORM TRADE-OFFS

The FMCW waveform parameters need to be carefully chosen. Basically, these parameters are frequency deviation, modulation period, and beat frequency.

Increasing frequency deviation improves resolution, lowers the radiated power spectral density, and increases the beat frequency. Reducing it enables us to better control linearity and go for higher output power.

Increasing modulation period yields better range and Doppler resolution and increases the coherent processing interval.

If we increase the beat frequency, the FM noise at carrier is less, causing higher sensitivity [21]. Decreasing the beat frequency makes for a reduced receiver bandwidth for each range bin filter for a given range resolution. The phase noise of the transmit and receive signals, may be more correlated. This acts to reduce phase noise effects. For high beat frequencies, a given Doppler shift corresponds to a smaller apparent shift in range. If we go in for too high beat frequencies, we run into problems for choosing a suitable AD converter with an adequate sampling rate. The converters also need to be relatively cheap, as we have multiple frequency channels and will require them in large quantities. In case such AD converters are not available, we will need to have an analog filter bank. We will also need to utilize a control loop to maintain a constant beat frequency across the analog range filter bank, by varying the modulation period in direct proportion to range. This increases hardware complexity.

8.10 CHANNEL ISOLATION AND GROUP DELAY PROBLEMS

It is seen from Figure 8–18, that the signal from the DDS source has to progress through six mixers and eight filters (9 if we take two F8 filters as will be necessary), before it enters the A/D converter for sampling. Normally, in any radar, we investigate the number of IM products that are permitted in the pass band. Based on this information, we select the filter bandwidths and orders. This is a complex process and is discussed in [13]. However, in the Pandora, our problem gets further complicated by the fact that we also need to control the group delay distortion due to the filters. This means that an ideally linear FM signal gets distorted at the stretch processor due to the filters *not* having a uniform group delay across the sweep bandwidth. This gives rise to a number of beat frequencies as shown in Figure 8–19.

We can see in Figure 8–19, that if there were no nonlinearities caused either due to group delays or due to the basic quality of the generated signal (in the figure, the nonlinearities are greatly exaggerated for the purposes of explanation), then the signal will vary as per the dotted line, yielding a beat signal proportional to range (round trip delay t_d), which is constant at frequency f1. However, in the presence of nonlinearities, we can see that we can get two frequencies f2 and f3. In reality there will be many more. These cause side lobes to appear, when ideally it should have been a line spectrum. Hence, it becomes necessary to precompensate this group delay distortion so that the received signal is as close to the dotted line as possible in quality. This is done using a

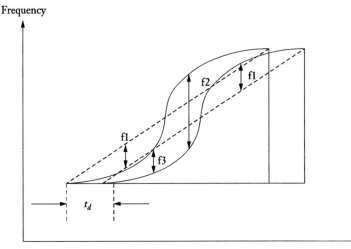

Figure 8–19 Towards the explanation of non-linear distortion

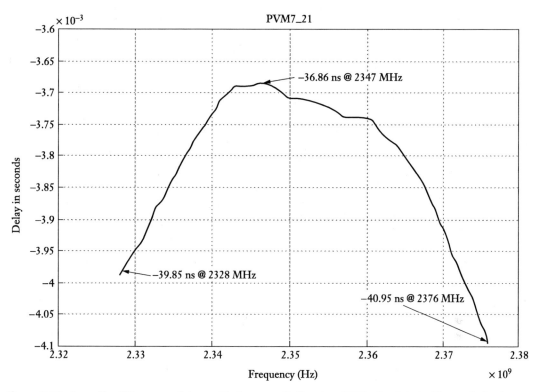

Figure 8–20 Correction filter group delay profile (measured). (Courtesy IRCTR: Reprinted with permission.)

correction filter. Correction filter only corrects for nonlinearities caused due to filter group delays. The remaining nonlinearity is controlled by using a pure signal source. Channel isolation also plays an important role toward curbing nonlinearities. A strong signal from an adjacent channel will distort the received waveform. Hence, we need to ensure proper channel isolation using steep baseband filters as discussed in Section 8.5.8. The S21 parameter is measured from the output of splitter PD3 (Figure 8–18) till the output of F8 filter. The group delay is then computed. Its inverse is the correction filter having a characteristic as shown in Figure 8–20.

An examination of the correction curve shows that the maximum group delay across the sweep bandwidth is not exceeding 4.1 nsec, that is, 4.1 nsec for a sweep frequency variation of 24 MHz. This has been the result of a careful selection of filter order and bandwidth. We now design a correction filter based on Figure 8–20, which gives practically a flat group delay response to within 1 nsec. We also examine the side lobe quality of the beat signal using a delay line of 80 μsec corresponding to a target at 12 km (see Appendix 'I' for details).

We next measure the beat signal with a view to determining its side lobe level. The measured side lobe level with Hanning weighting is shown in Figure 8–21. This measurement was carried out without any correction filter. We note that even without a correction filter, the result is excellent. This is explained by the fact that the RMS value of the correction required across the sweep bandwidth is 1 nsec. In accordance with antenna theory, near field distortion starts at about $\pi/8 < \Phi < \pi/4$, where Φ is the phase distortion. Similar logic applies to our problem. Klauder

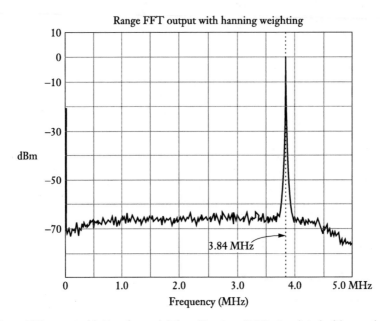

Figure 8–21 Range FFT output with Hanning weighting. (Courtesy IRCTR: Reprinted with permission.)

[22] showed that with no weighting, the side lobe structure gets distorted with a quadratic phase distortion as small as 22.5° or $\pi/8$. He particularly showed that for phase shifts between 22.5° and 90° the side lobe deteriorates by as much as 4 dB. In our case, from Figure 8–20, we note that the RMS value of the group delay is 1 nsec across the sweep bandwidth. This is equivalent to a 17° phase shift, which is less than Klauder's limit of 22.5°. Klauder [22] and Scheer [21] also showed that in order to combat quadratic phase shifts, if we resort to heavy weighting (for about −40 dB side lobes), the peak is attenuated by 4 dB, the 3 dB width increases by a factor of 2.3, but the side lobes remain better than 36 dB down for a 360° phase error. A 48 MHz sweep yields a 3.12 m range resolution. This figure will improve to 0.34 m at the output of the high resolution eight-point FFT allowing for Hamming weighting and taking into consideration the available bandwidth across eight channels of 776 MHz. Such a correction filter can be manufactured to within 1 nsec of the required curve in Figure 8–20.

8.11 AMBIGUITY ANALYSIS: PANDORA RADAR

The ambiguity function of an FMCW radar of this configuration has been discussed in [23]and is based on the work done by Jankiraman et al. [23]. The functioning of the Pandora radar at system level is described in Section 8.3. We examined the effect of the high resolution FFT, which yielded a range resolution of 0.19 m for a 48-MHz sweep. Effectively, we noticed that the achievable resolution of 3.12 m for a 48-MHz sweep was divided by eight (the number of channels) by the high resolution FFT yielding 0.19 m. If we correct this for Hamming weighting, we multiply it by a factor of 1.8 to yield 0.34 m. Hence, we have utilized the guard bands to achieve a higher range resolution than can be obtained by the individual sweeps themselves. The ambiguity function of the Pandora radar for the case of M channels and taking into account just one sweep is given by

$$\chi(T_R, f_d) = \frac{1}{MT} \sum_{m=0}^{M-1} \exp j\pi \left(f_d - 2mF_2 \right) T_R \left(T - |T_R| \right) si \left(\pi \left(f_d + \frac{T_R}{T} F_1 \right) (T - |T_R|) \right) rect \left(\frac{T_R}{T} \right)$$

$$\Rightarrow |\chi(T_R, f_d)| = \frac{\sin(M\pi F_2 T_R)}{M \sin(\pi F_2 T_R)} \frac{T - |T_R|}{T} si \left(\pi \left(f_d + \frac{T_R}{T} F_1 \right) (T - |T_R|) \right) rect \left(\frac{T_R}{T} \right)$$

(8.3)

where
T is the sweep time;
F_1 is the sweep deviation;
F_2 is the distance between channels;
M is the total number of parallel channels;
m is the channel number;
f_d is the target Doppler; and
T_R is the target round trip time.
To clarify the various terms, see Figure 8–22. The guard band is given by $F_2 - F_1$.

In equation (8.3), we assume that all the channels are equal i.e., they are calibrated for identical performance and contribute equally to the overall ambiguity function. Throughout this section we review cases where the guard band is absent i.e., $F_1 = F_2$, unless explicitly stated. Moreover, when we speak of BT product, we mean the BT product of each individual channel. The ambiguity function as defined by equation (8.3) is shown in Figure 8–23 along with its contour plot.

Inspection of equation (8.3) shows that the ambiguity function has a relative maxima if $f_d = -(F_1/T)T_R$. This is along the tilted line shown in the contour plot in Figure 8–23. The magnitude along this tilted line is, therefore, given by

$$\left| \chi \left(T_R, -F_1 \frac{T_R}{T} \right) \right| = \frac{\sin(M\pi F_2 T_R)}{M \sin(\pi F_2 T_R)} \frac{T - |T_R|}{T} rect \left(\frac{T_R}{T} \right)$$

(8.4)

Along this tilted line peaks exist if $\pi F_2 T_R$ equals $z\pi$, where z is an integer i.e., $T_R = z/F_2$. This accounts for the distance $1/F_2$ shown in Figure 8–23. If $M > 1$, the first term dominates and the width in time direction equals $1/MF_2$ and in the frequency direction equals F_1/TMF_2. This is clearly shown in Figure 8–24. Therefore, based on Figure 8–23, if $F_2 \to \infty$ or 0, the side lobe

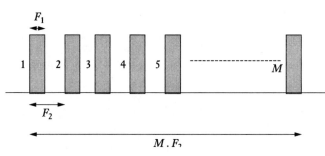

Figure 8–23 Toward an explanation of terms in equation (8.3).

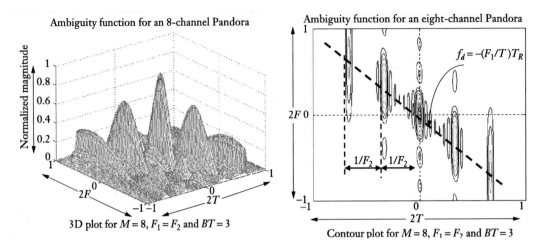

Figure 8–23 Ambiguity function for an eight-channel Pandora and contour plot.

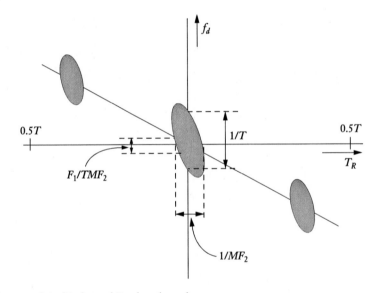

Figure 8–24 Contour plot of 8-channel Pandora (zoom).

due to F_2 will either tend to merge with the main lobe, or remain at infinity, i.e., the ambiguity diagram will reduce to that of a single chirp pulse as shown in Figure 8–25. A spinoff from this is that if we do keep a substantial guard band by having a large value of F_2, we gain advantage of an increased bandwidth. This is explained by the fact that normally if we have eight sweeps each of 48 MHz, we would achieve an overall bandwidth of 384 MHz with a Rayleigh range resolution of 0.39 m ($c/2 \times$ Bandwidth). If we retain a guard band of 50 MHz between the channels then our overall bandwidth will be 784 MHz corresponding to a Rayleigh resolution of 0.2 m. Hence, this is a trade-off between our desire for high resolution and the nuisance of a side lobe close to the mainlobe. A simulation was carried out for eight-channel Pandora radar with a 50 MHz gap

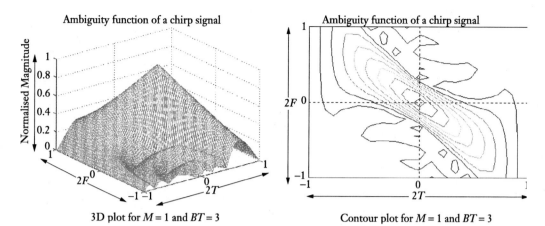

3D plot for $M = 1$ and $BT = 3$ · Contour plot for $M = 1$ and $BT = 3$

Figure 8–25 Ambiguity function of a chirp pulse and contour plot.

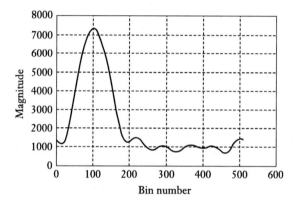

Figure 8–26 Two targets separated by a range of 0.1 m.

between channels, each channel having a 48-MHz sweep. Two targets were simulated, separated in range by 0.1 m. The result is shown in Figure 8–26. We note that there is no discrimination. However, if the separation were 0.2 m we can clearly discern two targets as shown in Figure 8–27. This is much better than the Rayleigh resolution of 0.39 m yielded by a total bandwidth of 384 MHz, where we need to use conventional FMCW radar.

Reverting back to equation (8.3), we note that the width along the frequency axis ($T_R = 0$) is given by

$$\left|\chi\left(0, f_d\right)\right| = si\left(\pi f_d T\right) \tag{8.5}$$

It follows that this width equals $1/T$ and is independent of the number of channels M (see Figure 8–24). We will now examine the behavior of the ambiguity function as the BT product changes and the number of channels change. It is better to analyze this using contour diagrams. We shall study various cases.

$$M = 1 \text{ and } BT = 3$$

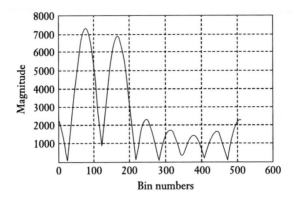

Figure 8–27 Two targets separated by a range of 0.2 m.

This is the ambiguity function of the classic chirp signal. It is based on equation (8.3) with the channel number $m = 0$ and number of parallel channels $M = 1$. This signal is basically a diagonal ridge. The diagonal nature of this ridge implies range–Doppler coupling. In equation (8.3), for $m = 0$ and $M = 1$ we obtain

$$\chi\left(T_R, f_d\right) = \frac{1}{T} \exp j\pi f_d T_R \left(T - \mid T_R \mid\right) si\left(\pi\left(f_d + \frac{T_R}{T} F_1\right)\left(T - \mid T_R \mid\right)\right) \mathrm{rect}\left(\frac{T_R}{T}\right)$$

$$\Rightarrow \exp j\pi f_d T_R \frac{\sin\left(\pi\left(f_d + \frac{T_R}{T} F_1\right)\left(T - \mid T_R \mid\right)\right)}{\pi\left(f_d + \frac{T_R}{T} F_1\right)T} \mathrm{rect}\left(\frac{T_R}{T}\right) \qquad (8.6)$$

In Figure 8–25, we detect a single peak at the origin.

$$M = 8 \text{ and } BT = 3$$

We now examine the contour plot for the case of an eight-channel Pandora. We note that the 3D plot in Figure 8–23 shows two side lobes $1/F_2$ apart. This is clearly shown in the contour plot in Figure 8–23. What we see in Figure 8–23 is the effect of increased bandwidth due to the 8-channel effect.

To summarize, it will be interesting to check out as to what happens if in an eight-channel Pandora, we have a high BT product. This is shown in Figure 8–28 and 8–29.

The ambiguity function looks like a multi-pronged candlestick, wherein the center prong is the main lobe and the other prongs are the side lobes $1/F_2$ apart. The implication here is that as the effective bandwidth is increased whether due to increasing BT of the individual channel or by adding guard band, in multi-channel radar side lobes appear at $1/F_2$ away from the main lobe. This result is the direct consequence of sampling a low resolution signal defined by F_1 at F_2 rate, i.e., due to the nature of the signal processing wherein we carry out a high resolution FFT based on the results of the low resolution FFT of each channel. Hence, if we seek to improve our range resolution by adding guard bands to the basic FMCW sweeps, in the Pandora configuration even if there is no guard band we need to deal with the side lobes which appear $1/F_2$ away from the main lobe and from each other. In the absence of a guard band, $F_1 = F_2$. Not having a guard band

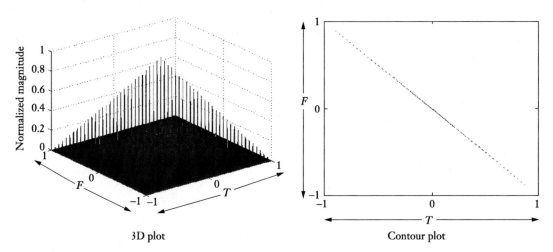

Figure 8–28 Pandora ambiguity function for a high *BT* product.

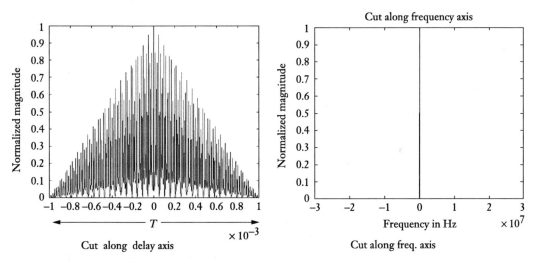

Figure 8–29 Pandora ambiguity function for a high *BT* product.

is, however, a theoretical concept as a guard band is necessary in order to ensure channel isolation. This means that each target return will have side lobes in the adjacent range bins. These side lobes can also be seen in the simulation in Figure 8–26 and Figure 8–27. Alternately, we can have a single-channel Pandora stepping through the entire bandwidth. Even in such a case, though the guard band is zero, the side lobes appear $1/F_2$ away wherein $F_1 = F_2$. This inconvenience due to the side lobes $1/F_2$ away is, however, not experienced when we carry out the signal processing of SFCW signals, which is a totally different approach as we shall see.

8.12 SFCW SIGNALS

During processing of SFCW signals, in the Pandora configuration, we cannot add guard bands and take advantage of the additional bandwidth provided by this approach to improve range resolution. This is because SFCW signals are intolerant of target Doppler. The addition of a guard band will be interpreted as such by the signal and it will suffer a range–Doppler coupling. This is illustrated in Figure 8–30.

Therefore, in order to eliminate the ill effects of range–Doppler coupling due to the presence of the guard band, it becomes necessary in the SFCW mode to laterally shift the entire spectrum of the radar to cover the gaps due to the guard band. This implies an additional measurement. The signal processing is then carried out collectively. Hence, a 32-step SFCW becomes actually a 64-step signal so as to also include the guard band gaps. This means that each sub-pulse will have a complex value stored in a RAM and we then carry out an IDFT on all the 64 sub-pulses. The basic advantage using this mode is that this is perhaps the fastest generator of SFCW signals in the radar world today. Consider a basic signal source like the DDS. Suppose we make it step through a sub-pulse bandwidth of 10 MHz and duration of 100 μsec (such a fast timing is possible for a DDS). In the normal course, such a DDS will step through 64 steps in 6.4 msec. But in the Pandora configuration, it needs only step through eight steps, by which time the remaining seven channels will have stepped through *their* eight steps, making 64 steps in all in just 800 μsec. Hence, if we include a second measurement to cover the guard band gaps, we need a total time of 1.6 msec for 128 steps. This is true of any other signal source other than a DDS. If due to phase locking considerations, and the signal source takes say, 1.5 msec per step, then we can use it as the driving signal source in place of the DDS and in the Pandora configuration complete 64 steps in a mere 12 msec achieving 8 times the basic bandwidth of the signal source for eight steps. The possibilities of this technology in the world of high resolution radars are staggering. The ambiguity analysis of such class of signals has been covered in Chapter 3.

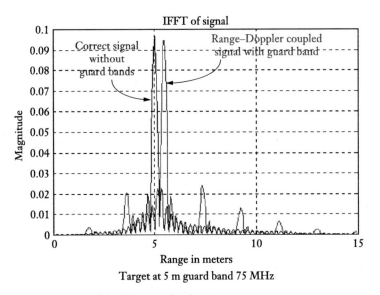

Target at 5 m guard band 75 MHz

Figure 8–30 Range–Doppler coupling with SFCW signals.

8.13 CONCLUSION ON AMBIGUITY ANALYSIS

In the preceding sections, we had briefly examined the design aspects of the FMCW radar Pandora. The massive resolution advantage gained by utilizing the guard bands in the FMCW mode was demonstrated. It was shown that this radar can operate in the SFCW mode if desired but without the attendant advantages of increased bandwidth due to the guard band as in the case of the FMCW signals. This is because the SFCW signal suffers from range–Doppler coupling effects due to the guard band. However, once we remove this constraint by a second measurement, we have demonstrated that the Pandora radar is perhaps the fastest SFCW signal source in the world today.

Figures 8–23 and 8–24 have been obtained with "*ambf.m*" file in the accompanying software. Figures 8–26 and 8–27 have been obtained with "*fmcw.m*" and its accompanying "*fmcw2.m*" files in the accompanying software. Figures 8–28 and 8–29 have been obtained using "*ambf.m*" and "*delay.m*" files in the accompanying software. Figure 8–30 has been obtained using "*sfcw.m*" and its accompanying "*sfcw2.m*" in the accompanying software.

References

1. Wehner, D. R., *High Resolution Radar*, Artech House, Inc., Norwood, MA, 1987.
2. Gjessing, D. T., "Matched illumination target adaptive radar for challenging application," in *IEE - 87 Radar Conference*, 1987.
3. Gjessing, D. T., *Target Adaptive Matched Illumination Radar*, Peter Peregrinus Press, Boca Raton, FL, 1995.
4. Taylor, J. D., *Introduction to Ultra-Wideband Radar Systems*, CRC Press, Boca Raton, FL, 1994.
5. Caputi, W. J., Jr., "Stretch: A time transformation technique," *IEEE Transactions on AES*, Vol. 7, No. 2, March 1971.
6. Nathanson, F. E., *Radar Design Priniciples*, McGraw-Hill, Boston, MA, 1991.
7. Levanon, N., *Radar Principles*, John Wiley & Sons, New York, 1988.
8. Skolnik, M. I., *Radar Handbook*, McGraw-Hill, Boston, MA, 1990.
9. Levanon, N. and Mozeson, E., *Radar Signals*, Wiley-Interscience, Hoboken, New Jersey, June 25, 2004.
10. Johnston, J. A., "Generating analogue FM pulses using a 1 bit digital technique," *IEE Proceedings-F*, Vol. 131, No. 4, July 1984.
11. Griffiths, H. D. and Bradford, W. J., "Digital generation of wideband FM waveforms for radar altimeters," in *IEE-87 Radar Conference*, 1987.
12. Eber, L. O. and Soule, H. H., "Digital generation of wideband LFM waveforms," in *IEEE-75 Radar Conference*, 1975.
13. Wessels, B. J., "Frequency optimization for the Pandora radar," in *IRCTR Book*, # 263, IRCTR, Delft University of technology, Delft,The Netherlands, 1998.
14. Liang, Xiao-Peng, Zaki, K. A., and Atia, A. E., "Channel expansion and tolerance analysis of waveguide manifold multiplexers," *IEEE Transactions on Microwave Theory and Techniques*, Vol. MTT-40, No. 7, July 1992, pp. 1591.
15. Rhodes, J. D. and Levy, R., "Design of general manifold multiplexers," *IEEE Transactions on Microwave Theory and Techniques*, Vol. MTT-27, No. 2, February 1979, pp. 111.
16. Müller, G., Bleißner, E., and Pfaff, R., "Zum Entwurf von Hohlleiter-Frequenzweichen," *AEÜ-Archiv für Elektronik und Übertragungstechnik—Electronics and Communication*, Band 34, Heft 3, 1980, p. 111.
17. Steenstra, H. T., *Haalbaarheidsstudie naar een power combiner voor Pandora*, Book number 68–220-A249-98, Delft University of Technology, Delft, The Netherlands, March 1998.
18. Matthaei, G. L., Young, L., and Jones, E. M. T., *Microwave Filters, Impedance-Matching Networks and Coupling Structures*, McGraw-Hill, Boston, MA, 1964.
19. Linz, E., Christensen, E. L., et al., "Review of the homodyne technique for coherent radar," in *IEEE-90 Radar Conference*, 1990.
20. Stove, A. G., "Linear FMCW radar techniques," *IEE Proceedings-F*, Vol. 139, No. 5, October 1992.

21. Scheer, J. A., et al., *Coherent Radar Performance Estimation*, Artech House, Inc., Norwood, MA, 1993.
22. Klauder, J. R., Price, A. C., Garlington, S., and Albersheim, W. L., "The theory and design of Chirp radars," *Bell System Technical Journal*, Vol. 39, July 1960, pp. 745–808.
23. Jankiraman, M., De Jong, E. W., Van Genderen, P., "Ambiguity analysis of PANDORA Multifrequency FMCW/SFCW radar," in *Radar Conference, 2000. The Record of the IEEE 2000 International*, pp. 35–41.

Implementation of the Single-Channel Pandora Radar and Other Issues

9.1 BLOCK SCHEMATIC OF THE SINGLE-CHANNEL RADAR

The group delay problem discussed in Section 8.10 required a detailed investigation by implementing a single-channel radar. This is just one of the eight channels used in the Pandora. The block diagram of the single-channel radar is shown in Figure 9–1.

This radar is divided into three sections: the digital sweep generator, sweep upconverter, and the receiver downconverter. We shall examine each of these in turn.

9.2 DIGITAL SWEEP GENERATOR

The block diagram of the digital sweep generator is shown in Figure 9–1. The signal source is a direct digital synthesizer (DDS) marketed by Stanford Telecom, USA, and operating at a clock rate of 1 GHz. A DDS was chosen, because of its high signal purity. This DDS generates a sweep extending from 200 to 248 MHz. This is up-converted via a local oscillator LO1 (2,128 MHz) to a sweep extending from 2,328 to 2,376 MHz. This signal is then amplified and fed to a two-way divider (PD3). One output of the divider constitutes the reference signal, while the other is the transmitted signal. Until now the channel is like it would be in the actual multifrequency radar. The difference lies in the sweep upconverter. In the actual radar, the signal is now split eight ways using an eight-way divider and fed to each channel as the basic signal. In our case, this signal is directly applied to mixer M2 of the sweep upconverter (see Figure 9–1).

9.3 SWEEP UPCONVERTER

The signal is routed via a correction filter. The drawing shows a power divider PD4 at the position it should have occupied, were this a multifrequency radar. In a multifrequency radar, the power divider PD4 is required to split the power 8 ways in order to cater to the eight channels of the radar. In the present instance, since this is a single channel radar, this is not necessary, but has been shown in Figure 9-1 for completeness. PD4 is the eight-way divider. This signal is then mixed with a local oscillator LO2 (3,640 MHz) to generate a sweep extending from 1,264 to 1,312 MHz. This is the actual sweep signal for this particular channel. Normally, it would have been applied to an eight-way power combiner PD5, along with the sweep signals from the other seven channels. But in this case, we ignore this aspect and feed the signal to a mixer M3 for the final upconversion

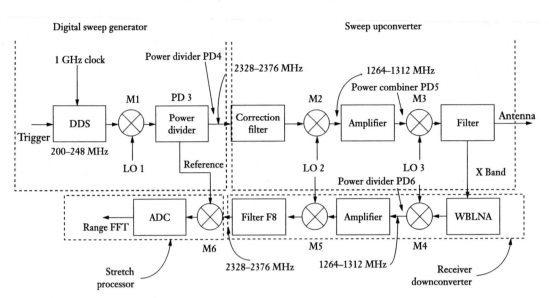

Figure 9–1 Schematic for single-channel Pandora radar. (Courtesy IRCTR: Reprinted with permission.)

to X-band. This is accomplished using a local oscillator LO3 (8,114 MHz). The output of the final filter, F5, is then applied to the antenna. This sweep spread is now a 48 MHz sweep in the X-band. But in our case, since we are investigating the group delay problem, we give the signal instead to the receiver downconverter. This is shown in Figure 9–1.

9.4 RECEIVER DOWNCONVERTER

The receiver downconverter is a mirror image of the sweep upconverter. The signal is down-converted from the X-band using oscillator LO3 at mixer M4, back to its original spread of 1,264–1,312 MHz. The signal is once again up-converted to 2,328–2,376 MHz sweep via mixer M5 and LO2. The output of filter F8 is then applied to the stretch processor.

9.5 LOCAL OSCILLATORS

We are using three local oscillators: LO1, LO2, and LO3 at 2,128, 3,640, and 8,114 MHz, respectively. Since LO2 and LO3 are required at the receiver for downconversion, the signal source is split and then applied both ways. This ensures phase synchronous local oscillator signals. Due to budget constraints, all the oscillators used were free running ones. In the actual radar, they need to be phase locked for reduced oscillator phase noise.

Details of the single channel are applied in Appendix H. The power levels at each point on the schematic are clearly indicated. We need to use attenuators in order to achieve the correct power levels at the mixers. These power levels are defined by the mixer specifications and need to be strictly adhered to in order to reduce IM distortion.

9.6 DEMONSTRATION MODEL

9.6.1 Pandora Verification Measurements

A single-channel receiver was constructed as per specifications in Appendix H. Thereafter, work was taken up to verify the performance parameters. The approach was to draw up a series of experiments, which will examine the various parameters in a phased manner. This radar was built through a judicious selection of the filters so that they fulfilled their specific tasks without overly compromising the group delays across them. Hence, it now remains to successfully compensate for the group delays that cannot be avoided due to the passage of the signal through the filters. We must also ensure that after the correction, the spectrum of the signal does not deteriorate. Hence, there is a need to monitor the spectrum of the signal stage by stage.

9.6.2 Measurement Procedures

Table 9–1 briefly outlines the purpose of each of the 12 principal experiments carried out to assess this radar.

Table 9–1 Twelve Principal Experiments

Experiment Number	Aim
1	Measure the performance of the DDS unit
2	Assess the performance of the M1 mixer
3	Assess the performance of M2 mixer, filter F3, and amplifier AMP3
4	Assess the performance of M3 mixer, filter F4, and filter F5
5	Verify the performance of the LOs
6	Assess the performance of amplifier AMP4, filter F6, and mixer M4
7	Assess the performance of amplifier AMP5, filter F7, filter F8, and mixer M5
8	Assess the performance of the stretch processor mixer M6
9	Assess the performance of the correction filter
10	Measurement of two tone IM distortion in mixers M3 and M4
11	Measurement of the receiver noise figure
12	Measure the level of adjacent channel interference at the output of mixer M6

9.6.3 Salient Results

The outcome of each experiment and the conclusions drawn are applied in detail in Appendix I. We shall only briefly examine the salient conclusions here.

1. The beat signal was measured for a delay line of 80 μsec corresponding to a target at 12 km. It can be seen in Figure 9–2 (Figure 8–20 is reproduced here for convenience) that the result is excellent even without a correction filter. We have used Hanning weighting. Hence, there is no need for a correction filter in this radar.
2. In case there is a need to correct, deterioration can be corrected using a correction filter whose group delay characteristic is shown in Figure 9–3 (Figure 8–21 is reproduced here for convenience).

The completed single-channel radar is shown in Figure 9–4 and Figure 9–5.

9.7 SUMMARY

An analysis of the multifrequency radar problem was carried out. Three key technological challenges were identified as the power combiner, power resolver, and the group delay problem.

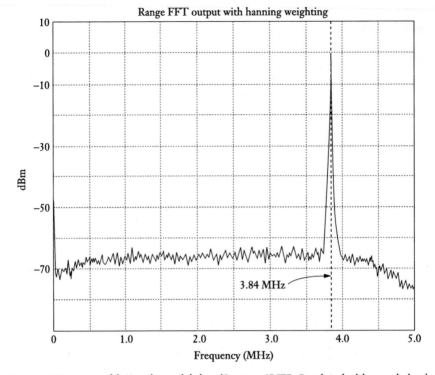

Figure 9–2 Range FFT output with Hanning weighting. (Courtesy IRCTR: Reprinted with permission.)

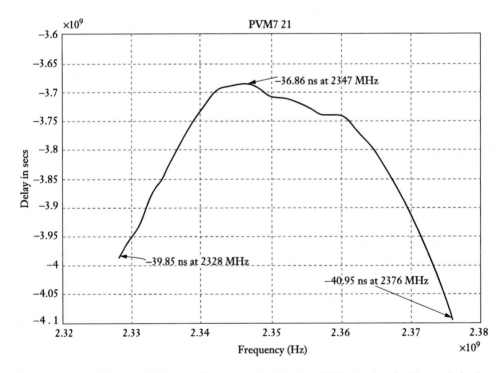

Figure 9–3 Correction filter group delay profile (measured). (Courtesy IRCTR: Reprinted with permission.)

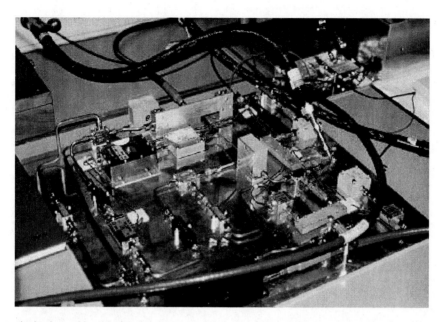

Figure 9–4 Single-channel board. The stretch processor (M6 mixer) is to the bottom right. The 3,640 MHz LO is located to the top left. (Courtesy IRCTR: Reprinted with permission.)

The various types of combiners were assessed and it was found that the passive Wilkinson combiner offers the best results, both in terms of accuracy and minimum parts count. The power resolver was identified as a bank of Chebyshev filters, which reject the adjacent channel interference by as much as −63 dB compared to the input signal level. A single channel of the radar was implemented and it was found that the side lobe levels were better than −70 dBc making a correction filter unnecessary. However, a correction filter was in any case designed so as to prevent side lobe deterioration at the output of the stretch processor. *In our analysis of the Pandora radar, not much emphasis was laid on the design of the antenna required for this application.* This is a specialized topic in its own right and we do not have enough space in this book to discuss these details. In the next chapter, however, we do discuss antenna configuration as it pertains to the specific application under question, viz., ground-penetrating radar for landmine detection.

The output of the high-resolution FFT experiences gaps due to the low sampling rate of the 776 MHz sweep. It is recommended that in order to obtain better performance, this problem can be solved by shifting the frequency spectrum of the entire radar by 50 MHz to cover the gaps which presently exist due to the 56 MHz guard band. This shift can be realized at the M3 and M4 mixers using frequency hopping techniques. Thereafter, we can multiplex the outputs in the time domain to remove the effect of the guard bands.

9.8 CONCLUSION

The Pandora radar can be classified as a new class of target recognition radars for target identification well beyond visual range and can be slaved to a tracking radar or used as a surveillance radar in its own right. It is pointed out that target recognition is the primary quest for ultra-wide band radars. As yet there are massive technical problems that need to be overcome in this area, especially in the

Figure 9–5 Bart Wessels (my colleague on this project) and I with the completed radar. (It took us just under 18 months. *Semper victrix.*) (Courtesy IRCTR: Reprinted with permission.)

designing of cheap, high-power, extremely narrow pulse generators, while the Pandora appears already commercially viable. Hence, as target recognition radar, it will remain competitive for many years to come.

It can also be used in a pulsed mode as a signal source for pulsed radar. This is possible, because of the parallel nature of the architecture. In such a mode, it can be used to generate large bandwidth step frequency signals or LFM many orders faster than is possible given the present state of technology. This will reduce *Doppler smearing* [1] when tracking fast targets. In pulsed mode or CW mode, the Pandora is capable of superior performance in the presence of high target Doppler or own platform Doppler. Once the own Doppler is nulled (ODN), or at least brought down to a low value, it is possible to use it on fast flying platforms to carry out mapping activities, for example, airborne surveillance/planetary probes. This is because of the Doppler resilience of the LFM waveform. Such performance is difficult to obtain with *relatively* poor Doppler tolerant phase-coded signals.

In the next chapter, we shall see the results of the completed multifrequency radar made expressly for profiling landmines. This is the first direct application of this new technology. In demining of landmines, it is imperative to recognize the type of mine deployed, as that determines the approach to be adopted toward disarming that mine. Till now this has been a major problem. Phase-coded radars cannot be used for this purpose, since phase-coded signals do not perform very well in a harsh propagation environment like sand and loamy soils due to the high amount of distortion that they would experience in such an unsuitable medium. Hence, such radars are unsuitable for ground mine detection and profiling. Target recognition FMCW radar based on range profiling offers the desired solution. In this application we have used a step frequency waveform in a parallel architecture with guard bands for channel separation between the channels.

This makes for high speed signal processing. The numbers on this radar are classified, but the calculations are on similar lines as already discussed. The radar uses guard bands to separate the channels. These guard bands (in addition to channel separation) help improve the range resolution when using LFM waveforms (by adding to the bandwidth). The radar is complemented with synthetic aperture techniques to raise the range resolution to even higher levels.

References

1. Skolnik, M. I., *Radar Handbook*, McGraw-Hill, Boston, MA, 1990.

10

Pandora Multi-Channel Radar*

10.1 INTRODUCTION

Radar for humanitarian demining must satisfy some requirements that are unique for radar: the resolution must be extremely high, the dynamic range must be very wide and the signal processing must rely on contrast between two or more different kinds of objects and/or media. Radar is one of the few sensors that is capable of positioning objects underneath the air/ground interface. Radar has problems in detecting surface laid or shallow buried objects, because of the very small effect of the little mines on the large reflection off the soil [1]. This very same property requires that ground-penetrating radar intended for detection of both deeply buried landmines and shallow buried or surface laid mines should provide both low frequencies (for ground penetration) and high frequencies (for having a wide band and sufficient resolution for the anti-personnel (AP) mines). An additional requirement on the radar is that it should provide features of the detected objects supporting their classification as well as eventual fusion with other sensors. Shape or size are examples of such features. The complexity imposed by the ultra-wideband (UWB) extends over all system components, from the low power signal generation, to the transmitter, to the receiver to the antennas. The International Research Centre for Telecommunications and Radar (IRCTR), Delft University of Technology, The Netherlands, have designed and built a stepped frequency continuous wave (SFCW) radar based on a detection capability for both deeply buried antitank mines with normally a significant metal content and surface laid antipersonnel mines, usually with very little metal content or even no metal at all. This chapter is based on various papers published by IRCTR and are given in the references to this chapter. A lot of information in this chapter has been reproduced from [1, 2] with permission from IEEE and IRCTR. Of course, this particular application of Pandora has nothing to do with LPI as there is no need for it here, but it is instructive to study the problems in Pandora radar design. After all, Pandora also can be used for anything and in any application which requires range profiling of a high order. The technology of Pandora is not confined only to CW radars, but it can also be used in pulse and chirp pulse radars, as the technology is not much different in terms of hardware and can be used to rapidly generate stepped frequency waveforms for very high resolutions.

This radar can be used for many applications, but in the first instance it has been designed for detecting land mines.

*This chapter has been jointly written with Piet van Genderen. Prof. Piet van Genderen, International Research Centre for Telecommunications and Radar (IRCTR), Delft University of Technology, The Netherlands. This project was sponsored and funded by the Dutch National Technology Foundation (STW), as part of its ongoing research program for investigating new technologies.

This chapter pertains to the design aspects of a multifrequency SFCW radar with an emphasis on the engineering problems involved in its implementation. This radar is an unmodulated SFCW radar [1]. The basic theory for such single-channel radar has been discussed in the previous chapters. Phase-coded signals were considered since there is no Doppler problem in this application. Since the platform carrying the antennas may move, the system should be robust against these Doppler effects. However, phase-coded signals do not perform well in ground radar due to poor propagation medium. Hence, the choice was either modulated (using LFM) or unmodulated. The design is identical for both and is switchable between these modes. The dimensioning of the radar is optimized for detection of surface laid and buried landmines with a maximum depth of 1 m and an initial requirement on range resolution of approximately 5 cm (since the size of the AP landmines is of the order of a few centimeters in vertical extension (height), the range resolution was set at approximately 5 cm). The depth of 1 m only concerns the burial depth, not the range of the radar. To obtain an acceptable data acquisition time, this radar generates eight separate SFCW signals, which are then additively mixed and radiated simultaneously. The target return also comprises a mixture of returns at these frequencies. The composite signal is then split into its constituents and upon completion of the signal collection over 128 frequencies, these are processed collectively to obtain an extremely high-resolution synthetic image of the target. The radiated power in each of the 128 frequencies can be chosen independently. The main area of application of this equipment will be in landmine detection. However, there are multiple possibilities for this design:

- To study waveforms of variable and flexible transmitted power spectra.
- To study UWB processing techniques, without the need to have a UWB instantaneous bandwidth.
- To study coherent, noncoherent, and mixed processing concepts.

The radar itself will belong to the class of LPI radars in the SFCW mode. The additional advantage here is that this radar can be built using essentially commercial items, that is, it does not require any radically new technology. We discuss some of the parameters of the design and the problems involved in the implementation of this radar.

One of the most important sensors currently considered for detection of AP landmines is the ground-penetrating radar. Considerable debate may be observed whether the preferred design for such a radar is the impulse radar, generating pulses of the order of 1 nsec length, or the SFCW radar, synthesizing a range profile from an acquired set of frequencies. The multifrequency SFCW radar under discussion in this chapter has been called Pandora radar and was introduced to the reader in Chapter 8. Recalling briefly, the technology of this multifrequency radar achieves wideband capability using multiple narrow band SFCW radars (in our specific case eight radars corresponding to 8 channels) *operating simultaneously*. All this has been achieved without the attendant engineering problems of UWB radars. In this chapter, we shall first review some of the basic parameters of the design of the SFCW radar and then examine the overall concept of Pandora, its working principle, and some of the key technologies that go into its design. We shall then discuss some of the experimental results achieved during their implementation.

10.2 BASIC PARAMETERS

The initial requirements that were put forward concerning the properties of the range profile are that it should be unambiguous over at least 1 m and that the resolution should comply with the size of small AP landmines, that is, 5 cm. Because of the large variety in types of soil and the associated variety in energetic ranges, a minimum unambiguous range of around 4.5 m was chosen.

In SFCW radar a range profile can be obtained by inverse Fourier transform of the complex signals achieved for a set of frequencies. With Δf, the interval between two successive frequencies and N_f, the number of frequencies, the total frequency band covered in discrete steps is $B = N_f \Delta f$.

The maximum unambiguous range that can be achieved in SFCW radar then is $R_{max} = c/2\Delta f$ and the resolution in the range profile is $\Delta R = c/2B$, where c is the speed of electromagnetic waves in the propagation medium.

Given that a typical value of the permittivity of soil is $\varepsilon_r = 4$, the speed of the electromagnetic wave will be half of that in vacuum. Taking the total band to be covered by the radar to be 400–4,800 MHz, so $B \cong 4,400$ MHz, the basic parameters of the waveform of $\Delta f = 35$ MHz and $N_f = 128$ will satisfy the initial requirement on range resolution of around 3 cm in air. The unambiguous range in free space will be $3 \times 10^8 / (2 \times 35 \times 10^6) \approx 4.3$ m. In soil this range will reduce according to the square root of the permittivity of the soil.

10.2.1 Overall System Approach

The reason for seeking an alternative to the frequently followed approach to apply network analyzers for generating and receiving the signals is, that the time it takes to collect data over all 128 frequencies is prohibitively long for performing significant field experiments. It is the objective of the Pandora approach to collect the dataset for one range profile in order of a few milliseconds.

The chapter now focuses on the Pandora design for SFCW radar [1] for the given application. The theory behind SFCW radars has already been discussed in Chapter 6 and is also discussed in [3].

The Pandora design comprises the following essential blocks:

1. SFCW waveform generator.
2. Power combiner block.
3. Wideband low noise amplifier.
4. Power resolver block.
5. Quadrature detection for each SFCW channel.
6. IFFT block
7. ADC and data acquisition system.

A system block diagram is shown in Figure 10–1. The SFCW source is based on a direct digital synthesizer (DDS). The output from the DDS is split eight ways and upmixed to impart bandspread. The final bandwidth extends from 400 to 4,845 MHz across eight channels. These outputs are then combined in a Wilkinson combiner after amplification. This output is then given to the antenna. On reception, the signals are routed through a wideband LNA as shown in the figure. The wideband LNA is fitted on the antenna. The signals are then split into the constituent channels based on bandpass filtering. They are then downmixed to a suitable intermediate frequency. These IF signals are then given to quadrature detectors. The final I and Q outputs from the detectors are then fed to the ADC. These signals are slowly varying DC (depending upon the target velocity, which is zero and antenna movement velocity, which is nearly zero), and hence, a low speed ADC will suffice. In Figure 10–1, we note that there are four delay lines, including a short. These are designed for calibration. This aspect will be discussed further down in this chapter. We now sample the signals to get one I and Q value per channel for all the eight channels. To ensure adequate channel isolation we need to have an adequate guard band between channels. To avoid any gaps caused due to the guard band, it was decided to use a two-stage approach. The two-phase approach is shown in Figure 10–2. The guard band gaps are covered in the second phase. This implies that each channel will step through 16 steps in two phases of eight

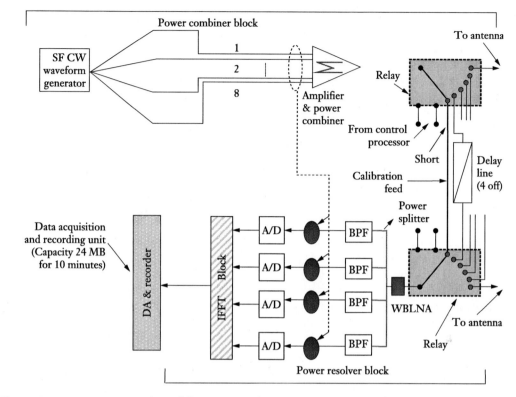

Figure 10–1 SFCW GPR system. (From [2], © IEEE 2001.)

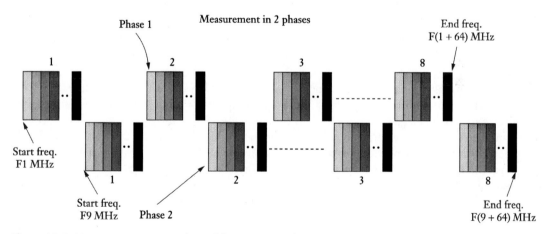

Figure 10–2 Measurement strategy. (From [2], © IEEE 2001.)

steps each, while overall the system steps through 128 steps. This can also be considered speed amplification in the measurement of a 128-step SFCW signal. *The time taken for data acquisition for 128 steps is effectively reduced to around 3 ms* [4]. This will yield a step frequency waveform without any guard band separation as if it were from a single-channel radar.

The radar operates on two antennas, one for transmission and the other for reception. This will have no impact on the performance of the combiner/resolver as these are buffered by amplifiers, the former by the RF amplifier and the latter by a wideband LNA. We ensure 60 dB of isolation between the antennas from the point of view of reduction of AM noise. The radar itself will belong to the class of LPI radars. The high resolutions obtained by this technique are comparable to those that can be achieved by UWB radar. The following paragraphs are taken from [1] and reproduced here with the permission of IEEE.

10.2.2 Design Procedure

10.2.2.1 Choosing the Frequency Band   The radar is intended for ground penetration. Hence, logically, it must provide low frequencies for deep penetration [1]. An overview of the absorption of electromagnetic fields in various types of soil is given in [5]. Based on this source and such similar sources, we can conclude that the lowest frequency should be 400 MHz. The resolution in air should be as small as a few centimeters to find the mines buried at or near the surface of the soil. It was felt that a resolution of 3 cm in air would suffice. The 3 cm would allow for estimating the size of the mine with reasonable accuracy. This requires a bandwidth of around 5 GHz. The reflection off the soil is a function of frequency; the higher the frequency, the stronger the reflection. Therefore, the bandwidth of 5 GHz should be obtained in the lowest possible bands.

In designing and manufacturing antennas, the ratio between the lowest and highest frequency transmitted and received becomes crucial. It can be seen that the higher the lower bound of the frequency band, given a fixed bandwidth of 5 GHz, the lower will be this ratio. For this reason, the lower frequency should not be too low and a compromise between the various effects must be chosen.

Based on these considerations, a lower frequency limit of 400 MHz was chosen. The upper frequency bound will be discussed in more detail. Taking into account the procedure for synthesizing the range profile, an upper limit of almost 4.9 GHz was selected.

10.2.2.2 Choosing the Frequency Step   In stepped frequency radar, the range profile can be synthesized using an IFFT, applied on a set of equidistant frequencies. The maximum unambiguous range in a synthesized profile is given by $R_{max} = c/2\Delta f$ where Δf is the frequency step and c is the velocity of light in the medium.

The resolution in range is $\Delta R = c/2B$ where B is the total bandwidth covered by the N frequency steps. The total range to be covered by the radar is not much; the distance between the antennas and the soil normally is less than 1 m. The depth of burial of landmines is typically less than 1 m. One must bear in mind that other effects may come into play; other objects than the surface and subsurface may reflect signal energy through the side lobes of the antennas, the cabling between the transmitter or the receiver and the antennas may produce internal reflections, and so forth. In general, we must anticipate on secondary reflections and reserve some room in the synthesized range profile to dimension the geometry such that aliased or irrelevant signals do not affect the area having the mines.

Given these considerations, an unambiguous range of at least 4 m was chosen. A frequency step of 35 MHz was chosen, yielding

$$R_{max} = \frac{3 \times 10^8}{2 \times 35 \times 10^6} = 4.28 \text{ m}$$

To achieve this range we require (5000/35)=142 frequencies. A total of 128 frequency steps were finally decided upon for ease of implementing the IFFT. Hence, the bandwidth of this radar becomes $(128-1)\times35 = 4445$ MHz. The lowest frequency being 400 MHz, the highest frequency will be 4,845 MHz and the ratio between the highest and the lowest frequency is 12:1. The corresponding resolution in air is 3.37 cm.

It should be noted that two effects with opposite impact would affect the resolution in the synthesized profile [1]:

- In the soil the relative dielectric permittivity will be higher than one. As an example, the permittivity of dry sand is of the order of three, so that the wavelength in sand equals 1.9 cm.
- The resolution in the synthesized profile is achieved by processing. Applying uniform weighting to the dataset prior to IFFT will yield large range side lobe levels of the order of −13 dB. Hence, at the cost of widening the main lobe, a Taylor weighting was chosen.

10.2.2.3 Dynamic Range   One of the major problems of ground-penetrating radar for detecting small AP landmines is the observability of the mine when it is either flush buried or laid on the soil surface. We will now derive the expression for the small signal dynamic range of the receiver [1].

Suppose the effective surface of the footprint of the antenna is A_a and the effective surface of the mine is A_m. Assume both the mine and the surface to be nonconducting. As an example, the soil might be dry sand with a relative permittivity $\varepsilon_{r,s} = 2.6$ and the mine might be a dielectric material with permittivity $\varepsilon_{r,s} = 4$. Then the reflection coefficient at the air/soil interface if it is only soil (no mine) is $r = -\frac{\sqrt{\varepsilon_{r,s}}-\sqrt{\varepsilon_{r,a}}}{\sqrt{\varepsilon_{r,s}}+\sqrt{\varepsilon_{r,a}}}$. Here, $\varepsilon_{r,a}$ is the dielectric permittivity of air. While $\varepsilon_{r,a} = 1$, it follows that $r = -\frac{\sqrt{\varepsilon_{r,s}}-1}{\sqrt{\varepsilon_{r,s}}+1}$.

Now if we let S_t be the power density incident on the soil, the back-scattered power will be $P_{r,1} = S_t \times A_a \times |r_s|^2$. If part of the soil is covered by another object, like the mine, then the total back-scattered power will be $P_{r,2} = (A_a - A_m)\times S_t + A_m \times |r_m|^2 \times S_t$.

It can readily be seen from this equation that there will be no difference between observation with and without the mine if the reflection coefficient of the mine is identical to that of the soil. It can also be seen that a difference as small as $\Delta P = P_{r,2} - P_{r,1}$ should be within the dynamic range of the radar in order to find the contrast between the areas with and without the mine. This is called small signal dynamic range DR_SS and can be computed from the above equations to be $\text{DR\_SS} = \frac{\Delta P}{P_{r,1}} = \frac{A_m}{A_a} \times \frac{|r_m|^2 - |r_s|^2}{|r_s|^2}$. If the antenna were to be elevated above the ground by some 0.7 m and if the beamwidth of the antenna is 90°, with the already given values of the permittivities of the mine and the soil and a mine of diameter 5 cm, one would find a small signal dynamic range of −28.9 dB [1].

These considerations concern the properties of the receiver. Due to this large number, it is a common practice in the processing of B-scan (scanning along a line) and C-scan (scanning a volume) images to subtract the background $P_{r,1}$. The terms B-scan and C-scan are peculiar to ground-penetrating radars and must not be confused with A-scope, B-scope, and C-scope in radar displays. In ground-penetrating radars, B-scan and C-scan pertain to the manner of searching for a hidden land mine, that is, along a line or searching a volume. Obviously, when the variation in the background itself is of the same order of magnitude as the variation in signal level due to the presence of a landmine, the distinction between the effects of surface roughness and presence of a mine is hard to detect. Signal processing accounting for coherent spatial integration and clutter decorrelation will then support this distinction [1].

Another property concerns the maximum signal level that may enter the receiver without saturating it. Once again the variations in the soil dominate the requirement. IRCTR have, therefore, chosen to equip the transmitter with the capability to modify the transmitted power such that the receiver never enters saturation.

10.2.2.4 Antenna Considerations   Until now we have observed that from the perspective of the small signal dynamic range, the elevation should be as low as possible above the soil. However, this triggers issues, for example, since the soil is in the near field of the antenna, the impedance of the antenna is affected. The impact of this effect depends upon the soil dielectric parameters and the frequency. In case two separate antennas are needed for isolation between the transmitting and receiving antenna, this isolation can significantly be reduced due to the vicinity of the soil. Spatial separation is a proper approach for achieving isolation. The reflection off the mines, soil, and other objects will then be due to bistatic scattering. The design of CW radar inherently concerns the problem of how to discriminate between the signal directly coupled from the transmitter to the receiver on one hand and the reflected signal on the other. Hence, IRCTR decided to choose a bistatic system with a strong isolation between transmit and receive antennas. The objective for the isolation in a fielded situation is to achieve better than 60 dB at all frequencies. These antennas were designed to cover a frequency band from 400 to 4,845 MHz. An antenna type with a footprint more or less independent of frequency is important for imaging algorithms. Since the phase evolution over the footprint will be used in the imaging, a smooth and symmetric pattern will ensure that the images will have resolution capabilities that are independent of the aspect of the mine. This property will be used in feature extraction from the images for the purpose of classification of the objects.

10.3 SYSTEM ANALYSIS

10.3.1 System Block Diagram and Schematics

The SFCW radar built by IRCTR generates a set of eight frequencies at the same time and in the receiver the mixture of signals in the echo signal is split into its constituent parts. After downmixing to baseband in a quadrature mixing scheme and analog-to-digital conversion, the data are stored in a buffer. Then another set of eight frequencies is transmitted and buffered. This procedure is repeated 16 times until a set of 128 frequencies covering the full band at an equidistant separation of 35 MHz is collected. IFFT synthesizes the range profile required for localizing the objects from the set of 128 frequencies. For the purpose of calibrating of the electrical components, the equipment has built in a number of delay lines of different lengths. They can be selected under the supervision of an operator control built in microprocessor. Also, for the same purpose a power equalizer function is built in. It enables attenuation of all 128 frequencies individually in steps of 0.25 dB. Through this feature, two requirements have been met: the transmitted power spectrum should be known and in this special case, it has additionally been equalized (pre-emphasis). Furthermore, the transmitted power can be adjusted so that the system can be adapted to the local soil condition in order to keep the receiver out of saturation. The system block diagram is shown in Figure 10–3.

The synthesized range profile is obtained after first calibrating the data and then performing an IFFT to all data collected from the 128 frequencies, buffered in the data acquisition board. This board is hosted by a PC. This PC is also used to control the radar and the scanning system. Details are given in [6].

The schematic for the pulse upconverter and receiver downconverter are shown in Figures 10–4, 10–5, and 10–6.

In Figure 10–4, the signal output from DDS via a filter F_1 (not shown). The DDS runs on a 1 GHz clock and steps through 115–360 MHz in eight steps at 35 MHz intervals. Each step is for 100 μsec duration. It is manufactured by Stanford Telecom, USA, and has been discussed in Appendix G. The DDS output signals are up-converted in two stages to 128 equally spaced frequencies. In the first stage the eight frequencies are up-converted to 16 frequencies using two mixers (not shown) [4]. Then these signals are applied to two eight channels, each of them having its own mixer M2 (with the mixing frequency ranging from 2,800 to 6,720 MHz) [4]. By two 8 channels we mean that in terms of hardware it is actually one 8 channel, but in each phase of

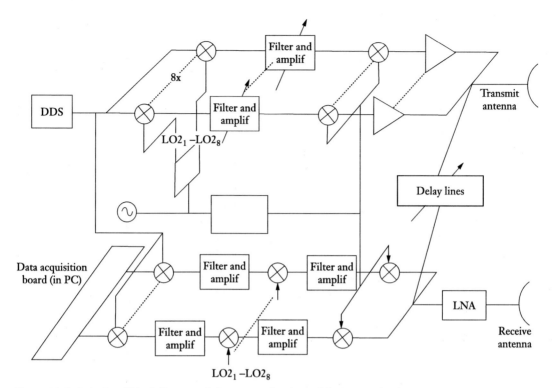

Figure 10–3 Functional block diagram of the SFCW radar. (From [1], © IEEE 2003.)

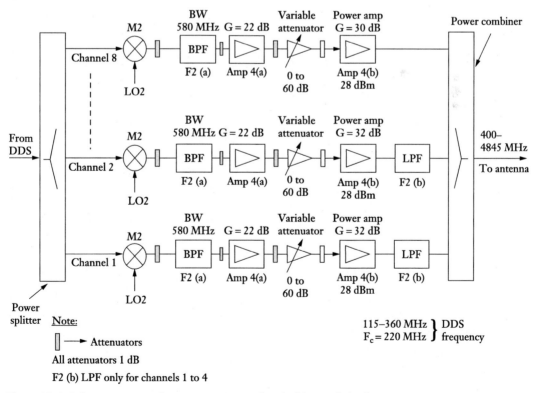

Figure 10–4 Pulse upconverter. (Courtesy IRCTR: Reprinted with permission.)

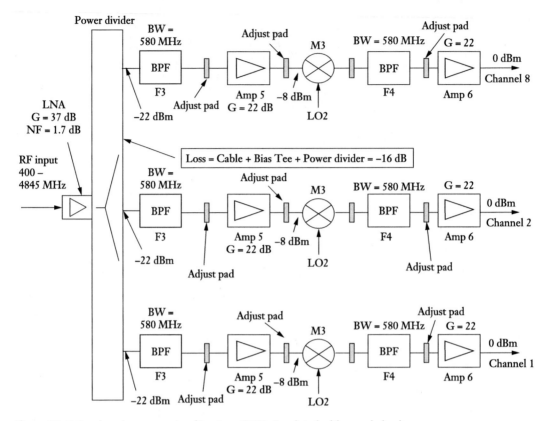

Figure 10–5 Receiver downconverter. (Courtesy IRCTR: Reprinted with permission.)

measurement (see Figure 10–6), the LO inputs to the DDS mixer (not shown) are correspondingly changed from Fr_1 to Fr_2. Therefore, if the base frequencies are changed, then correspondingly we have effectively made it as if we have two eight-channel radars (eight channels for phase one and eight channels for phase two). The M2 mixers impart bandwidth to the signal from the DDS. The filter bandwidths are chosen so that harmonics that can get into the pass band of the adjacent channel are suitably attenuated. Channels 1 to 4 were so low in frequency spread that they were requiring additional low pass filters to achieve this. This is because the F2(a) bandpass filters did not exhibit steep enough "skirts" at these low frequencies. There is no linearity constraint on the amplifiers, unless the signal is LFM. This application deals with unmodulated SFCW.

The receiver downconverter in Figure 10–5 splits the signal from the LNA using an eight-way power splitter. The loss here of 16 dB is due to the cabling from the antenna, Bias-Tee (not shown) for the calibration relays (see Figure 10–1), and the configuration loss (−9 dB) of the power divider. Note the LNA has a noise figure of 1.7 dB and a gain of 37 dB. Note the power levels of −22 dBm at the input of each channel. The final output has a power level of 0 dB.

The baseband processing shown in Figure 10–6 downmixes the IF to baseband using M4 mixer. This mixer has two frequencies switched into it. Fr_1 is for the first phase of measurement (see Figure 10–6) and Fr_2 the next phase. M5 is the stretch processor, which receives the reference signal. The ADCs have been provided a linearity margin of 4 dB. Adjust pads or attenuators are shown throughout for controlling the power levels into mixers and amplifiers.

The electrical block diagram is shown in Figure 10–7.

Transceiver: This unit comprises one transmitter and one receiver.

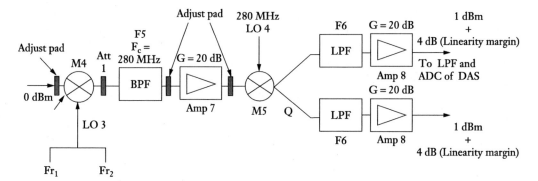

Figure 10–6 Receiver downconversion: Baseband (1 of 8). (Courtesy IRCTR: Reprinted with permission.)

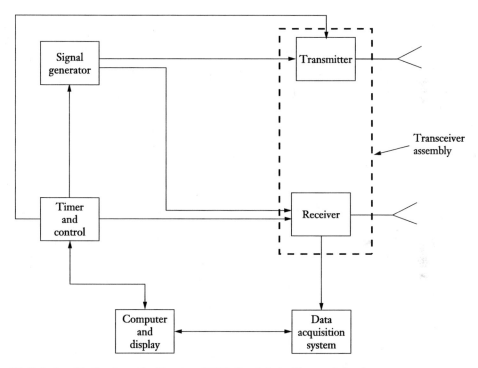

Figure 10–7 System block schematic. (Courtesy IRCTR: Reprinted with permission.)

Transmitter: The transmitter receives the input from the DDS and splits it into eight channels. It then imparts bandwidth coverage to these eight channels by dividing the frequency range of each channel. The bandspread of these eight channels range from 400 to 4,845 MHz at intervals of 525 MHz each. Each channel then feeds into a power amplifier via a digital attenuator. The attenuator is digitally manipulated by the control processor and is meant to control the quality (by controlling amplitude) of the transmitted spectrum across the entire bandwidth. A power combiner is located at the output of the individual power amplifiers for combining the power of each channel additively. Finally, the output of the transmitter is a single coaxial line leading to the antenna via a direction coupler. The direction coupler is meant to tap a portion of the transmitted power for calibration. The transmitter and signal generator are located away from the scanner and connected to the transmitting antenna via a 10-meter coaxial cable.

Receiver: The receiver is located away from the scanner via 10 meters of coaxial cabling. It is connected to the receiving antenna via a direction coupler, which feeds it an input from a calibration device. There is a wideband LNA at the input of the receiver. This is followed by a power divider that divides the input power into eight separate channels using bandpass filters. Thereafter, the signal is amplified and fed to the I/Q demodulator. The I/Q demodulators are fed with the corresponding signal from the transmitter local oscillators for downmixing. There are eight such demodulators. The output of each demodulator is one I- and one Q-signal. Both are baseband values. Together they compose a single complex number with a phase proportional to the range of the object and the frequency.

The entire system is controlled by a computer and display unit. This unit interacts with a timer and control unit and a data acquisition system (DAS). The timer and control unit is directly controlling the switching and measurement timings in the system. We can program this unit from the computer. The timer unit commands the signal generator to start stepping in frequency and at the same time prepares the transmitter for transmission. It also simultaneously prepares the receiver for reception. The transmitter and the receiver are located in a separate transceiver assembly. During the frequency stepping process, the timer switches in filters, amplifiers, and so forth, as the bandwidth increases. When the stepping process is complete, the timer informs the computer, which then commands the DAS to start downloading the received data. The DAS also formats the data and stores it on the hard disk of the computer. It also carries out various signal-processing activities using MATLAB®.

Timing and Control Unit: This is based on a microcontroller and it controls the switching and frequency stepping operation of the system. It interacts with the main computer and is easily programmable from the computer.

Data Acquisition System: The DAS is a 16-channel device. It can simultaneously sample 16 channels with a 16-bit resolution. The system is DAP 5216a marketed by Microstar Laboratories, Bellevue, Washington, USA. The processor is an AMD processor. The inputs are routed to the ADC via a changeover switch, which first samples the I channel and then the Q channel. The DAS incorporates switched capacitor anti-aliasing filters at its input so that the user can adjust the quality of frequency response of the anti-aliasing filters, on-line. The entire system can be controlled from the main computer. The DAS is located inside the main computer and directly interfaces to MATLAB® software on the main computer.

Computer and Display: This is a personal computer. It is the interactive user controlled Man Machine Interface (MMI) to the SFCW-GPR (Stepped Frequency Continuous Wave–Ground-Penetrating Radar) system. It not only processes the received data from the SFCW-GPR off-line using MATLAB, but also controls the entire radar system frequency stepping and data acquisition timing control. This is achieved via a control processor, which is connected to the computer. It uses the computer hard disks for data storage. We require a minimum of 24 MB for a 10-min sustained data recording.

This is computed as follows:

Data are recorded at each set repetition interval (SRI). There is one 16-bit I and Q channel (i.e., 4 bytes) to be recorded at 10 kHz set repetition frequency (SRF), which is equivalent to 100 μsec SRI. The term set repetition frequency (SRF) must not be confused with sweep repetition frequency (SRF) used in the preceding chapters. The former is a term peculiar to GPRs. There is effectively only one range bin as it covers the entire target extent of 4.5 meters, unambiguous range. This means a recording rate of 40 kB/sec (4 bytes, 1 range bin, 10 kHz). To record data for 60 seconds at this rate requires a data storage capacity of 2.4 MB. For 10 minutes this would imply 24 MB.

10.3.2 The Effect of Phase Noise

It was discussed in Chapter 6 that stability of an oscillator is paramount in using step frequency waveforms. It is no different here. Phase noise might limit the radar's high resolution. The radar's transmitter is based on frequency sources with a limited stability over time. Therefore, the transmitter and the receiver signals are distorted due to cumulative phase noise. Cumulative phase noise means the phase change accumulated over the time delay τ between transmission and reception of an echo signal. The noise free received signal would ideally have a phase $\phi(t) = 2\pi f_c t$, apart from an arbitrary initial phase. The actual reference frequency, however, is $f_c + \delta_f(t)$. As a result, the measured phase will deviate from the ideal value, leading to a noisy background disturbance. The background is expressed by the phase noise limited dynamic range (PNDR) [1]. The PNDR is defined as the ratio between the maximum squared of the IDFT output and the variance of the noise induced by the phase noise at ranges away from the object:

$$PNDR = \frac{(NC_f)^2}{NC_f^2(1 - C_f^2)} \tag{10.1}$$

Here N is the number of frequencies and C_f is the characteristic function of the effective random frequency, with

$$C_f = \exp[-(2\pi\tau\sigma_f)^2/2] \tag{10.2}$$

σ_f is the standard deviation of the effective random frequency. Bearing in mind that in this application $2\pi\tau\sigma_f << 1$ we can approximate C_f by

$$C_f = 1 - (2\pi\tau\sigma_f)^2/2 \tag{10.3}$$

Equation (10.1) then reduces to

$$PNDR = \frac{N}{(2\pi\tau\sigma_f)^2} \tag{10.4}$$

Taking these equations the length of the cables into account, it can be shown that the phase noise of the components is not a limiting factor to the dynamic range in this application [7].

The most critical component is the DDS that shows a phase noise power spectral density of approximately -90 dBc/Hz at higher offsets (see Appendix G). Bearing in mind that the other sources generate a phase noise as low as or better than -110 dBc/Hz over the receiver bandwidth, it can be concluded that PNDR is not limited by the phase noise of the signal sources.

10.3.3 The Ambiguity Function

The ambiguity function of the stepped frequency waveform was derived in Chapter 3. It was shown that the SFCW radar is sensitive to range–Doppler coupling. We now analyze this ambiguity function in order to evaluate the effect of Doppler due to antenna motion [2, 8].

The ambiguity function for the SFCW train of pulses can be derived to be (see Chapter 3):

$$\chi(\tau, f_d) = \frac{1}{N} \exp\left[j2\pi p T_R f_d\right] \chi_c\left(\tau - pT_R, f_d + p\Delta f\right).$$

$$\cdot \sum_{m=0}^{N-1-p} \exp\left[j2\pi m\left\{\Delta f\left(\tau - pT_R\right) + T_R f_d\right\}\right] \tag{10.5}$$

Here p is a parameter referring to the various phase and amplitude modified repetitions of the basic component ambiguity function χ_c. For negative values of p the expression would be symmetrical. T_R is the pulse repetition interval as was explained in Chapter 3.

The contour plot of this ambiguity function is shown in Figure 10–8 [1].

The ambiguity function of an SFCW signal exhibits a number of Doppler- and delay-shifted kernel functions. Figure 10–8 shows a kernel at the origin, that is, with delay and Doppler shift equal to zero. This is the one relevant for this application. The slope of the tilted line is $\Delta f / T_R$. Inspection of this figure shows us that the antenna may move at a speed of a few m/sec without significant distortion of the range profile.

The ambiguity function of a stepped LFM waveform was not derived in this book, but is available in [3]. If the guard bands are covered as discussed in Section 10.2.1 using the two-phase approach, then the ambiguity function will be exactly the same as for a single-channel stepped LFM radar.

10.3.4 The Antenna

10.3.4.1 Selection of the Design   IRCTR considered a number of types of antenna for this application. It is interesting to examine these in Table 10–1.

It can be seen from Table 10–1, that the TEM horn, although very well matching the bandwidth requirements, has an open structure, preventing the isolation between the transmitting and the receiving antenna to be at a high level. The same holds for the log periodic dipole array. The ridged horn is too narrow in bandwidth. Since the radar has to handle relatively low power

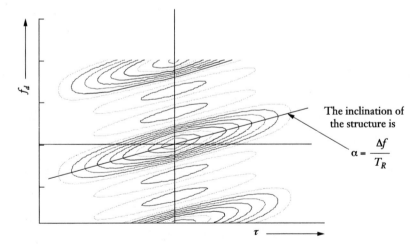

Figure 10–8 Ambiguity diagram near the origin. (From [1], © IEEE 2003.)

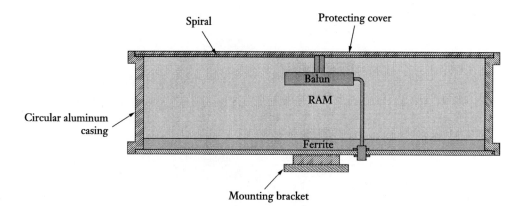

Figure 10–9 Sketch of the mechanical construction of the antenna. (From [1], © IEEE 2003.)

Table 10–1 Summary of the Properties of Some Relevant Antenna Designs

	Spiral	Log Periodic Dipole Array	Ridged Horn	TEM Horn
Bandwidth	20:1	20:1	5:1	12:1
Gain	2 dBi	7–15 dBi	6–18 dBi	2.5–15 dBi
Polarization	Circular	Linear	Linear	Linear
Power handling	Poor	Good	Reasonable	Adequate
Dispersion	Yes	Yes	Yes	No
Extra (T-R Isolation)	Closed structure	Open structure	Closed structure	Open structure
	No backlobe	Poor backlobe	Excitation problem	Excitation problem

From [1], © IEEE 2003.

(10 mW for each frequency), the poor power handling capability of a spiral antenna poses no problem. The fact that the structure of such a design is closed is well suited for good isolation. An important property of spiral antennas is that they transmit a circularly polarized field.

Consequently, IRCTR have chosen a set of two spiral antennas with opposite sense of rotation. A bistatic arrangement was preferred because of the isolation problem. The antenna design is illustrated in Figure 10–9.

Figure 10–9 shows that the spiral is cavity backed, while this cavity is filled with wide band absorbing material. The bottom of the cavity is plated with ferrite tiles in order to have the lower frequencies better absorbed. A Marchand balun is mounted in the cavity. Spiral antennas are basically quasi-frequency independent. This particular antenna is an Archimedean spiral with two arms. It is fed at the center point of the arms, after a balun. The usual power handling problems with both the balun and the narrow feed center, corresponding to the highest frequency, do not occur here because of the low power utilized in this application. Figure 10–10 shows the spiral antenna.

The transfer function of a spiral antenna is directly linked to the applied current. However, a time delay occurs as the signal travels from the feed point to the radiating section of the antenna.

Figure 10–10 Spiral antenna. (Courtesy IRCTR: Reprinted with permission.)

The lower the frequency, the more outward is the radiating section. This delay must be accounted for during calibration. This factor does lengthen any pulse. Therefore, this type of antenna is probably not the preferred solution for impulse radar.

10.3.4.2 Coupling   The directly coupled signal, measured in a time-gated mode in the antenna measurement chamber, shows very low values. The level of the coupling slightly depends upon the rotation of the spirals. The antennas are not fully symmetric due to the finite length of the arms, causing this dependency. Figure 10–11 shows two of these orientations, where one of the antennas is rotated over 90° along its boresight axis in one case and in the other case the two antennas are coincident. The dotted line represents the asymmetric geometry.

10.3.4.3 The Footprint of the Antenna   Figures 10–12 and 10–13 show the footprint of the antenna at a frequency of 1,100 MHz. The graphs show that the footprint of the amplitude is slightly asymmetric and tilted. This is common in such antennas. The higher the frequency, the more the initial circle will convert to an ellipse with a tilted angle. Note that the twin antennas have an opposite sense of rotation and that as a consequence the elliptic shape is different. The footprint of the phase is circular and (nearly) maintains this circularity also, for high frequencies.

10.3.4.4 Calibration of the Antenna   The antenna is fed at the center. In the feedpoint a current is imposed that propagates along the spiral arms toward the outer edge. The high frequencies are being emitted near the feed point. The low frequencies are being emitted in the proportionally larger part of the antenna. This effect makes the current in case of low frequencies

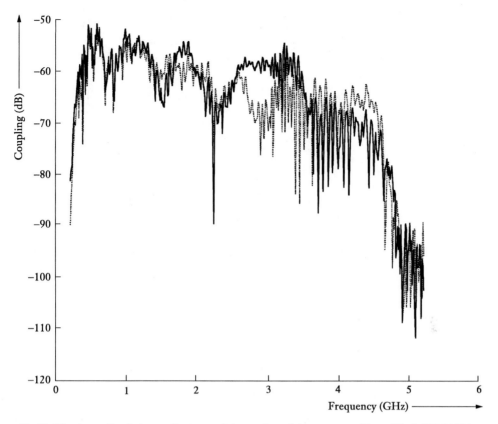

Figure 10–11 Direct coupling between the transmitting and receiving antenna. (From [1], © IEEE 2003.)

propagate along the spiral. This propagation takes time, estimated as around 3 nsec for each antenna. Hence, the antennas are dispersive. The delay in the antennas is a function of frequency and therefore, each frequency must be calibrated separately in order to create a focused synthetic profile. The procedure for this calibration is detailed in [9]. Basically the phase of each frequency is recorded in a well-defined scene, normally a metal plate (acts as a calibration target). Then the recorded signals are compared to the original ones and a complex correction coefficient is computed for each frequency and subsequently applied to the data. The results are illustrated in Figure 10–14.

In Figure 10–14, it can be seen that the sequence of successive peaks in the calibrated profile corresponds to a multiple reflection of the field emitted by the transmitting antenna. These multiple reflections occur because of multiple bounces between the metal plate and the receiving antenna. Therefore, during the calibration the antennas must be elevated high enough to let the multiple bounce from the early signal not interfere with the first reflection of the late signal. In short: the "pulse smearing interval" must be less than the propagation delay of the bounced signals. The first peak occurs at 1.27 m, which is the distance between the antenna and the metal plate. The second peak and third peak are at 2.54 and 3.81 m, respectively, with decreasing peak amplitude. The area of the flush buried or surface laid mines is directly near the first peak, where the side lobe level is down by more than 50 dB.

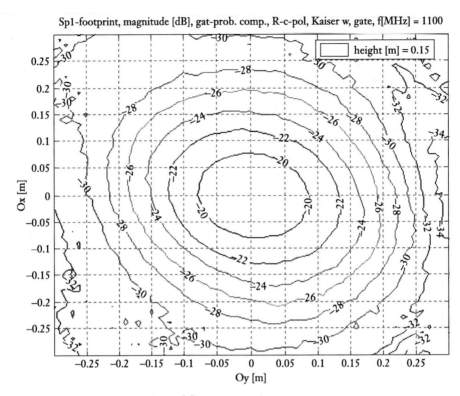

Figure 10–12 Footprint E-field power. (From [1], © IEEE 2003.)

The results discussed here are from an operational system and are therefore, representative of the system's capability.

10.4 EXPERIMENTAL SETUP

The setup for the measurements was basically a sandpit with buried mines. This is shown in Figure 10–15.

The sandpit is a large structure with the motorized antenna assembly capable of easily carrying out a C-scan of the area. The cables in the ground are power cables buried in the sand. They do not obstruct the experiment. We can see two cabinets in the background. That is the Pandora radar. One is a transmitter cabinet and the other the receiver cabinet. The assemblies are admittedly large. This is because a lot of EMI/EMC shielding was used, in view of the high frequencies involved. It will be appreciated that in the earlier discussions on Pandora radar in the previous chapters, we had dwelt extensively upon the need to isolate the eight radar channels using filters with steep skirts, and so forth. That entire work can get undone if we have poor shielding between the channels. This will cause one channel to superimpose its signal on the other channel causing the radar to fail. Therefore, extensive thought had gone into the engineering aspects of this radar. It is felt that future radars can be made more compact by extensively using MMICs.

The antenna assembly is shown in detail in Figure 10–16. The antenna itself was shown in Figure 10–10.

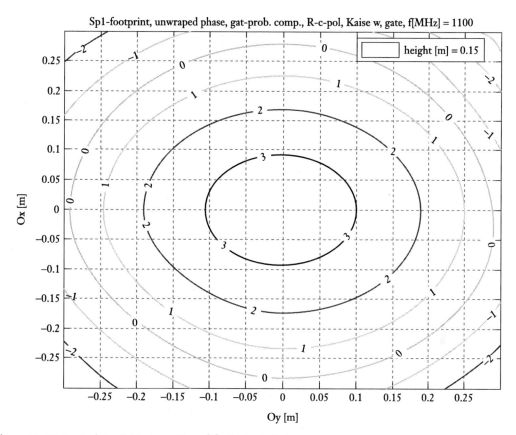

Figure 10–13 Footprint E-field phase. (From [1], © IEEE 2003.)

10.5 EXPERIMENTAL RESULTS

Two examples are presented here from [1] with the system discussed in this chapter. Figure 10–17 shows the ground truth.

Figure 10–18 shows the image produced as a direct intensity plot at the level of the surface, after subtraction of the average background. It can be seen that the image is unfit for detecting objects. This is mainly due to the effect of the poor small-signal dynamic range inherent in the large antenna footprint.

The second example concerns the very same dataset. But this time the processing was based on an SAR algorithm as described in [10]. This result is shown in Figure 10–19.

10.6 CONCLUSION

In this chapter we have analyzed the main parameters of the SFCW radar. We have investigated the considerations which went into formulating the numbers for this radar. We studied the radar at block diagram level, circuit schematic level, and at system level. We have seen as to how fast this radar is in taking a 128-step measurement in a mere 3.3 msec. This measurement has yielded a high-resolution of around 3 cm in air. The most remarkable properties of the Pandora radar are the fact that multiple frequencies are transmitted simultaneously, the very short data acquisition time of 3.3 msec, and the very wide bandwidth (400–4,845 MHz).

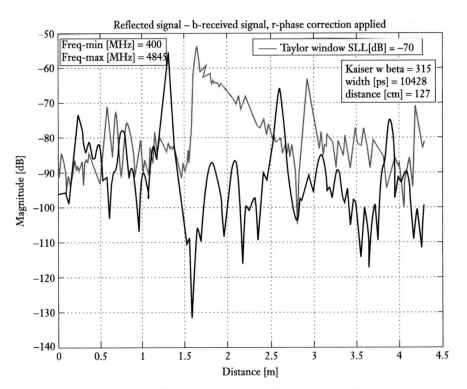

Figure 10–14 The synthesized range profile with and without calibration. (From [1], © IEEE 2003.)

Figure 10–15 Experimental setup for mine detection using Pandora. (Courtesy IRCTR: Reprinted with permission.)

Figure 10–16 Antenna assembly. (Courtesy IRCTR: Reprinted with permission.)

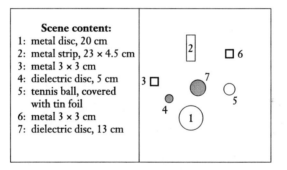

Scene content:
1: metal disc, 20 cm
2: metal strip, 23 × 4.5 cm
3: metal 3 × 3 cm
4: dielectric disc, 5 cm
5: tennis ball, covered
   with tin foil
6: metal 3 × 3 cm
7: dielectric disc, 13 cm

Figure 10–17 Ground truth of the data presented in Figures 10.19 and 10.20. (From [1], © IEEE 2003.)

10.7 FUTURE POSSIBILITIES OF PANDORA

The Pandora type of multifrequency architecture lends itself not only for CW applications but also pulse radar applications. This includes both unmodulated as well as chirp pulse radars. We have seen that due to the large bandwidths involved, generating linear chirp pulses is not easy nor is it fast enough. The Pandora architecture solves these issues admirably. It can, therefore, be applied for airborne survey and planetary survey from spacecraft in either pulsed mode or CW mode and also in a variety of roles like missile/aircraft tracking or profiling. Due to the wideband nature of the signal it can also be used against stealth aircraft as these aircraft have radar absorbers, which are narrow band. The signal processing can be step frequency waveforms or LFM. Above all, this radar can be built very cheaply based on components bought off the shelf (COTS).

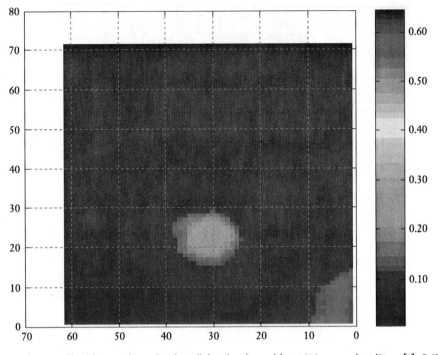

Figure 10–18 Five metallic objects and two simulate dielectric mines without SAR processing. (From [1], © IEEE 2003.)

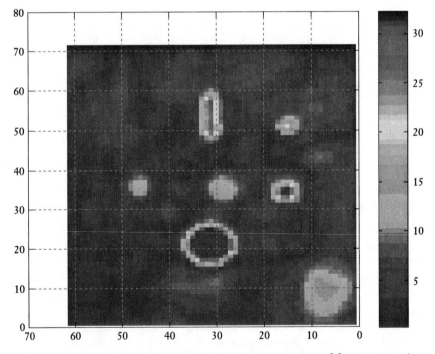

Figure 10–19 Same dataset as in Figure 10–18, but after SAR processing. (From [1], © IEEE 2003.)

References

1. van Genderen, P. and Nicolaescu, I., "System description of a stepped frequency CW radar for humanitarian demining," in *Proceedings of the 2nd International Workshop on Advanced Ground Penetrating Radar, 2003*, May 14–16, 2003, pp. 9–15.

2. Jankiraman, M. and van Genderen, P., "Multi-frequency SFCW radar for ground penetrating radar," in *Proceedings of the GRS2000*, pp. 431–435.

3. Levanon, N. and Mozeson, E., *Radar Signals*, John Wiley & Sons, New York, 2005.

4. Nicolaescu Ioan, van Genderen, P., Heijenoort, Joost van, "Stepped frequency Continuous wave ground penetrating radar," *International Radar Symposium*, Dresden, Germany, 2003

5. Daniels, D. J., "Surface penetrating radar," in *IEE Radar, Sonar, Navigation and Avionics Series 6*, ISBN 0 85296 862 0, 1996.

6. van Genderen, P., Hakkart, P., van Heijenoort, J., and Hermans, G. P., "A multi-frequency radar for detecting landmines: Design aspects and electrical performance," in *31st European Microwave Conference*, 0-86213-148-0, London, 2001, pp. 249–252.

7. van Genderen, P., "The effect of phase noise in a stepped frequency continuous wave ground penetrating radar," in *CIE International Conference on Radar IEEE*, 0-7803-7000-7, Beijing, China, 2001, pp. 581–584.

8. Jankiraman, M., de Jong, E. W., and van Genderen, P., "Ambiguity analysis of Pandora multi-frequency FMCW/SFCW radar," in *IEEE 2000 International Radar Conference*, Alexandria, USA, 2000.

9. Ioan, N., van Genderen, P., and Palmer, K., "Calibration procedures of a stepped frequency continuous wave radar for landmine detection," in *IEEE 2003 International Radar Conference*, Adelaide, Australia, 2003.

10. Ioan, N., van Genderen, P., van Dongen, K. W., van Heijenoort, J., and Paul, H., "Stepped frequency continuous wave radar-data preprocessing," in *Proceedings of the IWAGPR*, Delft, May 2003.

Appendix A: CW Radar Range Equations

1.1 AIM

The program "*cw.cpp*" calculates the range/power levels of FMCW and phase-coded CW radars given either of the parameters. It outputs the range in kilometers or the average power in watts. It is intended as a tool for quick performance evaluation of CW radars.

1.2 DESCRIPTION

This software covers two aspects of FMCW/PCCW radars: scanning mode and nonscanning mode. Certain assumptions are made in each of these modes.

1.3 FMCW RADARS

The FMCW radar range equation is

$$R^4 = \frac{P_{Av}G_TG_R\sigma\lambda^2}{(4\pi)^3 FkTL(SNR_o)SRF}$$

where
(*Note:* all dB values will be converted to absolute values before inserting in equation)

P_{Av}—Average transmitted power in watts
G_T—Transmitting antenna gain
G_R—Receiving antenna gain
σ—Target RCS in square meters
λ—Wavelength in meters
F—Receiver noise figure, typically 3 dB
k—Boltzmann's constant, 1.38×10^{-23} J/K
L—System losses
T—System noise temperature in K, typically 400 K
SNR_o—Output SNR, that is, desired at the detector for a given P_d and P_{FA}
SRF—Sweep repetition frequency

1.3.1 Scanning Mode

In this mode, the signal processing is essentially non-coherent. This is because the short dwell times typically encountered are suitable for detecting only extremely fast targets well in excess of Mach 12. Hence, only non-coherent processing is carried out. Please see Appendix B for a more detailed explanation.

Steps

1. We calculate the number of sweeps,

$$\text{Number of sweeps} = \frac{\text{dwell time}}{\text{sweep time}}$$

The number of sweeps then equals the number of noncoherent integration pulses.

2. We next calculate the integration loss. Hall [1] gives the integration loss for a Swerling 1 target for a P_d of 0.5 and P_{FA} of 10^6. The loss curve is essentially a straight line with a slope of 3.7542 and the number of pulses integrated on the x-axis varying logarithmically:

Integration loss $= 3.7452 \text{Log}_{10}(N)$ where N = number of noncoherent integration pulses.

3. The noncoherent gain is then given by

Noncoherent gain $= 10 \text{Log}_{10}(N) - \text{Integration loss.}$

The basic philosophy here is to calculate the gain, as if the pulses were coherently integrated and then correct for noncoherent integration by subtracting the integration loss. Some authors disagree with this philosophy, for example, Blake [2]. Blake, however, gives the reduction of signal-to-noise ratio in decibels as compared to purely coherent integration. These curves are very comprehensive and cover cases of Swerling 0–4 and can also be used as integration loss curves. It is pointed out that coherent integration can only be carried out for Swerling 0 and 1 type of targets, that is, nonfluctuating or slowly fluctuating targets from scan to scan.

4. The program also calculates the minimum required sampling rate needed to achieve the instrumented range given the sweep bandwidths. This is given by the following equation:

$$R_{\text{max}} = \frac{Nc}{4 \Delta f}$$

where N is the minimum sampling rate meeting the Nyquist criterion, c is the velocity of light, Δf is the sweep bandwidth, and R_{max} is the maximum instrumented range.

5. The program also keeps track of the sweep repetition frequency. It corrects this depending upon the sweep time. The user also has the option to override the SRF figure. This is a very important parameter and effectively influences the radar range by altering the dwell time. This is discussed in Appendix B.

In stretch processing, we have seen that there is a code compression gain. However, since our range equation is based on "average power × look time," this aspect of allowing for FFT gain is automatically taken into account. Therefore, code compression gain does not figure in the equation for net processing gain.

1.3.2 Nonscanning Mode

In this mode, we can carry out coherent processing.

Steps

1. The program asks you for Doppler filtering option. If we say "N," it will then carry out noncoherent processing. If we say "Y," it asks you for the maximum expected target

velocity in m/sec. It then calculates the definition of the Doppler FFT bank. This definition is given by

$$a = \frac{v^2}{R_{max}}$$

where a is the target radial acceleration during integration time, v is target velocity, and R_{max} is maximum instrumented range.

$$\therefore v = at = \frac{v^2 t}{R_{max}}$$

Doppler definition [3] [4],

$$\Delta f_d = \frac{2\Delta v}{\lambda} = \frac{2(\Delta v)^2 t}{R_{max}\lambda}$$

where t is the integration time.

This expression for Doppler definition is a useful one for our application, because it allows for target acceleration during look time. This means that targets which have a spread spectrum characteristic in Doppler, like helicopter blades which vary from zero at the beginning to 300 m/sec at the tip of the rotor can be seen *in one bin*.

2. The coherent processing interval is, therefore, the inverse of Δf_d.

$$\therefore \text{ The number of coherent sweeps } = \frac{\text{coherent processing interval}}{\text{sweep time}}$$

If, for example, the number of coherent sweeps is less than 2, the program asks you to input a new sweep time, so that at least two coherent sweeps are available for processing. It also corrects the SRF accordingly. It is, however, recommended that at least 30 sweeps be integrated.

3. The program next asks you to choose the nearest radix 2 FFT length, to the number of coherent sweeps. We will need to choose our parameters such that the number of coherent sweeps is equal to or at least one or two pulses less than the nearest radix 2 FFT length. This is to avoid signal loss. It then calculates the Doppler FFT gain.

$$\text{Doppler FFT gain} = 10\,\text{Log}_{10}(\text{FFT length}) - 3.5\text{dB}$$

The 3.5 dB loss is calculated on the basis of 2.5 dB for Doppler frequency and phase being unknown [1] and 1 dB for loss due to weighting. It is to be noted here that range FFT gains do not figure in our calculations due to reasons already discussed.

4. The noncoherent integration pulses are now given by

$$\text{Noncoherent integration pulses} = \frac{\text{integration time}}{\text{sweep time} \times \text{number of coherent s weeps}}$$

We should choose our parameters in such a way that at least a large number of pulses are obtained, for example, 128 pulses.

5. We now calculate the noncoherent gain

$$\text{Integration loss} = 3.7452 \, \text{Log}_{10} \, (\text{noncoherent integration pulses})$$

$$\therefore \text{noncoherent gain} = 10 \, \text{Log}_{10} \, (\text{noncoherent integration pulses}) - \text{integration loss}$$

The constant 3.7452 is the slope of the integration loss curve [1].

6. Therefore, the net processing gain

$$\text{Net processing gain} = \text{Doppler FFT gain} + \text{noncoherent gain}$$

The *power calculations* are just the inverse of the range calculations and therefore we shall not dwell upon it.

1.4 PHASE-CODED CW RADARS

No change is envisaged. It is the same as for FMCW radars.

1.4.1 Scan Mode

The argument here is the same as for FMCW radars in the scan mode. We cannot carry out coherent processing due to time constraints (because of low dwell times due to scanning). Please see Appendix B for details. Hence,

$$\text{Net processing gain} = \text{code compression gain} + \text{noncoherent gain}$$

It should be noted that in these phase-coded radars, it is more convenient to calculate code compression gain, due to the presence of matched filters (circular correlators). We subtract 1 dB from the code compression gain, due to the phase being unknown. This causes a loss of typically 1 dB in phase-coded signals ([5], p. 217).

$$\text{Code compression gain} = 10 \, \text{Log}_{10}(N) - 1$$

where N is the number of segments (chips). This is equal to the time-bandwidth product for phase coded signals.

1.4.2 Nonscanning Mode

Once again, the argument here is the same as for FMCW radars in the nonscanning mode. We can carry out coherent processing here.

$$\text{Net processing gain} = \text{code compression gain} + \text{Doppler FFT gain} + \text{noncoherent gain}$$

The PCCW *power calculations* are just the inverse of the range calculations and therefore, we shall not discuss it.

1.5 ERROR CORRECTIONS

The accuracy of this software depends upon the assessment of the errors and corresponding corrections. There are many corrections to the radar range equations. We shall, however, only dwell upon those that are peculiar to FMCW radars [5]. These essentially, fall into two categories.

1. *Errors affecting sensitivity:*

   (a) Thermal noise.
   (b) Antenna reflection.
   (c) Circulator leakage.
   (d) Mixer LO leakage.

In FMCW systems having separate transmit/receive antennas, points b, c, and d above do not apply and the coupling between the antennas takes the place of circulator leakage.

2. *Errors due to near-field clutter:* Far-off echoes compete with nearby clutter for detection. We need to eliminate this problem by employing clutter cancellers. In Appendix B, it will be shown that unless there is an MTI with an improvement factor of around 20 dB, it will drastically cut down the range of the radar. Alternately, we need to use pencil beams to avoid clutter returns from the ground altogether (but side lobe pick up will still be there) (see Figure A–1).

1.5.1.1 Thermal Noise   This is the standard correction in any radar. The thermal noise referenced to the antenna port is given by

$$N = kTBF$$

where k is the Boltzmann's constant $= 1.38 \times 10^{-23}$ J/K, T is the reference temperature $= 290$ K (typically), B is the receiver noise bandwidth (this is approximately equal to the ideal frequency resolution available, viz., the inverse of the sweep time (sweep repetition frequency)), and F is the system noise figure $= 3$ dB (typically).

This correction has already been applied in the range equations.

1.5.1.2 Antenna Reflection FM Noise Power Calculation   In systems having a common antenna for transmit/receive, we need to compute the FM noise power at the maximum beat frequency. The signal path is as follows:

signal source $\rightarrow$ isolator $\rightarrow$ coupler $\rightarrow$ circulator $\Leftrightarrow$ antenna port
$\qquad\qquad\qquad\downarrow$
$\qquad\qquad$ mixer
$\qquad\qquad\qquad\downarrow$
$\qquad\qquad$ mixer pre-amp

The power reflected by the antenna is a function of the VSWR of the antenna. The ratio of reflected power to incident power is equal to

$$\frac{P_r}{P_i} = \left(\frac{VSWR-1}{VSWR+1}\right)^2$$

where P_r is the reflected power and P_i is the incident power.

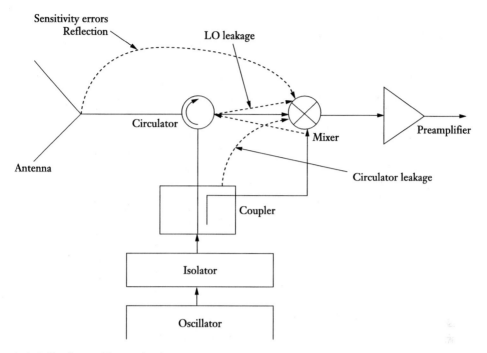

Figure A–1 Reflection and internal noise components.

Using the same source to transmit and receive a signal reduces the effect of the source's phase noise because the noise of the received signal is correlated to the transmitted signal. This phase noise correlation factor K^2 depends on the offset frequency from the carrier (beat frequency in the case of FMCW homodyne radars) and the delay between the transmit and receive signals. This delay is the difference between the source-to-mixer LO ports path and the source-to-antenna–reflection-to-mixer RF port path. This correlation is expressed as (see Appendix E)

$$K^2 \cong 4\left(\pi f_b t_r\right)^2$$

where t_r is the delay.

Lower beat frequencies result in both greater FM noise power density and greater phase noise correlation. The phase noise correlation factor increases with beat frequency, while the FM noise decreases with beat frequency. These may offset each other so that the phase noise is more or less a constant over the band of interest.

This approach has been incorporated in the "*cw.cpp*" program.

1.5.1.3 Circulator Leakage   The calculations here are identical to the antenna reflection problem. However, the principle is different. This case pertains to the leakage from the circulator to the mixer caused due to power that leaks in the opposite direction than the circulator polarity, directly into the mixer pre-amplifier input. The phase noise correlation factor will be poorer than for the antenna case, because the path length is less. This factor has also been taken into account in the "*cw.cpp*" program.

1.5.1.4 Mixer LO Leakage FM Noise Power Calculation   This concerns the leakage of the LO power to the output of the mixer pre-amplifier. The LO output leaks via the mixer to the circulator. It is then reflected by the circulator, depending upon the VSWR at the circulator, back to the mixer, along with the signal. This correction has also been incorporated in the program "*cw.cpp*" for single antenna systems.

1.5.2 Clutter Corrections

In this section, we will calculate the *C/N* ratio and hence the MTI improvement required for noise-limited performance. We will now correct the equation for detection of point targets in area clutter [6], [7]. The initial range equation is

$$R^4 = \frac{P_{Av}G_T G_R \sigma \lambda^2}{(4\pi)^3 FkTL(SNR_o)SRF} \tag{A.1}$$

Correcting for a single hit and looking at only the received power level from a point target:

$$P_R = \frac{P_{Av}G_T G_R \sigma \lambda^2}{(4\pi)^3 LR^4 SRF} \tag{A.2}$$

where P_R is the received power from a point target.

Similarly, for area clutter,

$$P_{Rclutter} = \frac{P_{Av}G_T G_R \sigma_A \lambda^2}{(4\pi)^3 LR^4 SRF} \tag{A.3}$$

where $P_{Rclutter}$ is the received power from an area illuminated on the ground by the beam and σ_A is the radar cross section for area targets.

Now the RCS of area targets is given by

$$\sigma_A = \sigma_0 A_I$$

where A_I is the area illuminated by the beam and σ_0 is the backscatter coefficient.

We now compute the value of the illuminated area.

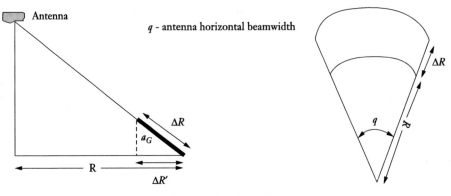

Computation of area clutter

Hence,

$$A_I = \frac{R\theta\Delta R}{\cos(\alpha_G)}$$

where α_G is the grazing angle, θ is the azimuth beamwidth, and ΔR is the range cell ΔR.

The expression $\Delta R/\cos(\alpha_G)$ denotes the projection of this range cell on the ground. Substituting in equation (A.3)

$$P_{Rclutter} = \frac{P_{Av}G_T G_R \sigma_0 \lambda^2 R\theta\Delta R}{(4\pi)^3 LR^4 SRF \cos(\alpha_G)} \tag{A.4}$$

Finally, we divide equation (A.2) by equation (A.4) to obtain,

$$\frac{P_R}{P_{Rclutter}} = \frac{S}{C} = \frac{\sigma \cos(\alpha_G)}{\sigma_0 R\theta\Delta R} \tag{A.5}$$

where S/C is the signal-to-clutter ratio.

The important thing to note here is that signal-to-clutter ratio is independent of transmit power, antenna gain, frequency, and losses. This is because the clutter echo power is affected by these factors exactly the same as point target echo power. The above equation gives the signal-to-clutter ratio for a single hit. For multiple hits, this is multiplied by the process gain. The process gain for point targets in clutter is the Moving Target Indicator Improvement factor (MTI_I). The final expression takes the form

$$\frac{P_R}{P_{Rclutter}} = \frac{S}{C} = \frac{\sigma \cos(\alpha_G)}{\sigma_0 R\theta\Delta R} MTI\_I \tag{A.6}$$

where S/C is the signal-to-clutter ratio.

Equation (A.6) is a vital equation and forms the basis of clutter effect calculations in the program.

The program "*cw.c*" gives you the option of calculating clutter-limited ranges. It will also suggest the level of MTI Improvement factor in dB to achieve the desired ranges. *It appears that using MTI is inescapable in a clutter environment as otherwise the radar ranges will be severely curtailed.* The range cell resolution in the clutter program is always the coarse range cell, since our fine range is synthetic.

References

1. Hall, W. M., "General radar equation," in Barton, D. K. (ed.), *Vol. 2: Radar: The Radar Equation*, Artech House, Inc., Norwood, MA, 1971.
2. Blake, L. V., *Radar Range Performance Analysis*, Artech House, Inc., Norwood, MA, 1986.
3. Wirth, W., "Omnidirectional low probability of intercept radar," in *IEE-89 Radar Conference*, 1989.
4. Nathanson, F. E., *Radar Design Principles*, McGraw-Hill, New York, 1991.
5. Scheer, J. A., et al., *Coherent Radar Performance Estimation*, Artech House, Inc., Norwood, MA, 1993.
6. Edde, B., *Radar, Principles, Technology, Applications*, PTR PrenticeHall, Englewood Cliffs, NJ, 1993.
7. Bussgang, J. J., et al., "A unified analysis of range performance of CW, pulse and pulse Doppler radar," *IRE Proceedings* 1959, pp. 1753.

Appendix B: The Design Process

We will now examine the procedure we need to adopt in the design of the eight-channel Pandora. This radar will essentially operate in two modes, the coarse mode and the fine mode. The initial detection of the target will be in the coarse mode, wherein only one channel will be used and the target recognition will occur in the fine mode wherein all eight channels will be used. We will examine these separately. The program *"pandora.cpp"* supplied with this book, gives you the initial option of starting with either of these modes. This is because the coarse mode also pertains to the design of a single-channel FMCW radar of the type as discussed in Chapter 7. The fine mode pertains to the multi-channel radar and is meant for range profiling. In our example in this appendix, we will design separately for this coarse mode as well as the fine mode.

1.1 THE COARSE MODE

In the Pandora configuration, the coarse mode is used for initial detection as well as measurement of target Doppler. Suppose we wish to detect a target, accelerating to 300 m/sec (Mach 1) during the observing interval, for example, a helicopter rotor which has a spread spectrum from 0 to 300 m/sec at the rotor tips. Our aim is to so choose the Doppler definition such that it comes to within one Doppler bin. We need to progress the design based on the following steps:

1. *Calculate the integration time:* We can either calculate the integration time based on a range cell resolution or vice versa. Suppose, we desire an integration time of say, 0.2 second. The target will traverse 60 m during the integration interval. If the azimuth beamwidth is 2°, then the cross range resolution of 60 m is achieved at 1.7 km, say, 2 km. Hence, our sweep bandwidth needs to have a resolution of at least 60 m (≥ 60 m). We select a maximum instrumented range $R_{inst} = 2$ km. This is the *first approximation.*

    The signal processing channel in the coarse mode, comprises the range FFT, followed by the Doppler FFT. Clutter in every range cell, will be mostly confined to the first filter or the zero Doppler filter. If we route the signal via an MTI to control the dynamic range of the Doppler filter bank, the clutter in the zero Doppler filter gets notched out. The clutter spread thereafter will be spread in Doppler only due to side lobe pickup and will not normally extend beyond the first filter (assuming own platform is stationary or at the most is a ship sailing at 25 knots).

    In the case of a ship sailing at 25 knots (12.5 m/sec) and at a frequency of 9.375 GHz ($\lambda = 0.032$ m)

$$\text{Maximum Doppler: } f_d = \frac{2v}{\lambda} = \frac{2 \times 12.5}{0.032} = 780 \text{ Hz}$$

We will see that this will fall entirely in the first and at the most second filters. We can then censor out these filters and have a clutter free display. Hence, we have no use for a Pandora type of configuration for this mode, but we use any one channel of the eight available channels for coarse detection via an MTI and Doppler filter FFT bank.

2. *Determining the sampling rate:* We need to design for 60 m resolution.
Sweep bandwidth based on 60 m range resolution

$$\Delta f = \frac{c}{2\Delta R} = \frac{3 \times 10^8}{2 \times 60} = 2.5 \text{ MHz}$$

If $N = 64$ is the number of samples/sweep,

$$R_{inst} = \frac{N}{2} \times \Delta R = \frac{64}{2} \times 60 = 1920 \text{ m}$$

This is the *second approximation.*

Therefore, we use a 64-point range FFT, with $R_{inst} = 1.92$ km, yielding 32 *real* range gates.

It is to be noted that the chosen value of R_{inst} (this is the instrumented range) is valid only if the calculated energetic range, R_{max}, is marginally less than this figure. Instrumented ranges should be marginally more than energetic ranges. The energetic range R_{max} is also the unambiguous range. At this stage we assume that $R_{max} = R_{inst}$.

3. *Determine the sweep time*: As discussed earlier, we determine the Doppler definition Δf_d for a target having a velocity of 300 m/sec and accelerating over the 0.2 sec observation (integration) period

$$\Delta f_d = \frac{2v^2 t}{R_{inst}\lambda} = \frac{2 \times 300^2 \times 0.2}{1920 \times 0.032} \approx 586 \text{ Hz}$$

This is the Doppler filter resolution. Hence, the clutter discussed earlier, falls essentially within the first filter and the second filter $f_d = 780$ Hz)

$$\therefore \text{ Coherent processing interval } = 1/\Delta f_d = 1.7 \text{ msec}$$

$$\text{Maximum target Doppler } f_{d_t} = \frac{2v}{\lambda} = \frac{2 \times 300}{0.032} = 18.75 \text{ kHz}$$

$$\therefore \text{ Number of Doppler filters required } = \frac{f_{d_t}}{\Delta f_d} = \frac{18.75 \times 10^3}{586} = 32 \text{ filters}$$

Therefore, we need to choose a sweep time that yields 32 coherent sweeps,

$$\text{Sweep time } = \frac{\text{coherent processing interval}}{\text{number of coherent sweeps}} = \frac{1.7 \times 10^{-3}}{32} = 53.3 \approx 50 \ \mu \text{sec}$$

Hence, we choose a 32-point Doppler FFT.

4. We now run the program with the following parameters, up to the end of the second approximation:

Stationary platform

Target velocity = 300 m/second
Number of Doppler filters = 32
Azimuth beamwidth = 2°
Integration time = 0.2 second
Average power = 5 W
SNR_0 = 12.8 dB for a P_d of 0.5 and $P_{FA} = 10^{-6}$ for a Swerling 1 target
N = 64 samples/sweep
Target RCS σ = 2 m²
λ = 0.032 m
G_T = 1 dB (short dipole)
G_R = 30 dB
Sweep time = 53.3 μsec
System loss = 10 dB
Noise figure = 3 dB
System noise temperature = 400 K

We obtain the following results:

Range = 0.64 km
Sampling rate = 64 samples/sweep for a 64-point range FFT

We note that the energetic range R_{max} is 0.64 km and it is much less than the instrumented range R_{inst} of 1.92 km, in the first approximation. This is not satisfactory as the instrumented range should be marginally superior to the energetic range but not so widely separated as this. We now round the instrumented range to the energetic range. We achieve R_{inst} = 640 m. This is the *third approximation*.

We once again determine the sampling rate for this R_{inst} using a sweep bandwidth of 2.5 MHz calculated earlier.

$$N = \frac{4\Delta f R_{inst}}{c} = \frac{4 \times 2.5 \times 10^6 \times 640}{3 \times 10^8} = 21.3 \text{ samples/sweep}$$

$$\therefore N = 32 \text{ samples/sweep (Nearest } 2^N)$$

The program next asks you for the sweep recovery time. Logically, it should be less than the sweep time. The program ensures this. It then calculates the round-trip time for the maximum energetic range. We need to ensure that the sweep time is at least 80% longer than the round-trip time to the maximum energetic range. This ensures that there is no deterioration in the attainable range resolution. In practice, best range resolution is attainable at near ranges and it progressively deteriorates at farther ranges due to the round-trip time. We can minimize the damage by ensuring that the sweep time is at least 80% longer than the maximum round-trip time. If the sweep time is less than the round-trip time, the program ensures that the sweep time is corrected so that

Sweep time = 5(round trip time + sweep recovery time)

It then recalculates the number of attainable Doppler filters for this new sweep time and the given coherent processing interval (CPI, which is dependant upon the target Doppler). If the number of filters is less than eight, there is no point in adopting the acceleration approach to identify helicopter rotors within the given system parameters. It, therefore, decides to adopt a simple Doppler filter bank whose bin width is decided by the user. It asks the user for the size of the

Doppler filter bank and then recalculates the CPI, based on the new sweep time and number of Doppler filters. Finally, the program ensures that the following equation is satisfied

$$T_{modulation} = T_{sweep} - T_{round\,trip} - T_{sweep\,recovery}$$

$$T_{modulation} >> T_{round\,trip}$$

This will ensure that in the range resolution equation based on the sweep bandwidth, the range resolution ΔR, is never negative.

$$\Delta R = \frac{c}{2\Delta f\left(1 - \dfrac{T_{Round\,trip}}{T_{modulation}}\right)} \tag{B.1}$$

The program now has the basic sweep time, CPI, and size of the Doppler filter bank that satisfies the range resolution requirements. It now recalculates the new energetic range based on these parameters and the given observation time.

This completes the design of the coarse channel. We can use any one channel for coarse tracking, so that the MTI and Doppler FFT bank is switchable. Hence, our Pandora radar will carry out initial detection of a 2 m² helicopter target at an instrumented range R_{inst} of 960 m and energetic range R_{max} (unambiguous range) of 640 m, using a 32 range FFT yielding 16 range gates and a 32-point Doppler filter bank. The sweep time is 53.3 μsec and the sweep bandwidth is 2.5 MHz.

It should be noted that we have allowed for a deterioration due to Hamming weighting and sweep recovery time of 10 μsec. The recovery time is essentially dictated by the settling time of the front end to settle to the LSB of the ADC. This is the impulse response of the low pass filter shown in Figure B–2. The target, however, stays within the same range cell during the integration time, since we have allowed for 60 m range resolution cell.

We note the following additional information.

The range FFT is completed every 53.3 μsec (sweep time).

$\therefore$ 53.3 μsec is the sampling rate for the Doppler FFT, that is, 18.76 kHz.

This means that there will be Doppler folding around the maximum target Doppler of 18.76 kHz. Basically, frequencies above 9.38 kHz will get folded.

1.2 FINE MODE

In this mode, we are not using target Doppler as we are interested only in target range profiling. We use the full eight channels in the Pandora configuration. Since the Pandora is a range profiling radar, it becomes necessary to improve the receiver beat frequency resolution (range bin resolution) to the extent possible. This yields very fine range bin resolutions, necessary for target profiling. This is the basic difference in the calculations as compared to the coarse mode.

We seek a range resolution of at least 40 cm. We calculate the sweep bandwidth for this resolution as (for number of channels $\Phi = 8$)

$$\Delta f = \frac{c}{2\Delta R \Phi} = \frac{3 \times 10^8}{2 \times 0.4 \times 8} = 46.875 \text{ MHz}$$

Suppose we choose a sweep bandwidth of 48 MHz (a round figure). This will yield a range resolution of

$$\Delta R = \frac{c}{2\Delta f \Phi} = \frac{3 \times 10^8}{2 \times 48 \times 10^6 \times 8} = 0.3906 \text{ m}$$

The program allows you to re-enter data based on this reasoning. The program also asks you to enter the wavelength as well as guard bandwidth and channel separation. The guard bandwidth, in this case, is the bandwidth required for filter roll-off. We adopt a value of, say, 6 MHz. The channel separation is the guard band as discussed throughout this book. We adopt a value of, say, 50 MHz.

The antenna bandwidth B is, therefore,

((Sweep bandwidth + guard band) × 8) + (Bandwidth separation × 8) = ((48 + 6) × 8) + (50 × 8) = 832 MHz

This is roughly 8% of the total antenna bandwidth, centered at 9.78 GHz and extending from 9.375 to 10.159 GHz.

The program next asks you to input the maximum target speed. We once again adopt a value of, say, 300 m/sec. It next asks you for sweep time. We will enter a value of, say, 1 msec. It is important that we choose a large sweep time, as we do not want the range resolution to deteriorate, which will be the case if the target round-trip time becomes comparable to the sweep time as is evident from equation (B.1). We can afford to choose our sweep time, because in the fine mode we do not have any need for target Doppler measurements, unlike in the coarse mode discussed earlier. Note that the sweep bandwidth has changed from 2.5 MHz in the coarse mode to 48 MHz in the fine mode. This has been dictated by the required resolutions. All entries need to be made using floating-point notation thus, $50e^{-6}$, and so forth. The program next asks for the maximum instrumented range. We will adopt an arbitrary value of, say, 1,000 m.

The following are the essential steps:

1. *Calculate the integration time*: We calculate the intergration time for a constant velocity target and not an accelerating one like helicopter blades as we are not using Doppler processing. Since, the coarse range resolution is approximately 3 m ($0.3906 \times 8 = 3.12$ m), our integration time must be such that the target remains within one tange cell during the integration time. Hence, for a 300 m/sec target

$$\text{Integration time} = 3.12/300 = 10.4 \, \text{msec}$$

The program does not ask you for this value, because it computes it as discussed above.

2. *Determination of sampling rate*: We choose the instrumented range as 1 km. We call this the *first approximation* of R_{inst}.

$$\therefore \text{first approximation of } R_{inst} = 1,000 \, \text{m}$$

$$N = \frac{4\Delta f R_{inst}}{c} = \frac{4 \times 48 \times 10^6 \times 1000}{3 \times 10^8} = 640 \, \text{samples/sweep}$$

We choose the nearest higher 2^N value

$$\therefore N = 1,024 \, \text{samples/sweep}$$

The coarse channel (low range resolution) for this mode is given by

$$\Delta R = \frac{c}{2\Delta f} = \frac{3 \times 10^8}{2 \times 48 \times 10^6} = 3.125 \, \text{m}$$

We, therefore, have a resolution of 3.125 m for a sweep bandwidth of 48 MHz.

$$R_{inst} = \frac{N}{2} \Delta R = \frac{1024}{2} \times 3.125 = 1,600 \, \text{m}$$

This is the *second approximation* of R_{inst}.

$$\text{Second approximation of } R_{inst} = 1,600 \text{ m}$$

3. We now run the following parameters for an eight-channel Pandora, in the *nonscanning* mode and *we do not exercise the Doppler option*:

Integration time = 0.0104 sec (this need not be entered, as the program already computes this as discussed earlier)
Desired range resolution = 0.3906 m
Guard band = 6 MHz (This allows for the filter "skirts")
Channel separation = 50 MHz
Maximum target velocity = 300 m/sec
Average power = 5 W
Sweep time = 1 msec
Sweep bandwidth = 48 MHz
Sweep recovery time = 10 μsec
Nonlinearity = 0.003% (see Appendix D)
SNR_o = 12.8 dB for a P_d of 0.5 and $P_{FA} = 10^{-6}$ for a Swerling 1 target
N = 1,024 samples/sweep
R_{inst} = 1,000 m
Target RCS σ = 2 m^2
λ = 0.032 m
G_T = 1 dB (short dipole)
G_R = 30 dB
SRF = 20 kHz (we round 18.76 kHz for convenience)
System loss = 10 dB
Noise figure = 3 dB
System noise temperature = 400 K

We obtain the following results:

Range R_{max} = 1.33 km
Sampling rate = 1,024 samples/sweep for a 1 K point range FFT

This is above the instrumented range of 1 km.
Hence, we increase the instrumented range to R_{inst} = 1,326 m
This is the *third approximation of* R_{inst}

$$\text{Third approximation of } R_{inst} = 1,326 \text{ m}$$

We now determine the new sampling rate

$$N = \frac{4\Delta f R_{inst}}{c} = \frac{4 \times 48 \times 10^6 \times 1326}{3 \times 10^8} = 848.64 \text{ samples/sweep}$$

Rounding to the nearest higher 2^N value:

$$N = 1,024 \text{ samples/sweep}$$

We now check whether this final sampling value satisfies the range resolution requirement. We assume a sweep recovery time of $10\,\mu$sec.

Basically, the range resolution due to the round-trip delay to the maximum coarse *energetic* range (we do not choose the instrumented range, because we are interested in the maximum energetic range for purposes of resolution) and the *range bin* resolution due to the beat frequency resolution (this is the convolution of the target beat-frequency spectral width and the receiver frequency resolution, see equation (4.55) are compared. These aspects have been discussed in Chapter 4. We then adjust the sampling rate till the range bin resolution becomes superior to the range resolution due to the round-trip delay, as we cannot do anything to control the latter. Hence, our endeavor will be to make the range bin resolution due to this factor, *less than or equal* to the range resolution due to the sweep bandwidth and round-trip delay. These aspects are shown and implemented in the algorithm in Figure B–1.

We do this by initially increasing the beat frequency (sampling rate) and then subsequently by reducing the nonlinearities as discussed in Chapter 4.

We find that it does not satisfy, but this condition is eventually satisfied by putting $N = 2,048$ and using a nonlinearity error of 0.003% and a sweep recovery time of $10\,\mu$sec

$$\therefore N = 2,048 \text{ samples/sweep}$$

$$\therefore R_{inst} = \frac{N}{2}\Delta R = \frac{2048}{2} \times 3.125 = 3,200 \text{ m}$$

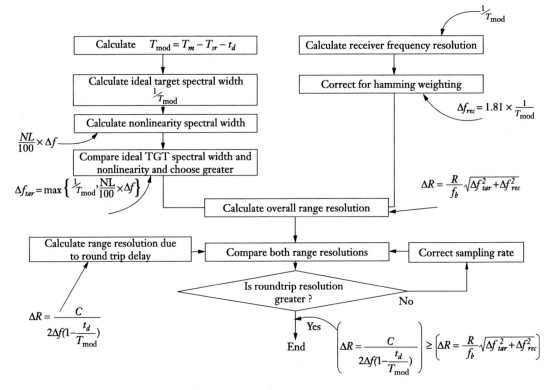

Figure B–1 The range resolution function.

This the *fourth approximation* of R_{inst}

Fourth approximation and final value of $R_{inst} = 3,200$ m

The sampling frequency becomes

$$f_s = \frac{\text{samples/sweep}}{\text{sweep time}} = \frac{2048}{1 \times 10^{-3}} = 2.048 \approx 2 \text{ MHz}$$

Steps Towards Calculation of R_{inst}
The Pandora Algorithm
↓
Adopt Initial Value   *First Approximation*
↓
Calculate Number Of Samples
↓
Choose Nearest 2^N Value   *Second Approximation*
↓
Calculate R_{inst} For This Value
↓
Calculate Energetic Range
↓
Adjust R_{inst} = Energetic Range   *Third Approximation*
↓
Check For Range Resolution
Adjust Sample Rate
↓
Calculate R_{inst} For This Value   *Fourth Approximation*
↓
End

It should be noted that we have already allowed for the deterioration of the coarse range resolution in this fine mode, due to the Hamming weighting by increasing the sampling rate/sweep to 2,048 samples/sweep. The target, however, stays within the same resolution cell during the integration time, since we have allowed for 3.125 m coarse range resolution cell, in our calculations.

This completes the design of the fine channel. Hence, our Pandora radar will carry out target recognition of a 2 m^2 target traveling at a constant velocity of 300 m/sec, at a range of 1.33 km. The sampling frequency is approximately 2 MHz and the range FFT will be 2 K.

We can run the coarse mode again using the fine channel parameters in order to calculate the initial detection *based on these parameters*. Alternately, we can calculate separate coarse mode parameters as we have done for initial detection. The radar should, therefore, be made switchable between these modes, that is, separate parameters for coarse mode and fine mode, respectively.

We have stated that the program determines the adequacy of the chosen sampling rate based on your chosen nonlinearity level. In case this sampling rate proves inadequate, it asks you to increase it. It then reiterates and if once again the sampling rate is inadequate, the program prompts you to increase it. In case you decline because the required sampling rate is unrealistic or you have ADC limitations, then there are two cases that will be examined:

1. If the range resolution calculated based on equation (B.1) is greater than the range bin resolution based on the beat frequency resolution, it will tell you that "irrespective of the nonlinearities you need to increase the sampling rate." This is because it is pointless to consider nonlinearities, unless the range bin resolution is at least superior or equal to the basic range resolution.
2. If condition 1 above is satisfied, the program calculates the required level of nonlinearity and informs you accordingly. Then the radar designer needs to take action by using superior signal sources, and so forth, to achieve this required nonlinearity. This is because the target spectral width is too high, such that merely hiking the sampling rate will not do.

These concepts are incorporated in the software "*pandora.cpp.*" The program prompts you through these steps. You can save the run details in two files, COARSE.DAT and FINE.DAT for later analysis. The files are saved in the local directory.

The program is supplied as a C file. It can be run on a C complier like Visual C++ in the Win32 mode as a console application. Alternately, an .exe file has been provided, which can be run on any PC with some minimum operating system. In such cases, C complier is not necessary, but the user cannot modify/edit the file.

Coarse Mode Results	
Parameter	**Units**
Integration time	2.00e − 01 sec
Average power	5.00 W
Wavelength	0.0320 m
Azimuth beamwidth	2.00 degrees
Sweep time	5.33e − 05 sec
Sweep bandwidth	2.50e + 06 Hz
Signal/noise ratio	12.80 dB
Target RCS	2.00 square meter
Target velocity	300.00 m/sec
Transmitting ant.gain	1.00 dB (short dipole)
Receiving ant.gain	30.00 dB
SRF	1.88e + 04 Hz
Doppler FFT length	32
Number coh.sweeps	32
Sweep recovery time	10 μsec
System loss	10.00 dB
Rx. Noise figure	3.00 dB
System noise temp.	400.00 K
Samples/sweep N	32
R_{inst}	0.96 km
Energetic range R_{max}	0.64 km
Desired target resolution	60.00 m
Achieved target resolution	118.95 m

This is the achieved target resolution at the *maximum energetic range* and allowing for Hamming weighting. We note that the achieved target resolution is 118.95 m (actually 118.95/1.81 = 65.7 4m, where 1.81 is the pulse broadening factor caused due to Hamming weighting) as compared to the desired 60 m. This is because our round-trip delay for maximum energetic range is 4.3 μsec as compared to the sweep time of 50 μsec. This causes a marginal deterioration in range resolution from the ideal.

Fine Mode Results

Parameter	Units
Integration time	1.04e − 02 second
Average power	5.00 W
Wavelength	0.0320 m
Azimuth beamwidth	2.00 degrees
Sweep time	10 μsec
Sweep bandwidth	4.80e + 07 Hz
Signal/noise ratio	12.80 dB
Target RCS	2.00 square meter
Target velocity	300.00 m/sec
Transmitting ant.gain	1.00 dB (short dipole)
Receiving ant.gain	30.00 dB
SRF	1 kHz
Number of channels	8
Antenna bandwidth	8.32e + 08 Hz
Nonlinearity error	0.0030%
Sweep recovery time	10 μsec
System loss	10.00 dB
Rx. Noise figure	3.00 dB
System noise temp.	400.00 K
Samples/sweep N	2,048
R_{max}	3.20 km
Energetic range	1.33 km
Desired target resolution	0.39 m
Achieved target resolution	0.39 m

We note that because of our large sweep time of 1 msec as compared to the round-trip delay time for maximum energetic range of 9 μsec, we have managed to retain our desired range resolution of 0.39 m. We can afford to do this in this mode, because we are not measuring the target Doppler (due to lack of time caused due to this large sweep time). Target Doppler, moreover, is not necessary in this mode, because we are only range profiling the target. It is to be noted that 0.39 m is the resolution at the maximum energetic range. This is the resolution already corrected for Hamming weighting. It is matching the permitted range resolution as given by the round-trip equation. This permitted resolution is maintained by increasing the sampling rate to a high value to cater for nonlinearities and receiver range resolution as discussed earlier. This technique was not used for the coarse mode, because the sweep time was not long enough, due to target Doppler measurement considerations. In reality, we are actually sampling at eight points the overall sweep bandwidth of 776 MHz. This yields a resolution of 0.19 m, which deteriorates to 0.34 m when we correct for Hamming weighting. *Hence, the actual range resolution at maximum energetic range will be somewhere between 0.34 m and 0.39 m.*

1.3 LOW PASS FILTER

The design of the low pass filter can now be carried out. The LPF needs to be basically an anti-alias filter. Since the sampling frequency is 2 MHz, frequencies above 1 MHz need to be attenuated as these will be aliased. Since our maximum instrumented range is 3.2 km, 1 MHz will correspond to this range (highest beat frequency $f_{max} = f_s/2$). However, our energetic range is around 1.33 km. Hence, we need the 3 dB cutoff to correspond to 500 kHz. A 6-order Butterworth fits these requirements. This will be an active filter (see Figure B–2).

The LPF also plays the role of cutting off the upper beat note. This is the note that occurs in the time interval $0 < t < T_d$ where T_d is the round-trip delay. This is at a higher frequency than our beat note of interest, that is, the lower beat note.

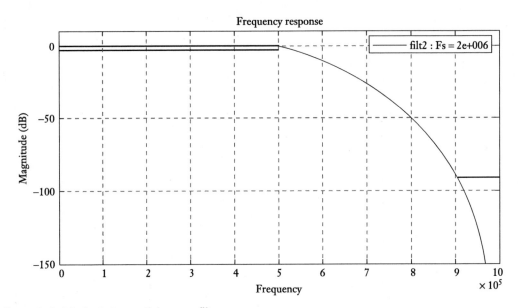

Figure B–2 6-Order Butterworth low pass filter.

C

Appendix C: A Hardware Solution to the Range Resolution Problem

1. INTRODUCTION

The following proposal envisages a hardware solution to the range resolution problem. We shall first examine the mathematical expression for FMCW radar returns.

This derivation is based on the paper by Stove [1]. For a linear frequency modulated sawtooth waveform that sweeps up in frequency, without taking into account nonlinearities, the transmitted frequency is given by

$$f_t(t) = f_0 - \frac{\Delta f}{2} + \frac{\Delta f}{T_m}t, \quad 0 < t < T_m$$

where f_0 is the RF center frequency, Δf is the total frequency deviation, T_m is the modulation period, and t is the time.

For zero initial RF phase, the transmitted signal $s_t(t)$ with amplitude a_0 is given by

$$s_t(t) = a_0 \sin 2\pi \left[\left(f_0 - \frac{\Delta f}{2} \right) t + \frac{\Delta f}{2T_m}t^2 \right], \quad 0 < t < T_m$$

The received signal $s_r(t)$ is the transmitted signal delayed in time by the round-trip propagation time to the target and back, t_d with reduced amplitude b_0.

$$s_r(t) = \frac{b_0}{a_0} s_t(t - t_d), \quad 0 < t < t_d$$

For a homodyne FMCW radar, the received signal is mixed with the transmitted signal so that the mixer output beat frequency signal $s_b(t)$ with amplitude c_0 is

$$s_b(t) = c_0 \cos \left\{ 2\pi \left[\left(f_0 - \frac{\Delta f}{2} \right) t + \frac{\Delta f}{2T_m}t^2 \right] - 2\pi \left[\left(f_0 - \frac{\Delta f}{2} \right)(t - t_d) + \frac{\Delta f}{2T_m}(t - t_d)^2 \right] \right\}$$

Simplifying,

$$s_b(t) = c_0 \cos 2\pi \left\{ \left(f_0 - \frac{\Delta f}{2} \right) t_d - \frac{\Delta f}{2T_m}t_d^2 + \frac{\Delta f}{T_m}t_d t \right\}$$

269

Since the transit time t_d is given by

$$t_d = \frac{2R}{c}$$

then

$$s_b(t) = c_0 \cos 2\pi \left\{ \left(f_0 - \frac{\Delta f}{2} \right)\left(\frac{2R}{c} \right) - \frac{\Delta f}{2T_m}\left(\frac{2R}{c} \right)^2 + \frac{\Delta f}{T_m}\left(\frac{2R}{c} \right)t \right\}$$

1.1 BEAT FREQUENCY FOR A CONSTANT VELOCITY TARGET

For constant relative velocity between the radar and the target, the range R is given by

$$R = R_0 + Vt$$

where R_0 is the initial range (at $t = 0$) and V is the relative velocity.

Substituting into the earlier expression,

$$s_b(t) = c_0 \cos 2\pi \left\{ \left[\left(f_0 - \frac{\Delta f}{2} \right)\frac{2R_0}{c} - \frac{2R_0^2}{c^2}\frac{\Delta f}{T_m} \right] + \left[\left(f_0 - \frac{\Delta f}{2} \right)\frac{2V}{c} + \frac{2\Delta f R_0}{T_m c}\left(1 - \frac{2V}{c} \right) \right]t \right.$$
$$\left. + \left[\frac{2\Delta f V}{T_m c}\left(1 - \frac{V}{c} \right) \right]t^2 \right\}$$

The constant phase term is approximately equal to the number of wavelengths in the round-trip propagation path to the target and back.

The beat frequency $f_b(t)$ is given by

$$f_b(t) = \left\{ \left[\left(f_0 - \frac{\Delta f}{2} \right)\frac{2V}{c} + \frac{2\Delta f R_0}{T_m c}\left(1 - \frac{2V}{c} \right) \right] + \left[\frac{4\Delta f V}{T_m c}\left(1 - \frac{V}{c} \right) \right]t \right\}$$

This beat frequency contains a constant component and a component linearly proportional to time. The constant beat frequency component includes both the Doppler frequency shift term that is linearly proportional to velocity and the FMCW equation term that is linearly proportional to range. Rearranging the constant term,

$$\left(\frac{2f_0}{c} - \frac{\Delta f}{c} - \frac{4\Delta f R_0}{T_m c^2} \right)V + \frac{2\Delta f}{T_m c}R_0$$

If we substitute the following values: $f_0 = 11.22$ GHz, $\Delta f = 25$ MHz, $T_m = 2$ msec then

$$f_b \cong 74.8\frac{\text{Hz}}{\text{m/sec}}V + 83.3\frac{\text{Hz}}{\text{m}}R_0$$

This equation depends on both the target range and velocity. To resolve this range–velocity ambiguity, the radar must use different frequency sweep slopes and then process the two resulting beat frequencies or as in our case, control the ambiguities to within one range gate. The nominal beat frequency for 3 km range is 250 kHz. For a maximum approaching relative velocity of –300 m/sec, the maximum Doppler frequency shift is −22.4 kHz. The linear frequency, quadratic phase beat-frequency component is 166.6 Hz/(m/sec) Vt. This term is 100 Hz for the maximum velocity of 300 m/sec at the end of the sweep when $t = T_m$. This component accounts for the beat frequency chirp induced by the target velocity during the frequency sweep.

It can be easily seen that the beat signal is a function of both range and velocity. However, it tells us nothing about the resolution. This is given by the following expression:

$$\Delta R = \frac{c}{2\Delta f \left(1 - \frac{T_d}{T_m}\right)}$$

where T_d is the round-trip propagation time and T_m is the sweep time.

Hence, we cannot control the resolution in the frequency domain, but only in the time domain by controlling the round-trip delay. This is because before performing the range FFT, we need to wait for the return from the maximum range. *This eats into the sweep time and consequently reduces the sweep bandwidth and therefore resolution.*

We, therefore, need to develop a system which caters to this need. Essentially, we need to delay the transmitted signal by the round-trip delay time before giving it to the mixer as reference.

We will illustrate the problem with an example. Suppose we are analyzing the PANDORA with the following parameters in the fine mode. These are the final parameters for a 32-channel radar.

Integration time = 20 msec
Average power = 5 W
Sweep time = 1 msec
Sweep bandwidth = 25 MHz
SNR_o = 10 dB for a P_d = 0.25 and P_{FA} = 10^{-6} for a Swerling 0 target
N = 2,048 samples/sweep
R_{max} = 6,144 m
Target RCS σ = 2 m^2,
λ = 0.032 m
G_T = 1 dB
G_R = 30 dB
SRF = 20 kHz
System loss = 10 dB
Noise figure = 3 dB
System noise temperature = 400°K

We obtain the following results:

Total number of coherent sweeps = 20
Range = 2.77 km
Sampling rate = 2,048 samples/sweep for a 2 K point range FFT = 2 MHz

If the instrumented range (it is more convenient for these calculations to work with instrumented range, rather than the energetic range) is 6,144 m, the round-trip time is 40 μsec. Hence, the loss of range resolution due to this is given by

$$\Delta R = \frac{c}{2\Delta f \left(1 - \frac{T_d}{T_m}\right)} = \frac{3 \times 10^8}{2 \times 25 \times 10^6 \left(1 - \frac{40 \times 10^{-6}}{1 \times 10^{-3}}\right)} = 6.25 \text{ m}$$

The ideal range resolution should have been 6 m, but it has now deteriorated to 6.25 m. *However, this pertains to only the last range bin, since, the resolutions progressively improve as we tend toward the reference, till it becomes ideal at zero range bin.*

The following facts also emerge:

The scale factor is given by

$$\text{scale factor} = \frac{T_m c}{2\Delta f} = \frac{1\times10^{-3}\times3\times10^8}{2\times25\times10^6} = 6 \text{ m/kHz}$$

We will obtain 1,024 range bins, each with a resolution of 6 m, totaling 6,144 m.

The maximum beat frequency is given by

$$f_{\max} = \frac{f_s}{2} = \frac{2.048\times10^6}{2} = 1.024 \text{ MHz}$$

We need to evolve a methodology to overcome this problem. The recommended method is to delay the transmitted sweep given as reference to the homodyne mixer, by the round-trip delay time, in this case 40 μsec. However, the price we pay is that the closer ranges will be forgotten as there is no transmitted reference for their returns.

In our approach, we subdivide the overall range delay of 40 μsec into a number of subsectors, say, 10 sectors, each of 4 μsec duration. This implies that we generate 10 references each delayed by 4 μsec with respect to each other. Hence, for *each transmitted reference*, the furthest range gate will be 4 μsec away. If we now compute the resolution,

$$\Delta R = \frac{c}{2\Delta f\left(1-\dfrac{T_d}{T_m}\right)} = \frac{3\times10^8}{2\times25\times10^6\left(1-\dfrac{4\times10^{-6}}{1\times10^{-3}}\right)} = 6.024 \text{ m}$$

This means that no range gate will have a resolution worse than 6.024 m. There are, however, certain hardware problems that we need to address (see Figure C–1).

The different delayed sweeps need to be generated by the FMCW generator itself. It is convenient to do this in the digital mode, rather than in the analog, because the delays need to be accurately adjusted. This process is difficult using analog delay lines and also bulky. It is, however, convenient for us as we in any case intend to use digital FMCW generators for the sake of signal purity. The output of each mixer is given to a bank of filters. It is pointed out that, since each beat frequency will have the same relationship with respect to its reference, the beat frequencies will be identical in each channel. Hence, though each filter in every one of the 10 channels will have similar frequency response as compared to its counterpart in the adjacent channels, it will in fact represent a different range bin in each channel. Thus, we will need to designate the range bin number for each filter.

These filters can be implemented as range FFTs. Hence, it becomes similar to the normal stretch processing but with the only difference, that instead of one mixer for a channel for the purposes of stretch processing, we now have 10 mixers. The quantity of these mixers depends upon tour desired range resolution. *More mixers mean higher resolution. In the limiting case, if the number of mixers equals the number of range bins, we achieve ideal resolution!* It is easy to see that the output of the 10 range FFTs is exactly identical to what would have come out of a single mixer/channel configuration, but with the proviso that now the resolution is controlled so as not to exceed 6.024 m for each range bin. A point to be noted here is that the sampling rate of the A/D converters will now be 1/10 of

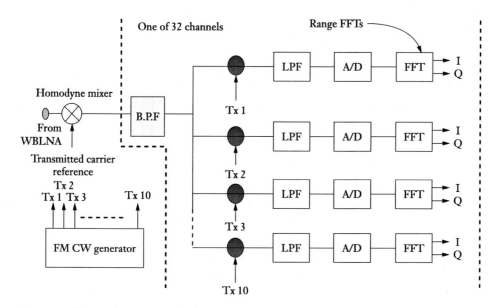

Figure C–1 Toward improving range resolution.

the original value, that is, 200 samples/sweep instead of the original 2,048 samples/sweep. Actually, it will be the nearest radix 2 value, that is, 256 samples/sweep yielding 128 range bins per channel. Another point to be noted here is that the receiver frequency resolution problem will improve since T_d does not exceed $4\,\mu$sec as compared to $40\,\mu$sec earlier.

The I and Q values will be fed to the high-resolution FFT. The output of this FFT will then be given to a magnituder, followed by noncoherent integration and then to a detector and finally to a display.

Reference

1. Stove, A. G., "Linear FMCW radar techniques," *IEE Proceedings-F*, Vol. 139, No. 5, October 1992.

Appendix D: Nonlinearity in FM Waveforms

It was discussed in earlier chapters that the nonlinearities in the FM waveform need to be controlled to ensure against deterioration of range resolution.

To illustrate the problem, let us examine the results of the fine mode analysis given at Appendix B. We reproduce the results here for convenience.

Fine Mode Results	
Parameter	**Units**
Integration time	10 msec
Average power	5.00 W
Wavelength	0.0320 m
Sweep time	1 msec
Sweep bandwidth	48 MHz
Signal/noise ratio	12.8 dB for a $P_d = 0.25$ and P_{FA} $= 10^{-6}$ for a Swerling 0 target
Target RCS	2.00 m²
Target velocity	300.00 m/sec
Transmitting ant.gain	1.00 dB
Receiving ant.gain	30.00 dB
SRF	1 kHz
Number of channels	8
Antenna bandwidth	8.32e + 08 Hz
Nonlinearity error	0.0030%
Sweep recovery time	10 μsec
System loss	10.00 dB
Rx. noise figure	3.00 dB
System noise temp.	400.00 K
Samples/sweep N	2,048
R_{max}	3.2 km

Energetic range = 1.33 km for a 2 m² target
Sampling rate = 2 MHz

Maximum beat frequency = $f_b = \dfrac{f_s}{2} = \dfrac{2 \times 10^6}{2} = 1$ MHz

We note that the beat frequency is given by

$$f_b = \frac{\Delta f \tau}{T_m}$$

where τ is the round-trip propagation time for a maximum instrumented range of 3,200 m = 21 μsec, T_m is the sweep time, and Δf is the sweep bandwidth.

$$\therefore f_b = \frac{48 \times 10^6 \times 21 \times 10^{-6}}{1 \times 10^{-3}} = 1 \text{ MHz}$$

(Half the sampling frequency as discussed above)

This yields 1 MHz/3,200 m or 312.5 Hz/m beat frequency to range ratio. Thus, a 3.12 m range resolution (coarse resolution in the fine mode) requires 975 Hz receiver frequency resolution and consequently frequency sweep linearity less than 1 kHz. This is 0.002% of the 48 MHz frequency deviation. This is, therefore, the linearity required to achieve our desired fine range resolution of 0.4 m. This is an ideal case as we have designed the Pandora at Appendix B using a nonlinearity of 0.003%. This discrepancy can be explained by the fact that in order to function with a nonlinearity of 0.003%, we have gone in for a higher beat frequency than otherwise necessary, as explained below. However, it can be verified that a higher nonlinearity of say, 0.04% does not satisfy our requirement. Hence, if we can achieve a level of 0.003% nonlinearity, we can expect to achieve the desired range resolution.

The logic behind this reasoning is explained as follows. If the frequency sweep linearity exceeds 1 kHz (in our example), then the target spectral width increases beyond the ideal figure given by $1/T_{mod}$ we had estimated in the ideal case, in Chapter 4. This will lead to loss of receiver frequency resolution. In such a case, one way of countering it is to yet increase the sampling rate further, that is, increase f_b. This is evident from inspection of equation reproduced here for convenience

$$\Delta R = \max \left\{ \frac{R}{f_b} \sqrt{\Delta f_{tar}^2 + \Delta f_{rec}^2}, \ \frac{c}{2\Delta f \left(1 - \dfrac{T_d}{T_m}\right)} \right\}$$

This will, however, unnecessarily increase the instrumented range, when the energetic range is only 1.33 km. There is no sense in unnecessarily increasing the sampling rate. The better method will be to control the nonlinearities in the waveform. Piper [1] has shown that the nonlinearities can be reduced by time gating the minimum and maximum points of the frequency sweep, since the greatest nonlinearities occur in these regions. For example, blocking 5% at each end of the sweep period reduces effective signal power by less than 0.5 dB. But this also yields a 2.1 dB decrease in the standard deviation of the nonlinearity. This makes a case for the employing of the digital FMCW generator, discussed elsewhere in this report. Such generators will not have discontinuities at the ends of the sweep, as they are gated.

The software "*pandora.cpp*," employs nonlinearity correction on request. It compares the target spectral width (inverse of the sweep time minus round-trip delay to maximum range in the ideal case) to the percentage nonlinearity with respect to sweep bandwidth and takes the greater of the two to compute the receiver frequency resolution.

Reference

1. Piper, S. O., "FMCW linearizer bandwidth requirements," *IEEE-91 Radar Conference*, 1991.

E

Appendix E: Transmitter Noise Leakage

1. INTRODUCTION

The problem of transmit–receive noise leakage is known as one of the most severe problems facing the FMCW radar designer. The FM sidebands at a deviation of ω_m like AM (amplitude modulation) noise sidebands at the same frequency give rise to noise also, but only at an IF of ω_m. There is thus a simple relationship between the transmitter's noise spectra and the spectrum of the detected noise in the receiver's IF. Moreover, in the case where the frequency deviation is independent of frequency, that is, it is the same at all frequencies of interest, then it will be shown that the detected noise is the same at all IF frequencies, which is the same as the case for white AM noise. This derivation is based on the paper by Stove [1].

2. AM NOISE

The AM noise on a carrier in a narrow frequency band about $\pm\omega_m$ from a carrier at a frequency ω_0 is quasi-sinusoidal and so can be expressed by the following well known expression:

$$E = E_0 \left[(1-\alpha) + \alpha \cos\omega_m t \right] \cos\omega_0 t \tag{E.1}$$

where $\alpha/(1-\alpha)$ is the modulation index and the peak signal level is E_0. This can be reformulated to separate the carrier and the sidebands:

$$E = E_0 (1-\alpha)\sin\omega_0 t + E_0 \frac{\alpha}{2}\cos(\omega_0 - \omega_m)t + E_0 \frac{\alpha}{2}\cos(\omega_0 + \omega_m)t \tag{E.2}$$

Similarly, an FM signal can be expressed as

$$E = E_0 \sin(\omega_0 + \Delta\omega \sin\omega_m t)t \tag{E.3}$$

which can be expanded as a series of Bessel functions of the deviation ratio $\Delta\omega/\omega_m$.

If $\Delta\omega$ is the noise in 1 Hz bandwidth, it is typically of the order of 1–10 Hz and ω_m is typically of the order of many kHz or MHz. The FM noise thus typically gives rise to narrowband modulation, that is, $\Delta\omega << \omega_m$, in which case the Bessel expansion can be reduced to

$$E = E_0 \sin\omega_0 t + \frac{E_0}{2}\frac{\Delta\omega}{\omega_m}\sin(\omega_0 - \omega_m)t - \frac{E_0}{2}\frac{\Delta\omega}{\omega_m}\sin(\omega_0 + \omega_m)t \tag{E.4}$$

It will be noted that in the expression for AM noise, the two sidebands have the same sign whereas in the expression for FM noise, they have opposite signs. Therefore, equations (E.2) and (E.4) can, therefore, be used to completely characterize the noise sidebands of the oscillator at a given frequency from the carrier.

2.1 ANALYSIS IN TERMS OF POWER RATIOS

We will now analyze the above expressions in terms of power ratios as the final results will be in terms of power. We define characteristic impedance Z_0 of the system, so that the voltages can be converted into powers, using the general formula

$$P = \frac{E^2}{2Z_0}$$

where P is the power and E is the peak voltage of the sinusoidal signal. The single sideband AM and FM noise to carrier power ratios are thus, respectively

$$\eta_{AM} = \frac{\left[\dfrac{E_0^2 \alpha^2}{8Z_0}\right]}{\left[\dfrac{E_0^2 (1-\alpha)^2}{2Z_0}\right]} \tag{E.5}$$

$$\eta_{FM} = \frac{\left[\dfrac{E_0^2 \left(\dfrac{\Delta\omega}{\omega_m}\right)^2}{8Z_0}\right]}{\left[\dfrac{E_0^2}{2Z_0}\right]} \tag{E.6}$$

If α is assumed small,

$$\eta_{AM} = \frac{\alpha^2}{4} \tag{E.7}$$

$$\eta_{FM} = \frac{\left(\dfrac{\Delta\omega}{\omega_m}\right)^2}{4} \tag{E.8}$$

If $\Delta\omega$ is a constant at all frequencies of interest, then in the FM case, this corresponds to a sideband level $(\Delta\omega/\omega_m)^2/4$, which falls at 20 dB/decade as m increases. This then counteracts the fact that, for small $\omega_m \delta t$, the degree of cancellation is reduced by 20 dB/decade as the modulation frequency rises.

The noise in the leakage signal will be detected by the receiver mixer and will degrade the receiver noise figure. The IF from an input $E = E_s \sin \omega t$ is

$$E' = kE_s \cos\left[(\omega - \omega_0)t + \phi\right] \tag{E.9}$$

where k is the mixer conversion loss and ϕ is the phase of the signal relative to the local oscillator. The power in the signal is,

$$P_s = \frac{E_s^2}{2Z_0}$$

The power in the IF signal is,

$$P_{IF} = \frac{E_s^2 k^2}{2Z_0}$$

and the mixer conversion loss is

$$L = \frac{P_{IF}}{P_s} \tag{E.10}$$

2.2 DETECTION OF AM NOISE

If the AM noise on the local oscillator signal is assumed to be substantially suppressed by the use of a balanced mixer, then the local oscillator can be considered to have the following expression,

$$E_L = E_{LO} \sin(\omega_0 t - \phi) \tag{E.11}$$

so the low frequency component of the AM noise at the mixer output (the IF noise) will be

$$E_{IF} = \frac{\alpha E_0 k}{2 \left[\cos(\omega_m t + \phi) + \cos(-\omega_m t + \phi) \right]} \tag{E.12}$$

The ratio of detected IF AM noise power to s.s.b. AM noise power is thus,

$$\eta_{AM} = 4k^2 \cos^2 \phi = 4L \cos^2 \phi \tag{E.13}$$

The mean power level is $2L$, that is, the RF double sideband power level is multiplied by the conversion loss.

3. DETECTION OF FM NOISE

In the case of FM noise, the noise on the local oscillator cannot be neglected as it is not affected by the balanced mixer and the local oscillator can be considered to have the following expression:

$$E_L = E_{LO} \left[\sin \omega_0 t + \frac{\Delta \omega}{2\omega_m} \sin(\omega_0 - \omega_m) t - \frac{\Delta \omega}{2\omega_m} \sin(\omega_0 + \omega_m) t \right] \tag{E.14}$$

The RF signal, in general, suffers a time delay relative to the local oscillator. This is defined by δt. It is the time delay caused due to the *path length difference* between the transmitter and the mixer via the leakage path and the transmitter and the mixer via the local oscillator path. This is similar to the phase shift ϕ between the RF signal and the local oscillator in the case of AM noise. Hence,

$$E = E_0 \left\{ \sin \omega_0 (t - \delta t) + \frac{\Delta \omega}{2\omega_m} \sin \left[(\omega_0 - \omega_m)(t - \delta t) \right] - \frac{\Delta \omega}{2\omega_m} \sin \left[(\omega_0 + \omega_m)(t - \delta t) \right] \right\} \quad \text{(E.15)}$$

When the RF and the LO signals are mixed and the difference extracted, we obtain,

$$E_{IF} = \frac{\Delta \omega E_0 k}{2\omega_m} \left\{ \begin{aligned} &-\cos \left[(\omega + \omega_m)(t - \delta t) - \omega_0 t \right] \\ &+\cos \left[(\omega - \omega_m)(t - \delta t) - \omega_0 t \right] \\ &-\cos \left[(\omega + \omega_m)t - \omega_0 (t - \delta t) \right] \\ &+\cos \left[(\omega - \omega_m)t - \omega_0 (t - \delta t) \right] \end{aligned} \right\}$$

We can write this as,

$$E_{IF} = 2k \left(\frac{\Delta \omega}{\omega_m} \right) E_0 \times \sin (\omega_0 \delta t) \cos \left(\frac{\omega_m \delta t}{2} \right) \sin \left[\omega_m \left(\frac{t - \delta t}{2} \right) \right] \quad \text{(E.16)}$$

The IF power level is then,

$$P_{IF} = \frac{2k^2 \left(\dfrac{\Delta \omega}{\omega_m} \right)^2 E_0^2}{Z_0 \sin^2 (\omega_0 \delta t) \sin^2 \left(\dfrac{\omega_m \delta t}{2} \right)}$$

The detected power relative to the s.s.b. RF level is then,

$$\eta_{FM} = 16 L \sin^2 (\omega_0 \delta t) \sin^2 \left(\frac{\omega_m \delta t}{2} \right) \quad \text{(E.17)}$$

If we put $\phi = \omega_0 \delta t$, we obtain,

$$\eta_{FM} = 4 L \sin^2 \phi \times 4 \sin^2 \left(\frac{\omega_m \delta t}{2} \right) \quad \text{(E.18)}$$

The first term in the above expression is the phase sensitive term. It is equivalent to the expression in equation (E.13), except that the FM noise is detected with the local oscillator in quadrature with the RF leakage signal, whereas the AM noise is detected when the two are in phase. This is a consequence of the fact that the upper and lower AM sidebands are in phase whereas the FM sidebands are in anti-phase.

The second term in equation (E.18) above pertains to the FM noise cancellation. If the time delay δt is much less than the modulation rate $(\omega_m / 2\pi)$ then, because the variations are the same on both the leakage signal and on the local oscillator, they are strongly correlated and no noise output is observed at the IF. As the time delay increases, the degree of correlation for a given modulation rate decreases and more noise is seen at IF. Basically, if there is a strong correlation between the leakage signal and the local oscillator, the mixing action wherein we take the difference frequency, will ensure that the noise cancels out. Strong correlation can be ensured by making the path length from the transmitter to the mixer via the leakage (the leakage can be the spillover in case of separate antennae or the reflected leakage due to the antenna reflection in case of an integral antenna) the same as the local oscillator path length from the transmitter to the mixer. Furthermore, the limit to the degree of cancellation which can be achieved will either be due to incidental FM to AM conversion (caused by imperfections in the frequency responses of the various components in the RF path) or to the presence of multiple leakage sources. The FM noise cancellation effect counteracts the fact that, for a typical oscillator, the FM noise power is much higher than the AM noise power and yet, because of the cancellation effect, it is less important than the AM noise contribution. This is provided the FM noise has a low enough level and is well correlated.

A corollary, is that target returns having a round trip delay greater than δt (which is always true!), will always be decorrelated with respect to the phase noise. This is the principal source of "noise" we have to contend with for such targets. But targets at short range need to be dealt with as discussed above.

Specifically, if we compare equations (E.18) and (E.13),

$$AM = 4L \cos^2$$

and

$$FM = 4L \sin^2\phi \sin^2\left(\frac{\omega_m \delta t}{2}\right)$$

we can identify

$$C = 4 \sin^2\left(\frac{\omega_m \delta t}{2}\right) \text{ as the differing term.}$$

For small values of $(\omega_m \delta t / 2)$,

$$C = 4\left[\frac{\omega_m \delta t}{2}\right]^2 \tag{E.19}$$

Equation (E.19) is the formula for FM noise cancellation. Using this equation, we can determine the delay δt required to reduce FM noise. This delay translates to a proportionate gap between the transmitter and receiver antennae (in case of a separate antenna system).

4. EXAMPLE NOISE CALCULATION

In the case of the Pandora, with a transmitted power of 5 W, let us assume a value of −160 dBc/Hz as AM noise level and −115 dBc/Hz as FM noise level, both at a carrier offset of 2 MHz. The radar requires maximum sensitivity at the higher end of the sweep corresponding to maximum range. This corresponds to a maximum IF (beat frequency) of 2 MHz. We need to calculate the isolation required when we operate at this maximum beat frequency. The assumed AM and FM noise levels are typical in the X-band (Scheer et al. [2]) and in fact it is even lower in some cases. We take the receiver noise figure as the measured value of 2.25 dB (Appendix I, Experiment 11).

If the variation of transmitter noise with distance from carrier is assumed to be reasonably well behaved, then the critical case will be at 2 MHz IF, corresponding to maximum range, where the receiver sensitivity is most needed. If the AM and FM noise levels are to be allowed to make equal contributions to the overall noise figure, then the FM noise at 2 MHz *must be cancelled by* 45 dB. From equation (E.19), this means that δt must be less than 448 psec. This corresponds to a path length error of nearly 0.13 m in air.

If the AM and FM noise levels are equally significant, then from equations (E.13) and (E.18) above, we can see that the effective noise level will be -154 dBc/Hz $(-160 + (10 \log 4 = 6) = 154$ dBc/Hz), referred to the receiver input. If the transmit/receive isolation is 60 dB, achieved by using two antennae, then the leakage level will be -23 dBm $(37 - 60 = -23$ dBm), so the total leakage power seen in the receiver will be -177 dBm/Hz (for a 5 W power, that is $(37 - 154 - 60) = -177$ dBm). If we correct for receiver noise, this becomes -174.75 dBm/Hz. This is the noise power due to the leakage of the transmitter sidebands. If we add to this, the thermal noise power of -174 dBm/Hz, we obtain a sum of -171 dBm/Hz. This means that the influence of the noise sidebands does affect the sensitivity of the FMCW radar, but it is not too serious. Another aspect is that the leakage power is -23 dBm. This is within the limit of most LNAs from the point of view of linearity of operation. Hence, there is no risk of saturation. Furthermore, even target returns from close-in targets are less than this figure, so FM noise reflected back from close-in targets will also not saturate the LNAs.

During design calculations leading to determination of isolation values, we check for both AM and FM noise.

If we assume the following parameters:

P_t—transmitter power in dBm $= 37$ dBm for 5 W
η_{AM}—oscillator AM noise level in dBc/Hz s.s.b $= -160$ dBc/Hz
η_{FM}—oscillator FM noise level in dBc/Hz s.s.b $= -115$ dBc/Hz
R—transmit–receive leakage in dB
F—receiver noise figure in dB at the input to the receiver $= 2.25$ dB
N_{ther}—Thermal noise $= -174$ dBm/Hz (for 289 K temperature)

Then in the case of AM noise, if we allow AM noise to be 9 dB below the thermal noise level,

$$P_t + \eta_{AM} + R \le -174 + F - 9$$
$$37 - 160 + R \le -174 + 2.25 - 9$$
$$\therefore R \le -37 + 160 - 174 + 2.25 - 9$$
$$\therefore R \le -57.75 \text{ dB}$$

For FM noise a similar inequality must be satisfied, but with an additional term to take account of the FM noise cancellation.

$$P_t + \eta_{AM} + R + C \le -174 + F - 9$$
$$37 - 115 + R - 45 \le -174 + 2.25 - 9$$
$$\therefore R \le -37 + 115 + 45 - 174 + 2.25 - 9$$
$$\therefore R \le -57.75 \text{ dB}$$

Hence, we can see that the isolation required is of the order of 58 dB.

Broadly, an isolation of 60 dB will ensure that AM phase noise does not pose a problem. However, it is a different situation with regard to FM phase noise. To ensure a cancellation of 45 dB, we need to have a transmitter/receiver separation not exceeding 0.13 m. This is not possible in our proposed configuration of two separate parabolic antennas using pencil beams to promote isolation. We can, however, add a suitable delay line in the reference signal path

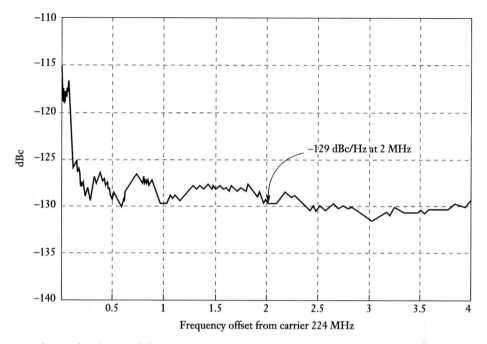

Figure E–1 Phase noise characteristic.

to the stretch processor whose delay corresponds to the antenna separation distance. This will cancel direct FM phase noise pick up. However, close range clutter will create a lot of FM noise in the system and so will close range targets. Clutter interference will, therefore, need to be eliminated in Doppler if required, while close range targets are of no interest to us. However, FM phase noise will be spread throughout the range bins. Ultimately FM phase noise will have to be controlled by using excellent signal sources, since distant targets with weak returns will have to compete with the FM phase noise level in the system to be detected.

In our DDS, for example, the phase noise curve is shown in Figure E–1. This curve is a combination of both FM and AM phase noises.

It can be seen that at 2 MHz offset (our highest beat frequency), the phase noise is -129 dBc/Hz. The individual AM and FM phase noises were not measured, but since FM phase noise is usually 30 dB higher than AM phase noise, it can be seen that -129 dBc/Hz is close to the FM phase noise value and the AM phase noise should in that case be around -160 dBc/Hz which is adequate. This makes our DDS suitable for this radar, from the point of view of phase noise.

References

1. Stove, A. G., "Linear FMCW radar techniques," *IEE Proceedings-F*, Vol. 139, No. 5, October 1992.
2. Scheer, J. A., et al., *Coherent Radar Performance Estimation*, Artech House Inc., Norwood, MA, 1993.

F

Appendix F: Pandora Receiver Channel— Basic Design

1. INTRODUCTION

We shall now discuss the basic design considerations of the Pandora radar receiver system. The discussions pertain to one of the eight receiver tracts and the signal input is assumed to come from one receiver beam.

The simplified block diagram is shown in Figure F–1. The signal input comes from one beam of the receiver antenna to a wideband low noise amplifier. It is thereafter fed to the first mixing stage of the heterodyne mixer M4 where it is downmixed with the carrier of 8.14 GHz. It then acquires a sweep bandwidth pertaining to that particular channel. The signal is now split using an eight-way power divider. It is then fed to the second mixing stage where it is upmixed with the

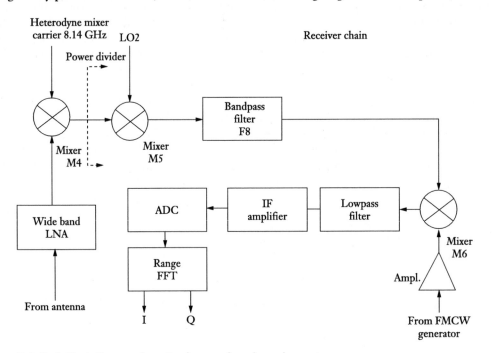

Figure F–1 Basic block diagram of *one* Pandora receiver channel.

basic LO2 signal pertaining to *that* channel. At this stage all channels acquire a standard sweep bandwidth extending from 2,328 to 2,376 MHz. The signal now enters the bandpass filter bank F8. The aim behind this concept is to take advantage of the high frequency of the F8 filters to ensure steep skirts and good adjacent channel rejection. Another advantage is to ensure commonality of filter design, that is, all the filters are identical. The output is then fed to the stretch processor mixer M6, where it is correlated with the reference signal. This output is then low pass filtered (anti-aliasing) and then given to an IF amplifier, followed by an A/D converter. Finally, it is given to a range FFT for spectrum analysis. *The entire electronics up to and including the AD converter is located in the antenna mounting in the interests of noise control.* Only the digitized signal comes out of the pedestal. We shall now discuss each block.

2. WIDEBAND LOW NOISE AMPLIFIER

Receivers generate internal noise that masks weak echoes being received from the radar transmissions. This noise is one of the fundamental limitations on the radar range. The analysis of radar sensitivity is facilitated if the noise contribution of each element of the system is expressed as a noise temperature. For radar receivers [1],

$$T_e = T_0\left(F-1\right)$$

where T_e is the equivalent noise temperature of the receiver (it accounts for the noise power above KTB added by the receiver), T_0 is 290°K, and F is the receiver noise factor.

A direct compromise must be made between the noise temperature and the dynamic range of the receiver. The introduction of an RF amplifier in front of the mixer necessarily involves raising the system noise level at the mixer to make the noise contribution of the mixer itself insignificant. Even if the RF amplifier itself has more than adequate dynamic range, the mixer dynamic range gets compromised.

We shall examine the various terms in receiver [2] with respect to the diagram at Figure F–2.

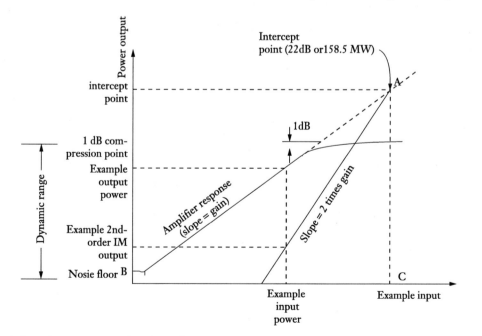

Figure F–2 Basic concepts: Receiver dynamic range.

Dynamic range is defined as the ratio of the largest signal a system can handle to the smallest signal that same system can handle. A measure of the dynamic range of the receiver is its 1 dB compression point, which is the output power at which the gain of the receiver is reduced by 1 dB due to saturation. In Figure F–2, the slope of the power-out/power-in curve is the receiver's gain and the 1 dB compression point is shown. We should not operate at this point due to excessive distortion that results, but in the linear region. However, the 1 dB compression point is a useful measure for comparing receivers. If the input signal power is excessive, the receiver operates in the nonlinear region. This gives rise to intermodulation distortion, which creates intermodulation products. The frequencies of the intermodulation products are

$$f_{out} = mf_1 + nf_2$$

where f_{out} is an output frequency, f_1 is one of the input frequencies, f_2, is the other input frequency, and m and n are integers (positive or negative).

Normally in radars, narrow bandwidth bandpass filters are involved, wherein only third-order and higher-order intermodulation products exist within the passband and appear as interfering signals.

$$\text{IM order} = |m| + |n|$$

However, in the Pandora, even second-order IM products appear in the passband of the higher filters. Hence, we have no room for allowing any IM distortion in the system.

The amplitude of the intermodulation products can be found from the intercept point specification of the system. The intercept point is plotted on the receiver's gain curve. From the intercept point, a line is drawn having a slope equal to the intermodulation order times the linear gain of the receiver. For example, in our case, since we are interested in the amplitude of second-order IM products, a line having a slope equal to twice the linear gain is drawn from the intercept point; for third-order products the slope is three times the linear gain and so on. For a given input power, the linear gain line gives the output power at the input frequency; the twice slope line gives the power out in second-order IM products.

We shall now examine the diagram at Figure F–2 with respect to a commercial LNA. We choose model number AMP-15 from "Mini-Circuits," Brooklyn, New York. The amplifier has the following characteristics:

Frequency range: 5–1,000 MHz
Noise figure: 2.8 dB, Gain: 13 dB
Output at 1 dB compression point: 8 dBm
Intercept point: 22 dBm
VSWR: 2:1 at both input and output

We shall now plot the parameters pertaining to triangle ABC in Figure F–2. This is shown in Figure F–3.
We have converted all parameters to absolute units. Hence,

Intercept point = 22 dBm = 158.5 MW
Output power = 8 dBm = 6.31 MW
Input power (output power in dBm – gain) = –5 dBm = 0.32 MW

$$\text{Slope} = \text{gain} = \frac{6.31}{0.32} = 20 = 13 \text{ dB}$$

$$\therefore \theta_{i_1} = 87°$$

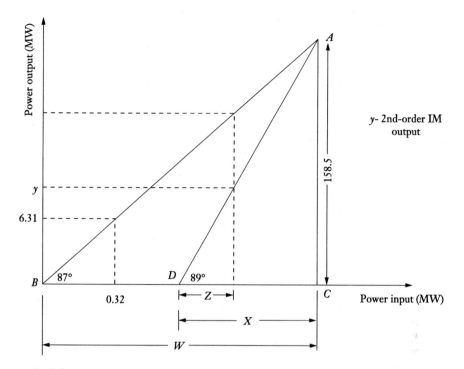

Figure F–3 Calculation of second-order *IM* output.

Second-order IM products
Slope = 2 × gain = 40

$$\therefore \theta_{i_2} = 89°$$

Intercept on x-axis = X

$$\therefore X = \frac{158.5}{\tan 89°} = 2.77 \text{ MW}$$

Now,

$$Y = \frac{158.5}{\tan 87°} = 8.31 \text{ MW}$$

$\therefore$ Intercept on x-axis = $8.31 - X = 8.31 - 2.77 = 5.54$ MW
$\therefore$ If signal at input crosses 5.54 MW, we obtain second-order IM distortion products.
  Suppose input signal is 6 MW.
  Then second-order IM products will have power given by

$$\text{Tan } 89° = \frac{Y}{Z} = \frac{Y}{0.46}$$

where $Z = 6 - 5.54 = 0.46$ MW

$$\therefore Y = \text{output power of second-order IM products}$$
$$= 0.46 \times \tan 89° = 26.4 \text{ MW} = 14 \text{ dBm}$$

The amplifier is capable of sustaining an input signal of as high as 13 dBm without any damage.

Therefore, the inference here for Pandora is that we need to select an LNA which has as high an intercept point as possible. This yields very low IM products. For example, if (as a hypothetical case) we had an intercept point of say, 25 dBm. In such a case, Intercept point = 25 dBm = 316.2 MW

$$\therefore X = \frac{316.2}{\tan 89°} = 5.52 \text{ MW}$$

$$Y = \frac{316.2}{\tan 87°} = 16.6 \text{ MW}$$

$$\therefore \text{ Intercept on } x\text{-axis} = 16.6 - 5.52 = 11.05 \text{ MW}$$

This is higher than the 5.54 MW calculated earlier. Hence, in this case in order to generate second-order and higher IM products, the input signal needs to cross 11.05 MW. Therefore, we have more safety. *This is the only option for us, as we cannot do anything to counter IM products existing in higher filters.*

The LNA needs to configured in a cascade of multiple stages, yielding a dynamic range of 72 dB. This is a typical value usually required between the weakest signal and near range clutter return. The response of the LNA should be flat over the bandwidth.

We have already considered the heterodyne mixer, the stretch processor mixer, and the baseband filter elsewhere in this report. Hence, we shall now study the design requirement of the IF amplifier and AD converter.

3. IF AMPLIFIER

We choose a typical model marketed by the same firm "Mini-Circuits" viz., ZFL-1000 GH. The specifications are given below:

Frequency range: 10–1,200 MHz
Noise figure: 15 dB, Gain: 24 dB
Output at 1 dB compression point: 13 dBm
Intercept point: 25 dBm
VSWR: 2:1 at both input and output

The same arguments discussed for LNA hold here except for the frequency response of the amplifier. The frequency response of the amplifier should have a rising characteristic like 6 dB/octave or 12 dB/octave so that high frequency beat signals coming from the maximum ranges should be amplified more than low frequency beat signals coming from near ranges. Another important aspect here is that this amplifier has a variable gain for AGC. The amplifier has a response time of 25 μsec for a change between 10 and 90%. The control range is 30 dB and the control voltage ranges from 0 to 5 V. The AGC might not be necessary as weak signals from farther ranges come in at higher beat frequencies. Hence, they will undergo higher amplification due to the IF amplifier's rising characteristic. This has been the experience with Philips [3] FMCW radar for navigation. However, it is useful to cater for the contingency of a strong transient signal.

4. ANALOG TO DIGITAL CONVERTER

Since our sampling rate is only 4 MHz there are many converters available that are suitable for our requirements. A suitable choice is ADC 614 marketed by Burr-Brown. The chip is an integrated chip complete with an ADC, sample/hold amplifier, voltage reference, timing and error correction circuitry in a 46-pin hybrid DIP package. Briefly, the essential parameters are as follows:

Number of bits: 14
Dynamic range: 90 dB
Sample rate: 5.12 MHz
SNR: 78 dB
Sample/hold bandwidth: 60 MHz
Jitter: 9 psec rms
Overload recovery time: 205 nsec

References

1. Edde, B., *Radar, Principles, Technology, Applications*, PTR Prentice-Hall, Englewood Cliffs, NJ, 1993.
2. Nathanson, F. E., *Radar Design Principles*, McGraw-Hill, New York, 1991.
3. Barret, M., et al., "An X-band FMCW navigation radar," *IEE-87 Radar Conference*, 1987.

G

Appendix G: Direct Digital Synthesis

1. INTRODUCTION

The discussion on DDS in this report is based on the paper by Kroupa [1] and company manuals from Qualcomm and Stanford Telecom [2–4] and reproduced here with permission.

2. DIRECT DIGITAL SYNTHESIS (DDS)

Direct digital synthesis (DDS) can be practically defined as a means of generating highly accurate and harmonically pure digital representations of signals. This digital representation is then reconstructed with a high-speed digital-to-analog (DAC) converter to provide an analog output signal, typically a sinusoidal tone or sequence of tones. The quality of the DAC plays an important role on the level of spurious signals generated by the DDS. DDS techniques offer unique capabilities in contrast to other synthesis methods. Although limited by the Nyquist criterion (up-to 1/2 the frequency of the applied clock reference), DDS allows frequency resolution control on the order of milli-hertz step size and can likewise allow milli-hertz or even nano-hertz of phase resolution control. Additionally, DDS imposes no settling time constraint for frequency changes other than what is required for digital control. This results in extremely fast switching speeds, on the order on nanoseconds. All frequency changes are completed in a phase continuous fashion, that is, a change to a new frequency continues in-phase from the last point in the previous frequency. Since the signal is being generated in the digital domain, it can be manipulated with exceptional accuracy. This allows precise control of frequency or phase and can readily accommodate frequency and phase modulation, that is, FSK or PSK.

A DDS works on the principle that a digitized waveform of a given frequency can be generated by accumulating phase changes at a higher frequency. Sampling theory requires that the generated frequency be no more than 1/2 the clock frequency. In practice, this is limited to 40% of the clock frequency. Figure G–1 shows the phase accumulation of a generated sine wave whose frequency is equal to 1/8 of the clock frequency.

The circle shows the phase accumulation process of $\pi/4$ at each clock cycle. The dots on the circle represent the phase value at a given time and the sine wave shows the corresponding amplitude representation. This phase to amplitude conversion occurs in the sine lookup. Note that the phase increment added during each clock period is $\pi/4$ radians, which is 1/8 of 2π. The phase value stored in an input frequency register is added to the value in the phase accumulator once during each period of the system clock. The resulting phase value (from 0 to 2π) is then applied to the sine lookup once during each clock cycle. The lookup converts the phase information to its corresponding sine amplitude as shown in Figure G–1. The digital word is then output from the

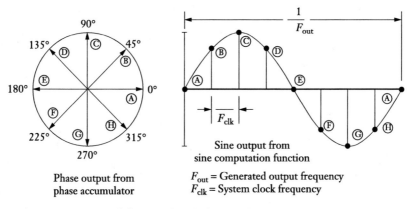

Figure G–1 Sample sine wave. (From [2], © Reprinted with permission)

DDS to the DAC. The phase increment value is given by

$$\Delta\phi = \frac{2^N \times F_{OUT}}{F_{CLK}}$$ where N is the number of bits in the phase accumulator

For example, for a required frequency of 3.75 MHz (1/8 of clock frequency as in Figure G–1 above) and clock frequency of 30 MHz with a 32 bit phase accumulator,

$$\Delta\phi = \frac{2^{32} \times 3.75 \times 10^6}{30 \times 10^6} = 2^{29} = 5386870912 \tag{G.1}$$

Strictly speaking $\Delta\phi$ is an increment value but *not* the phase increment value. The $\Delta\phi$ value is a mere number and by itself has no meaning. We need to examine its value in the context of the largest number in the phase accumulator, which is 2^{32} and represents 2π. In relation to this number, the value of $\Delta\phi$ is *equivalent* to a phase increment of $\pi/4$ radians or 1/8 of 2π as required in Figure G–1 above.

Conversely, if we adjust the clock frequency to a suitable value, we can generate exact decimal frequencies. For example, if clock frequency is 2^{25} Hz, the frequency resolution is given by

$$\text{Frequency resolution} = \frac{F_{CLK}}{2^N} = \frac{225}{232} = \frac{1}{27} \text{ Hz}$$

This means that if,

$$\Delta\phi = 2^7, F_{OUT} = 1 \text{ Hz}$$
$$\Delta\phi = 2^8, F_{OUT} = 2 \text{ Hz}$$
$$\Delta\phi = 2^9, F_{OUT} = 4 \text{ Hz}$$

and so on.

The block diagram of a typical DDS is shown in Figure G–2.

In Figure G–2, the frequency control word k constitutes the phase increment $\Delta\phi$. It is given to an adder where modulus-2 addition is carried out on each clock cycle. Then $f_X = (X/Y) f_{clock}$. Here $k = X$ in the expression. Henceforth, we shall use X as the phase increment required for a given frequency f_X as discussed above. Frequency f_X will be called generated frequency or carrier.

2.1 SPECTRAL PURITY OF DDS

The spectral quality of a DDS system is dependent upon a number of factors including the phase noise of the clock source, the number of phase bits applied to the sine lookup function (i.e., phase truncation) and the number of bits output from the lookup (i.e., amplitude truncation).

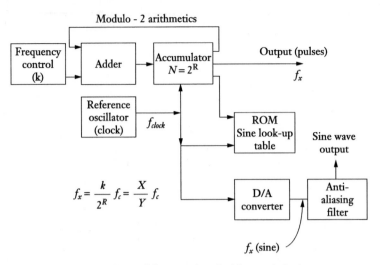

Figure G–2 Typical DDS block diagram. (From [2], © Reprinted with permission)

The specification of the DAC, that is, its resolution, LPF design, that is, how many harmonics are let through and circuit card design also affect the quality of the converted sine wave. The linearity and glitch energy specifications of the DAC are especially important for spectral purity of the sine wave. We shall examine each of these factors in the succeeding paragraphs.

2.2 CLOCK SOURCE.

The phase noise of the DDS output will show an improvement over the phase noise of the clock source, as per the relationship $20\text{Log}(F_{CLK}/F_{OUT})$. The frequency accuracy of the clock is also propagated through the DDS. *Therefore, an accurate clock source with low phase noise is essential.*

2.3 PHASE TRUNCATION

Suppose the phase accumulator has 32 bits, phase truncation implies selecting only a portion of it, usually the most significant bits (MSB). Truncation is essential, as otherwise the ROM lookup tables will become enormous if all 32 bits are used as address. The truncation of the least significant bits (LSB) is a loss of phase information and contributes errors. If R is the number of bits in the phase accumulator and W is the number of bits truncated, then $B = R - W$ bits introduces a phase modulation. This is given by [1]

$$\sin\left[2\pi \frac{2^B}{Y}\text{int}\left(\frac{mX}{2^B}\right)\right] = \sin\left[2\pi \frac{mX}{Y} - \varphi(m)\right]$$

$$= \sin\left[2\pi \frac{mX}{Y}\right] - \varphi(m)\cos\left[2\pi \frac{mX}{Y}\right]$$

(G.2)

where m is an integer and $\varphi(m) = 2\pi \dfrac{2^B}{Y} s(m) = 2\pi \times 2^{-W} s(m) << 1$

where $s(m) = \dfrac{mX}{2^B} - \text{int}\left(\dfrac{mX}{2^B}\right) = m\dfrac{P}{2^B} - \text{int}\left(\dfrac{mP}{2^B}\right)$

where P is an integer smaller than 2^B. This is the integer that determines the position of the spur due to phase truncation on the spectrum.

In the above equation, $Y = 2^R$ is the largest number in the accumulator and X is the phase increment that gives rise to the desired frequency as in Figure G–1. The integer P is the *residual* number which remains in the truncated register B and determines the frequency of phase modulation. If we can *a priori* determine the value of P, we can predict the position of the spur caused due to phase truncation. The phase modulation index is, however, small, that is, it is not so severe. Hence, it is generally much below carrier. The expression for $\varphi(m)$ indicates that it is a periodic function repeating after each 2^B clock periods with the first term on the right-hand side being a sawtooth wave and the second one a mere integer. We can, therefore, expand $\varphi(m)$ as

$$\varphi(m) = 2 \times 2^{-W} \sum_{r=1}^{2^B-1} \frac{1}{r} \sin\left(2\pi m \frac{rP}{2^B}\right) \tag{G.3}$$

If we substitute equation (D.3) in equation (D.2), then the first term represents the carrier and the second term the spurious sidebands of the rth harmonic. This expression for spurious sideband can be rearranged and is given by

$$\frac{2^B}{rY}\left\{\sin\left[2\pi m\left(\frac{rP}{Q} + \frac{X}{Y}\right)\right] + \sin\left[2\pi m\left(\frac{rP}{Q} - \frac{X}{Y}\right)\right]\right\} \tag{G.4}$$

where $Q = 2^B$.

The expression in equation (G.3) shows that the waveform comprises staircase values of spurious sine waves around the carrier. The waveform shown in Figure 8–9 of the main report, for multi-bit generation, corresponds to equation (G.2). A plot of equations (G.2) and (G.4) is shown in Figure G–3.

Note the relatively small amplitude of the rth harmonic ($r = 1$) as given by equation (G.4). This is a sine wave signal. We can determine the level of the spurious signal from the amplitude of equation (G.4)

$$SP_{phase} = 20\log\left(\frac{2^B}{rY}\right) = 20\log\left(\frac{2^{R-W}}{r2^R}\right) = 20\log\left(\frac{2^{-W}}{r}\right) \pm 3 \text{ dB} \tag{G.5}$$

In equation (G.5), there is an error margin of 3 dB. This is the level of the spurious signal caused due to phase truncation, W. An inspection of equation (G.4) will indicate that the spurious harmonic is located $rPY/2^B$ away from the carrier frequency (note that the generated frequency f_x is given by $f_x = (X/Y)f_c$, where f_c is the clock and X is a constant phase increment ($\Delta\phi$) required for this frequency f_x), that is

$$n_r = \left|X + \left(r\frac{PY}{2^B} + sY\right)\right| < \frac{Y}{2}, \text{ where } (s,r = ..., -2, -1, 0, 1, 2, ...) \tag{G.6}$$

s is the number of the harmonic which exceeds the passband of the LPF of $Y/2$ and are folded back (aliased). Inspection of equation (G.4) suggests that for $|r|$ varying from 1 to $2^B/2$ while simultaneously P varies from $2^B/2$ to 1, the inequality is satisfied (i.e., if $r = 1$, $P = 2^B/2$, the highest

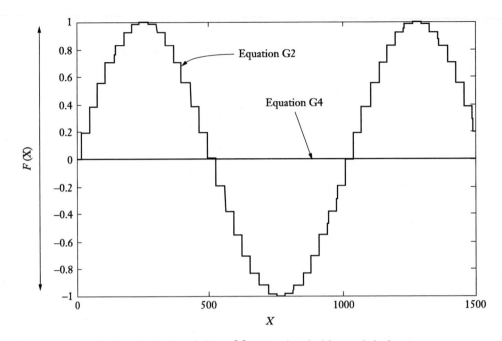

Figure G–3 Equations (G.2) and (G.4) plotted. (From [2], © Reprinted with permission)

spur closest to the clock). This means that for r varying in the range 1 to $2^B/2$, the spurious signal level can vary from (based on equation (G.5)),

$$-6W \text{ till } -6(W+B-1) = -6(R-1)\,\text{dB} \tag{G.7}$$

Equation (G.7) tells us the maximum level of variation of the signal caused due to phase truncation. The location of the phase truncation spur on the spectrum is given by (using equation (G.6))

$$f_p = \left| f_x + \left(r\frac{P}{2^B} + s \right) f_c \right| \tag{G.8}$$

The problem with equation (G.8) is that we cannot know the frequency f_{sp} *a priori*, because we do not know the value of P *a priori*, as it is also corrupted by noise. In the absence of noise, P is just a number, some power of 2, which gives rise to this spur. However, since the harmonic is a periodic signal, the value of P is a *constant increment* and the spurious signal consequently builds up that position on the spectrum. This aspect is discussed in greater detail with respect to Figure G–10. Finally, using equation (G.5) for a 32 bit phase accumulator ($R = 32$), with an 18 bit truncation ($W = 14$, $B = 32 - 14 = 18$) we note that the largest spurious signal caused due to phase truncation is -84 dB. This is a typical figure for most DDS having 32 bit accumulators. This is, however, not reflected at the output, because the quality of the DAC is the inhibiting factor, as we shall see.

2.4 AMPLITUDE TRUNCATION

Amplitude truncation is carried out at the sine lookup table in order to match the resolution of the DAC. The reduction is to S bits, where S is usually less than W. This introduces an error

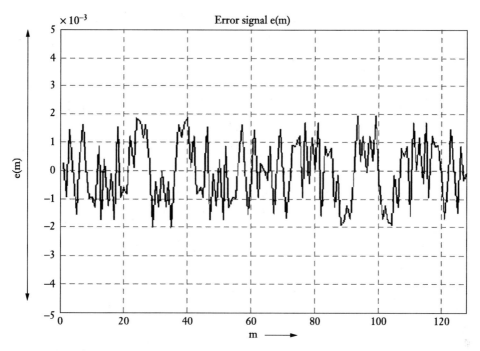

Figure G–4 Error signal $e(m)$.

signal [1]

$$e(m) = \sin\left(2\pi m \frac{X}{Y}\right) - \frac{1}{2^s}\operatorname{int}\left(2^s \sin\left[2\pi \frac{2^B}{Y}\operatorname{int}\left(\frac{mX}{2^B}\right)\right]\right) \tag{G.9}$$

where m is an integer.

Note that the sine term on the right-hand side of equation (G.9) is the result of the phase truncation given by equation (G.2). From equation (G.9), we find that

$$|e(m)| < \left|\frac{1}{2^s}\right| \tag{G.10}$$

A plot of equation (G.9) is shown in Figure G–4. This is for $X = 207$, $Y = 1024$, $S = 11$, $B = 0$. The DAC is 8 bits ($A = 8$), discussed below.

Inspection of Figure G–4 confirms the validity of equation (G.10). According to Kroupa, computer simulations have revealed that the spurious signal due to phase truncation will not exceed that due to amplitude truncation if,

$$W - 3 \le S \le W - 2 \tag{G.11}$$

2.5 DIGITAL-TO-ANALOG CONVERTERS

Digital-to-analog converters (DACs) cause degradation of the output spectral purity by finite bit resolution, nonlinearity, and transient effects or glitches. Present day technology is confined to mostly 8 to 12 bit DACs. If a DAC has D bits for generation of a sine wave, then only A bits are used, where $A = D - 1$ bits. If $S > A$, we need to replace S in the preceding equations with A, as A is the deciding factor for spurious levels. In fact, this is what has been done to obtain the graph in Figure G–4, where we have taken $A = 8$, in lieu of S.

By assuming an ideal DAC with a resolution of D bits in the range of V_{pp} (peak-to-peak) volts, the LSB voltage difference is

$$V_{LSB} = \frac{V_{pp}}{2^D} \tag{G.12}$$

and will introduce a random voltage error which may occupy the range from $-(1/2)$ LSB to $+(1/2)$ LSB. Consequently, its variance will be (assuming uniform distribution)

$$\sigma^2_{DAC} = \frac{1}{12}(V_{LSB})^2 = \frac{1}{12}\left(\frac{V_{pp}}{2^D}\right)^2 \tag{G.13}$$

and the ideal signal-to-noise ratio will be

$$\frac{S}{N} = \frac{\frac{1}{2}\left(\frac{V_{pp}}{2}\right)^2}{\frac{1}{12}\left(\frac{V_{pp}}{2^D}\right)^2} = \frac{3}{2}2^{2D} \tag{G.14}$$

or in dB,

$$\frac{S}{N} = 1.76 + 6.02D \text{ dB} \tag{G.15}$$

Due to the manufacturing processes in DACs, there exists an additional nonlinearity between the input and output of a DAC. By assuming an error increase due to this, by ± 1 LSB, the SNR ratio would be reduced by 4 dB.

Finally, we examine the transient effects or glitches. Glitches occur during major code transitions, for example, from 0111 to 1000. A typical glitch is shown in Figure G–5. Glitches are usually expressed as voltage–time area $Vg\tau$. If we assume uniform distribution, as glitches are encountered at every DAC switching, we obtain

$$V_g\tau \approx \frac{V_{pp}}{2^D}T_c, \text{ where } T_c = \frac{1}{f_{clock}} \tag{G.16}$$

Usually, SNR decreases by 3 dB due to glitches. Application of sample-and-hold circuits at DAC output reduces the glitch problem considerably.

Figure G–6 gives the overall picture of the source of spurious signals and the equations governing them.

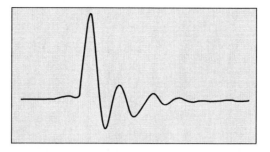

Figure G–5 Glitch in a system.

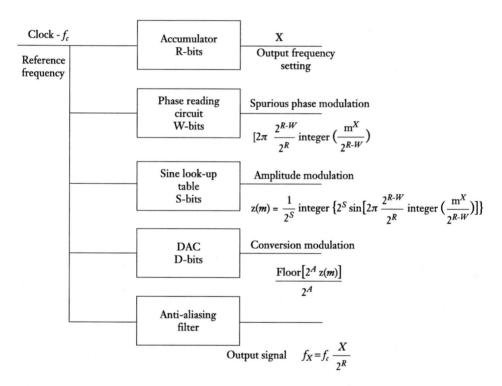

Figure G–6 Overall picture of spurious signals.

2.6 LOW PHASE ANTI-ALIASING FILTER

The low pass filter is designed to cut off all frequencies in excess of $f_{clock}/2$. This is required because the nonlinearities in the DAC give rise to IM products (image frequencies) of the clock and the generated frequencies. The effect of DAC nonlinearities is shown in Figure G–7.

In some cases, the LPF is replaced with a BPF for filtering higher frequencies in the ranges from $f_c/2$ to f_c, from f_c to $1.5 f_c$, and so forth.

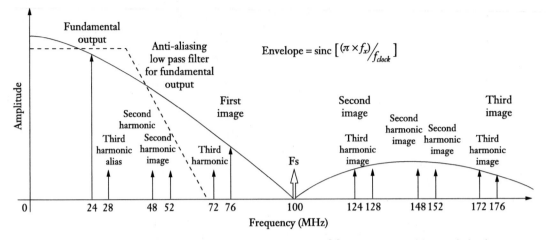

Figure G–7 DDS output showing effect of DAC nonlinearities. (From [2], © Reprinted with permission.)

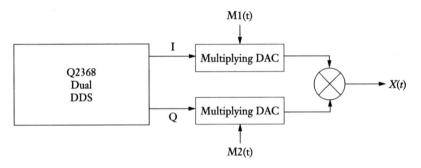

Figure G–8 QAM modulator.

2.7 BACKGROUND PHASE NOISE

The background phase noise due to the clock signal is reduced by the division ratio of $\frac{X}{Y} = \xi_X$. This noise is added to the noise from the DDS electronics. Kroupa [1] gives the estimate of this phase noise as

$$S_{DDS}(f) = S_{clock}(f) \times \xi_X^2 + \frac{10^{-10\pm2}}{f} + \frac{2^{-2D}}{f_c} \tag{G.17}$$

where f is the specific generated frequency.

In the preceding paragraphs, we have analyzed the sources of spurious noise in DDS systems. Certain manufacturers like Qualcomm and Stanford Telecom have developed patented algorithms to further reduce the spurs caused due to phase and amplitude truncation. In fact, the above estimates are baseline values. Qualcomm have claimed that their noise reduction algorithms eliminate the disadvantages of using an 8 bit DAC! Generally, DDS units are marketed in dual configuration, that is, there are two DDS chips in one board. The idea behind this is to carry out quadrature modulation like QAM, and so forth. A configuration for QAM modulation is shown in Figure G–8 and demodulation in Figure G–9. This schematic uses a Qualcomm Q2368 chip.

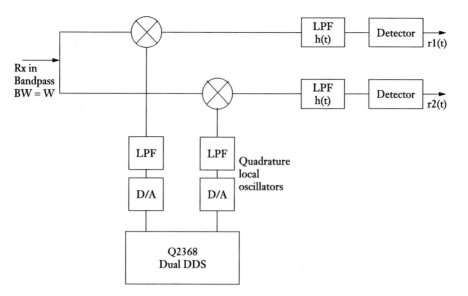

Figure G–9 QAM demodulator.

The advantage of using a DDS becomes apparent in Figure G–9, because we need not have an elaborate circuit to extract the carrier for QAM demodulation. Since the source is digital, we can generate carrier frequencies synchronous with that of the transmitter. It is for similar reasons that in radars, if we use a DDS source, the receiver need not be close to the transmitter, as the receiver mixers are driven by a DDS source which guarantees a local oscillator signal which is phase synchronous with that of the transmitter.

We shall now examine two different DDS chips now in the market, with a view to consolidating our ideas.

3. QUALCOMM Q2334 DUAL DIRECT DIGITAL SYNTHESIZER

The following are the salient features of this chip:

- Two complete DDS functions on one chip
- Processor interface for control of phase and frequency
- Patented algorithmic sine look-up function
- Patented noise reduction circuit
- Synchronous PSK and FSK modulation inputs
- Three maximum clock speed versions, 20, 30, and 50 MHz (This means that the highest generated frequency will be 20 MHz from 50 MHz clock)
- Can be paralleled for higher clock rate operation
- Frequency resolution: 0.005 Hz at 20 MHz clock rate

The specified features require no further discussion except for the following.

Patented algorithmic sine look-up function: The Q2334 DDS implements a patented technique to generate a sine wave lookup. This algorithm takes the 16 MSB from the phase accumulator to generate a 12 bit sine wave value (a maximum of 12 bit DAC can be used with this chip). Using this high precision lookup function, the phase truncation noise of the sine wave output is kept

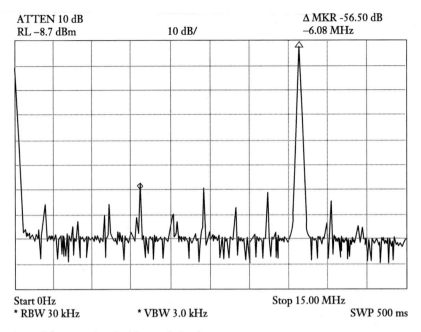

ATTEN 10 dB   Δ MKR -56.50 dB
RL −8.7 dBm   10 dB/   −6.08 MHz

Start 0Hz   Stop 15.00 MHz
* RBW 30 kHz   * VBW 3.0 kHz   SWP 500 ms

Figure G–10 (From [3], © Reprinted with permission.)

below 76 dB. This technique differs considerably from the traditional method of using an ROM lookup function. The firm claims that this patented technique of theirs, provides highly accurate and precise sine wave generation.

Patented noise reduction circuit: Noise due to amplitude quantization is often assumed to be random and uniformly distributed. In reality, this is not the case as the sine wave function is periodic. At certain output frequencies, amplitude quantization errors become highly correlated, causing spurs. These spurs can be reduced by enabling the noise reduction circuit (NRC). The NRC distributes the noise energy evenly across the frequency band, thus reducing the amplitudes of peak spurious components.

The firm has claimed that this circuit drastically improves the performance of 8 bit DACs making it unnecessary to go in for higher DACs. However, since the maximum generated frequency is 20 MHz using a 50 MHz clock, 12 bit DACs are available for such frequencies. In Figure G–10, we see the effect of a synthesized frequency of 10.8 MHz with 30 MHz clock and NRC disabled. In Figure G–11, the NRC is enabled. The improvement is evident.

In Figure G–10, we can see a large number of spurs, but they are all better than 56 dB down. This is further improved in Figure G–11, using NRC to better than 64 dB down. In Figure G–10, we cannot have any control over the number of spurs or their location in relation to the carrier signal, because we cannot control the value of P in equation (G.4) to relocate it in the spectrum well away from the carrier. The implication here is that though P is theoretically a constant, which can be precalculated, in reality it is also corrupted by noise. Therefore, instead generating a pure sine wave like in equation (G.4), it also produces noise spurs. By noise spurs, we mean spurs other than the actual noise spur generated due to the value of P. In Figure G–10, the marker is located on the actual spur while the rest are noise spurs. Even though P can be precalculated, we cannot

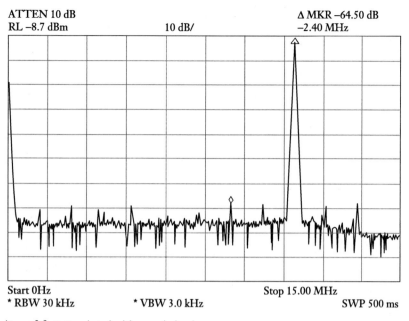

ATTEN 10 dB   Δ MKR −64.50 dB
RL −8.7 dBm   10 dB/   −2.40 MHz

Start 0Hz   Stop 15.00 MHz
* RBW 30 kHz   * VBW 3.0 kHz   SWP 500 ms

Figure G–11 (From [3], © Reprinted with permission.)

do anything about it, as it is a result of bit truncation. Hence, relocation of the spur in relation to the carrier is not possible. Therefore, NRC seems to be the only way out, to at least control the level. Using equation (G.5) wherein we substitute D for W, we can calculate the level of spurs to be expected for a 10 bit DAC. It works out to −60 dB ± 3 dB ($r = 1$) exactly as in Figure G–10 (without NRC). Figure G–12 shows the typical performance of the Q2334 DDS when operating with a 10 bit DAC with NRC disabled and no LPF. This figure shows a 5 MHz output generated from a 20 MHz system clock frequency and the image at 15 MHz. This 15 MHz spur results from the two image frequencies, 15 MHz and 25 MHz folded around the 20 MHz clock frequency. This image would normally be filtered by an LPF at the output of the DAC.

Parallel mode for higher clock rate operation: In this mode, we load the phase accumulators of both the DDSs, so that one takes over when the other has finished. This means that we can generate a maximum frequency of 40 MHz with a 50 MHz clock. This is effectively doubling the clock rate. But the price is that *effectively* there is only one DDS!

Finally, we examine the basic block diagram of the Q2334 DDS in Figure G–13.

We shall not discuss this figure in detail regarding each input/output signal. The reader is advised to consult the manual in this regard [2]. There are distinctly two DDSs. The phase accumulator of each is controlled by the system clock. The external multiplex control is required for FSK modulation wherein phase increment register A determines one frequency and register B determines the other. If extremely fine phase control is desired, we load the basic frequency increment in register A and the fine phase increment in the most significant *byte* of register B. We then add the contents of the most significant *bytes* of A and B. We then append the LSB of A (24 bits) to make a 32 bit word. This is then given to a phase accumulator. Since the fine phase increments are defined by the MSB of register B (8 bits), the overall phase increment in the accumulator is (360°/256) = 1.4°. This is the maximum definition of phase increments in Q2334 DDS.

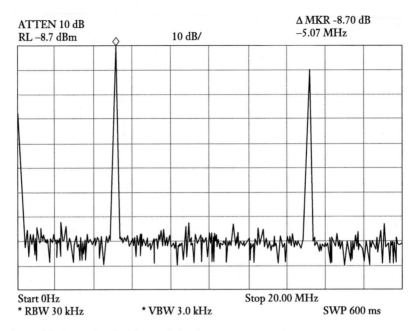

Figure G–12 (From [3], © Reprinted with permission.)

4. STEL 9949 DIRECT DIGITAL CHIRP SYNTHESIZER

The following are the principal features of this DDS marketed by Stanford Telecom:

- 1 GHz clock rate.
- 0–400 MHz output range.
- Generates chirp and CW signals.
- There is no phase control using a processor like in Q2334.
- Chirp duration 1.65 sec max.
- 32 bit frequency resolution (0.23; Hz steps; Compare this to 0.005 Hz in Q2334).

The DDS itself is shown in Figure G–14.

The spurious response of this DDS is shown in Figure G–15.

Figure G–16 is self-explanatory. The auxiliary DDS operates at 1/8 of the basic clock frequency and is intended as a reference. This means that its upper frequency limit is 1/8 of the limit of the main DDS, that is, 50 MHz. This is intended as a reference. The spurious content in Figure G–15 is −50 dBc for a frequency range of 120–320 MHz. Compare this to the figures of −60 dBc using a 10 bit DAC in Q2334. The problem with STEL 9949 is that since the upper frequency limit is 400 MHz, the firm has used an 8 bit DAC. Presently, 12 bit DACs are not available at these frequencies. Using (G.5), with $A = 8$, the spur level works out to −46 dBc ± 3 dB ($r = 1$). This is in accordance with Figure G–15. The firm has not mentioned using an NRC like in Q2334. Hence, the baseline values for the spurs are high. In Figure G–17, we see the spurs obtained for a carrier CW signal of 224 MHz. The spurs are at −63 dBc which is better than the specification and is as good as that obtained from a 10 bit DAC. This appears to be a matter of chance rather than design.

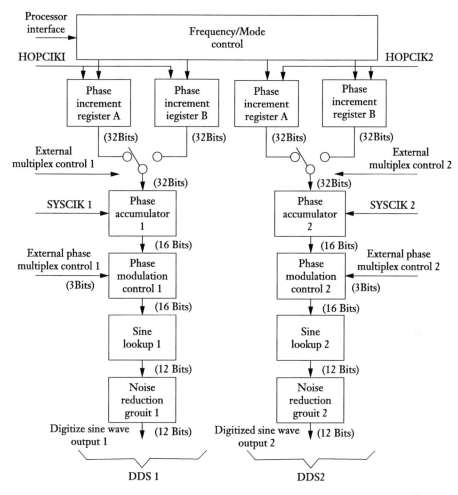

Figure G–13 Q2334 Dual DDS block diagram. (From [3], © Reprinted with permission.)

Figure G–14 STEL 9949 direct digital chirp synthesizer. (From [4], © Reprinted with permission.)

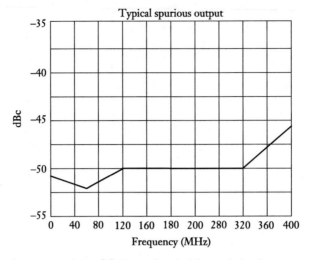

Figure G–15 Typical spurious output. (From [4], © Reprinted with permission.)

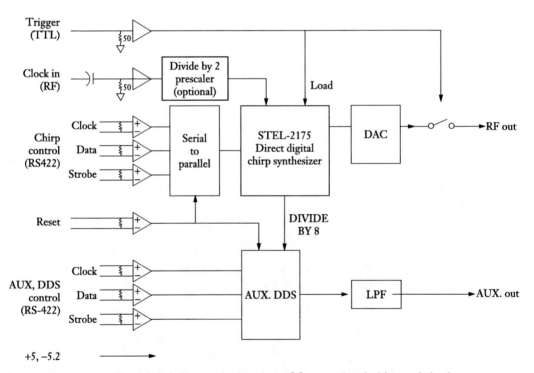

Figure G–16 STEL 2715 direct digital chirp synthesizer. (From [4], © Reprinted with permission.)

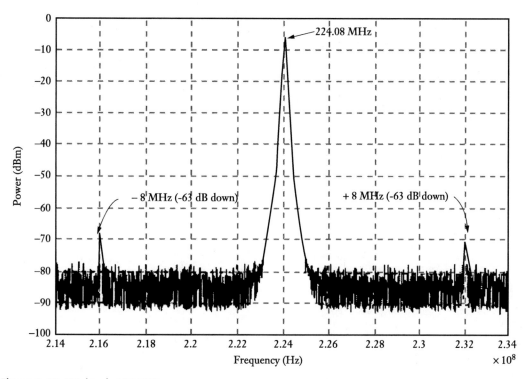

Figure G–17 CW signal at 224 MHz.

Finally, in order to see the internal workings of the DDS itself, let us examine the STEL2375-A DDS chip in Figure G–18.

In Figure G–18, the frequency input latch plays the role of register A in Q2334 chip as discussed above. The delta frequency input latch plays the role of register B for fine phase increments. The only difference is that the user has no control over the delta increments, as the chip has no phase control interface except for setting the initial phase. This is unlike the arrangement in the Q2334 chip which can, therefore, be used for nonlinear modulations. The 32 bit word is then formed in the 32 bit digital frequency accumulator just like in Q2334. This is then given as one incremental word to the phase accumulator. A 10 bit phase truncation is carried out ($W = 10$). The phase modulation block is intended for the initial phase setting if desired and of course, phase modulation. These 10 bits are then used for sine lookup. The S value in this case is 18 bits. This does not satisfy equation (G.11), but this equation applies to ROM-based sine lookup. In this case, the firm is using a patented sine function instead of an ROM. Hence, theoretically the chip can drive a 16 bit DAC. The DAC used is 8 bits. Hence, the entire fine signal processing before the DAC is wasted! This is one of the enduring problems with DDSs.

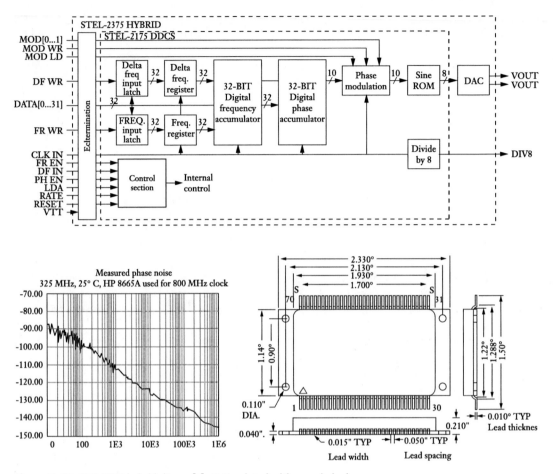

Figure G–18 STEL 2375 hybrid. (From [4], © Reprinted with permission)

References

1. Kroupa, V., "Direct digital frequency synthesizers (DDS) in microwaves," *Microwave Engineering Europe*, December 1998/January 1999, pp. 23–33.
2. Qualcomm, *User's Guide for Q2334 Direct Digital Synthesizer Evaluation Kit (Part Number Q0310-1)*, VLSI Products Division, 10555 Sorrento Valley Road, San Diego, CA, USA.
3. Qualcomm, *Dual Direct Digital Synthesizer, Technical Data Sheet*, VLSI Products Division, 10555 Sorrento Valley Road, San Diego, CA, USA.
4. Stanford Telecom, *STEL-9949 Direct Digital Chirp Synthesizer*, 59 Technology Drive, Lowell, MA, USA.

Appendix H: Implementation of the Single-Channel Radar

1.1 INTRODCTION

This appendix contains the drawings leading to the implementation of the single-channel radar.

1.2 UPCONVERTER

Note the correction filter inserted at the input to the M2 mixer. This is the only correction filter at the transmitter end. The other correction filter is located at the receiver end, viz., at the output of the F8 filter and just before the M6 mixer (stretch processor). This correction filter caters to the distortions caused by filters F3 through F8 (see Figures H–1 and H–2).

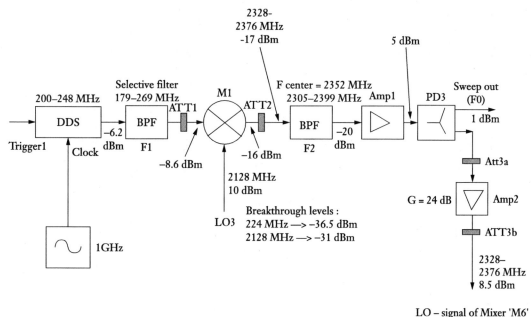

Figure H–1 Digital sweep generator (DSG).

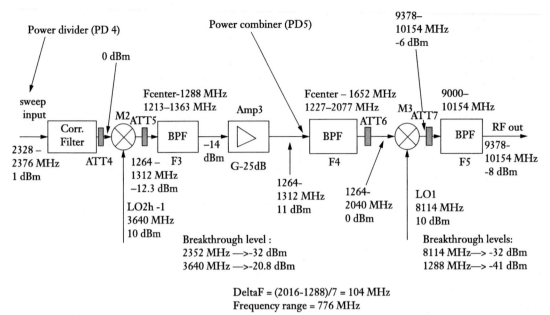

Figure H–2 Sweep upconverter.

1.3 DOWNCONVERTER

A correction filter (not shown) is to be inserted at the output of the F8 filter. These correction filters in the transmitter and receiver are not mandatory. There are merely additional tools available to the radar designer to improve the side lobe quality. If the side lobes are already considered satisfactory, then these correction filters are unnecessary (see Figure H–3).

At mixers "M1," "M3," and "M5," the input and output RF and IF are interchanged with each other. Triple-balanced mixers are unnecessary. They were used because of the unavailability of double-balanced mixers.

Mixers

Mixers	Type Number	Type	Make	LO Power (dBm)	RF Range	LO Frequency	IF Range	Conversion Loss (dB)
M1	ZFM-15	Double balanced	Mini-Circuits	+10	200–248	2,128	2,328–2,378	8.5
M2h	TB0218LW2	Triple balanced	MITEQ	+10	2,328–2,376	3,640–4,368	1,264–2,040	7
M3	M77C	Double balanced	Watkins–Johnson	+10	1,264–2,040	8,114	9,378–10,154	5
M4	M79HC	Double balanced	Watkins–Johnson	+17	9,378–10,154	8,114	1,264–2,040	6.5
M5(=M)	TB0218LW2	Triple balanced	MITEQ	+10	1,264–2,040	3,640–4,368	2,328–2,376	7
M6	MY84C	Double balanced	Watkins–Johnson	+10	2,328–2,376	2,328–2,376	0–F (max. beat freq.)	6

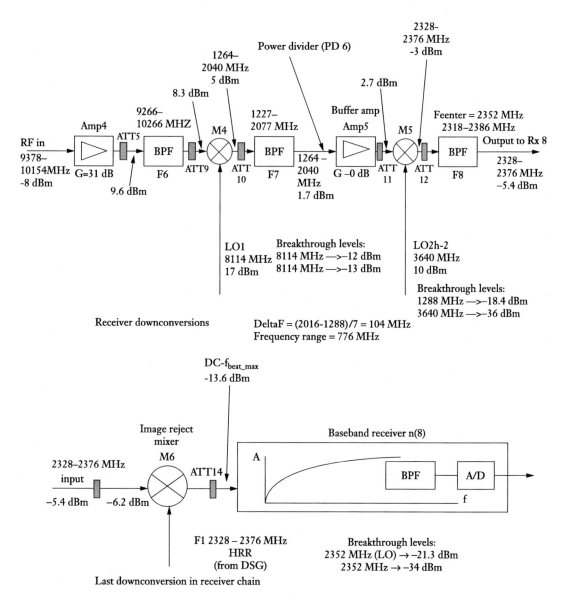

Figure H–3 Top: Receiver downconversions. Bottom: Last downconversion in receiver chain. *Note*: There is signal + noise in one sideband and only noise in the other sideband. This noise has to be suppressed, as otherwise the sensitivity decreases by 3 dB.

Filters

Filter	Type Number	Make	Type	BW (3 dB) (MHz)	F (Center) (MHz)	% BW	I. Loss (dB)	Sections	F (Spur) (MHz)
F1	81B42-224/ T90-0/0	K&L	BPF	90	224	40	1.7	8	400
F2	8MC19-2352/ T94-0/0	K&L	BPF	94	2,352	4	4.5	8	2,528

F3-h	6MC10-1288/T150-0/0	K&L	BPF	150	1,288	11.65	0.7	6	1,112
F4	6IB10-1652/H850-0/0	K&L	BPF	850	1,652	51.45	0.2	6	2,528
F5	7FV10-9577/T1154-0/0	K&L	BPF	1,154	9,577	10.24	0.3	4	10,642
F6	4FV10-9766/T1000-0/0	K&L	BPF	1,000	9,766	10.24	0.3	4	
F7	6IB10-1652/H850-0/0	K&L	BPF	850	1,652	51.45	0.2	6	2,528
F8-h	5C45-2352/T68-0/0	K&L	BPF	68	2,352	2.89	1.1	5	2,456

Note: Correction filter is inserted *after* the F8 filter. It was not used in this single-channel radar because the performance was very good even without this filter. This will become evident when we see the results in Appendix I for the single-channel radar. We reiterate that it is not mandatory to use correction filters in every such radar. Correction filters are merely an additional option resting with the radar designer if poor side lobe quality so warrants it.

Amplifiers

Amplifier	Type Number	Type	Make	Frequency (MHz)	Gain (dB)	NF (dB)
Amp 1	ZEL-1724LN	LNA	Mini-Circuits	2,328–2,376	20	1.5
Amp 2	ZEL-1724LN	LNA	Mini-Circuits	2,328–2,376	20	1.5
Amp 3	ZEL-1217LN	LNA	Mini-Circuits	1,264–1,312	20	1.5
Amp 4	JCAB12-300	LNA	JCA	9,378–10,154	24	2.2

Note: Amplifier "Amp 5" is not needed in the one full-chain concept.

Attenuators

Attenuator	Make	Type Number	Frequency (MHz)	Attenuator (dB)
ATT1	Mini-Circuits	SAT-1	200–248	1
ATT2	P.E.	PE7005-1	2,328–2,376	1
ATT 3a	P.E.	PE7005-10	2,328–2,376	10
ATT 3b	P.E.	PE7005-6	2,328–2,376	6
ATT 4	P.E.	PE7005-6	2,328–2,376	6
ATT5	Mini-Circuits	SAT-1	1,264–1,312	1
ATT6	Suhner	6810.19.A	1,264–2,040	10
ATT7	P.E.	PE7005-1	9,378–10,154	1
ATT8a	P.E.	PE7005-2	9,378–10,154	2
ATT8b	Suhner	6810.19.A	9,378–10,154	10
ATT9	P.E.	PE7005-1	9,378–10,154	1
ATT10	P.E.	PE7005-1	1,264–2,040	1

ATT11	P.E.	PE7005-1	1,264–2,040	1
ATT12	P.E.	PE7005-1	2,328–2,376	1
ATT13	P.E.	PE7005-1	2,328–2,376	1
ATT14	Mini-Circuits	SAT-1	DC-1	1
ATT15a	Suhner	6810.19.A	8,114	3
ATT15a	Suhner	6810.19.A	8,114	3

Power Dividers

Power Divider	Type Number	Type	Make	Frequency (MHz)	I. Loss (dB)
PD 1	IZY2PD-86	2-way	Mini-Circuits	8,114	3 + 0.5
PD 2h	ZN2PD-9G	2-way	Mini-Circuits	3,640	3 + 1.4
PD 3	ZFSC-2-2500	2-way	Mini-Circuits	2,328–2,376	3 + 1.5

Isolators

Isolator	Type Number	Make	Frequency (MHz)	I. Loss (dB)	Isolation (dB)
ISO1-1	PE8303	Pasternack Enterprises	8,114	0.6	20
ISO1-2	PE8303	Pasternack Enterprises	8,114	0.6	20
ISO2h-1	PE8301	Pasternack Enterprises	3,640	0.6	18
ISO2h-1	PE8301	Pasternack Enterprises	3,640	0.6	18

Local Oscillators

Local Oscillator	Type Number	Make	Frequency (MHz)	Power (dBm)
LO1	DRO-F-08114-ST	MITEQ	8,114	+13
LO2h	DRO-E-03640-ST	MITEQ	3,640	+13
LO3	LP-2128-B-0-15P	MITEQ	2,128	+13
CRO-DDS	CRO	CTI	1,000	+5

Note:

1. LO2 must provide a power of +13 dBm, so that it can be split by a two-way power divider. It is required to provide +10 dBm at two points in the schematic.
2. LO3 has to provide power at only one point in the schematic of +10 dBm.
3. LO1 has to provide two points, +10 dBm and +17 dBm, in the schematic. This is achieved by taking it as a +13 dBm source and splitting it. One output is then amplified to +17 dBm and the other is amplified to +20 dBm split again and then attenuated to +10 dBm (this ensures isolation) (see Figure H–4).

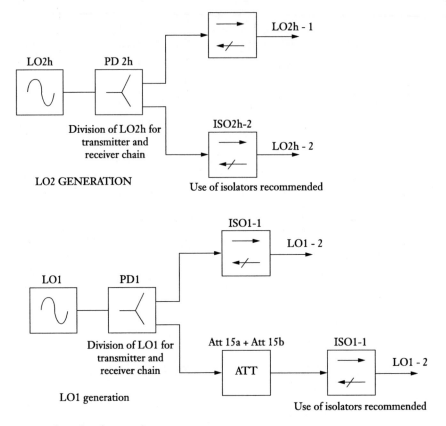

Figure H–4 Generation of LO frequencies.

The reader can verify that this single-channel radar has indeed been implemented as **Commercial Off The Shelf (COTS)** as was projected in Chapter 8. In achieving this, the **cost of the single-channel as well as the subsequent multi-channel radar was kept very low so as to make it economically viable**.

Appendix I: Pandora Radar Performance Verification Measurements

1.1 INTRODUCTION

The aim of this appendix is to assess the Pandora radar performance through measurement. The measurements are necessary to confirm that the radar is behaving as predicted and that there is no deterioration in its performance. This radar has basically the following technological challenges:

1. Power combining
2. Power splitting
3. Group delay in filters

The first two aspects have been successfully solved [1]. It now remains to minimize the group delays through the filters and successfully compensate for the residual group delays that exist due to the passage of the signal through the filters. We must also ensure that during this process, the spectrum of the signal does not deteriorate. Hence, there is a need to monitor the spectrum of the transmitted signal stage by stage. Toward this end, a single tract of this radar had been constructed based on the schematic at Appendix H.

1.2 STRUCTURE

This appendix is defined as a series of experiments based on the single tract schematic at Appendix H. Due to the lack of space, we cannot go into the details of each experiment. Interested readers are requested to refer to [2]. Instead, we only present the salient results. In some cases, where it is required, we shall discuss the experiment. In these experiments, we are not measuring the group delays across the individual filters. These have already been measured earlier and found to conform to company standards. Instead, we track the behavior of the signal spectrum as it progresses through the tract. We will, however, measure the collective group delays of filters F3 through F8, as this delay will determine the exact parameters of the correction filter, as well as the group delay from the input of F1 filter right upto and including the power splitter PD3 which determines the group delay pattern of the reference signal.

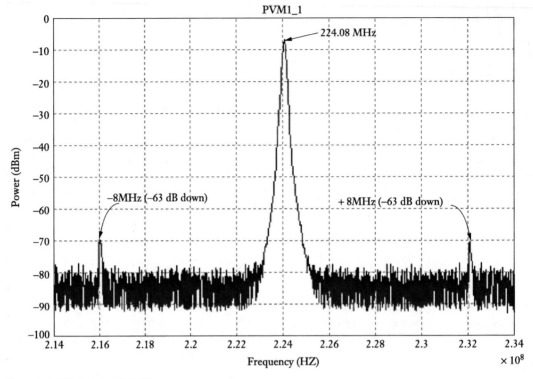

Figure I–1 CW signal at 224 MHz.

Center frequency: 224 MHz
Frequency span: 20 MHz
Video bandwidth: 300 kHz
Resolution bandwidth: 100 kHz
Sweep time: 20 msec
Marker frequency: 224.08E6 Hz
Marker amplitude: −6.90 dBm
2nd harmonic distortion: −56 dB
3rd harmonic distortion: −55 dB

Note: Spurs at ±8 MHz offset.

1.2.1 Observation

Figure I–1 shows the quality of the DDS signal. The spurs do not bother us as it is 63 dB below the maximum signal value. The performance of the DDS has been verified.

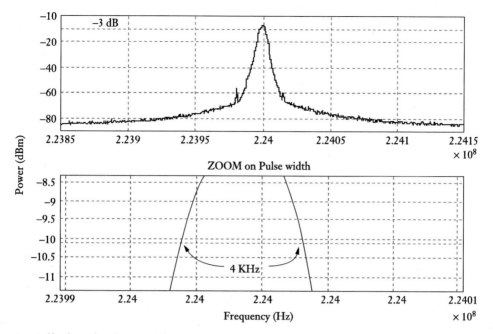

Figure I–2 Drift of CW signal measured over a 10-minutes period.

Center frequency: 224 MHz
Frequency span: 0.305 MHz
Video bandwidth: 3 kHz
Resolution bandwidth: 3 kHz
Sweep time: 100 msec
Marker frequency: 224.0043E6 Hz
Marker amplitude: −7.20 dBm

1.2.2 Observation

In this experiment, we have measured the drift of the DDS output signal over a period of 10 minutes (see Figure I–2). This drift of 4 kHz is caused solely due to the 1 GHz clock as this signal is measured at the DDS output and, therefore, there is no other cause for this drift.

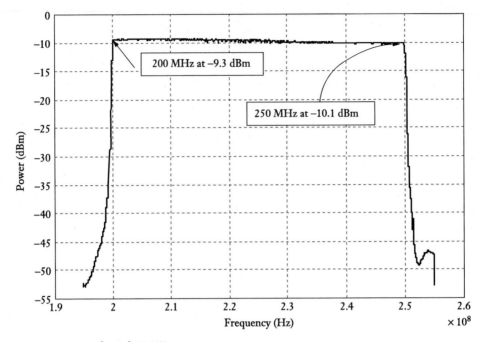

Figure I–3 FMCW sweep through 50 MHz.

Center frequency: 224.9 MHz
Frequency span: 60 MHz
Video bandwidth: 300 kHz
Resolution bandwidth: 100 kHz
Sweep time: 20 msec
Marker frequency: 224.90E6 Hz
Marker amplitude: −9.70 dBm

1.2.3 Observation

A perfectly normal spectrum and no anomalies were detected.

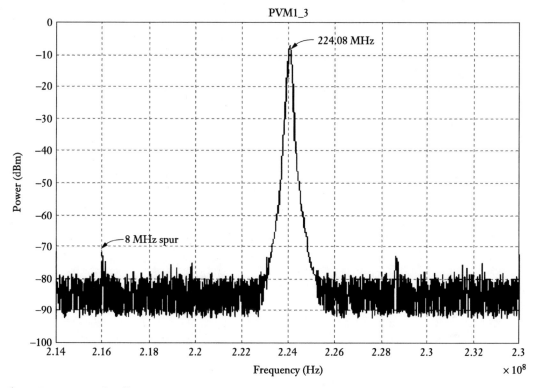

Figure I–4 Output of F1 filter.

Center frequency: 224 MHz
Frequency span: 20 MHz
Video bandwidth: 300 kHz
Resolution bandwidth: 100 kHz
Sweep time: 20 msec
Marker frequency: 224.08E6 Hz
Marker amplitude: −7.40 dBm

1.2.4 Observation

Note that the spurs still exist.

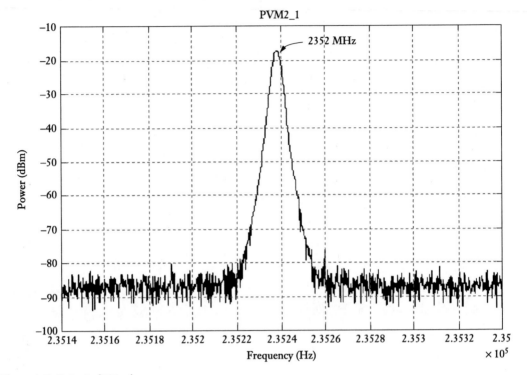

Figure I–5 Output of M1 mixer.

Center frequency: 2352.4 MHz
Frequency span: 2 MHz
Video bandwidth: 100 kHz
Resolution bandwidth: 30 kHz
Sweep time: 20 msec
Marker frequency: 2.352400E9 Hz
Marker amplitude: −19.60 dBm

1.2.5 Observation

The "grass" is not too much. The spurs are outside this bandwidth. The signal has been "upmixed" to 2,352 MHz. The IM products of this mixing can be seen in Figure I–6. The bandwidth shown here is too narrow to see these.

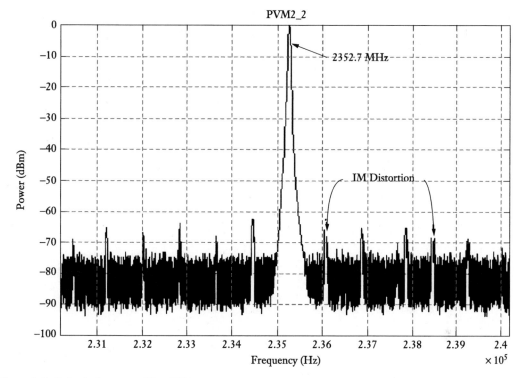

Figure I–6 Output of power splitter PD3.

Center frequency: 2,352 MHz
Frequency span: 100 MHz
Video bandwidth: 1,000 kHz
Resolution bandwidth: 300 kHz
Sweep time: 20 msec
Marker frequency: 2.3527E9 Hz
Marker amplitude: −0.60 dBm

1.2.6 Observation

The signal has passed through M1. Hence, we see IM products. The products cannot be eliminated, but they can be controlled. The levels shown here are not serious enough to merit consideration. This output is passed on for reference to M6 as well as to the next stage. The correction filter, if required, is fitted at the output of PD3 before passing the signal onto the next stage.

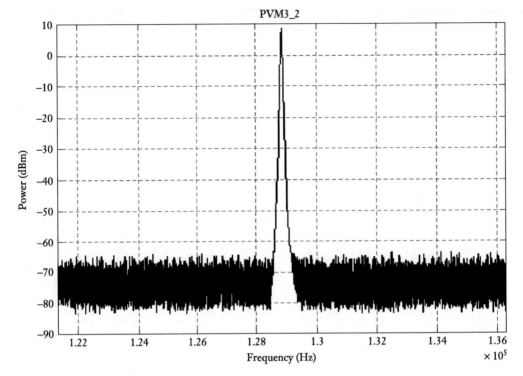

Figure I–7 Output of Amplifier 3.

Center frequency: 1,286 MHz
Frequency span: 150 MHz
Video bandwidth: 1,000 kHz
Resolution bandwidth: 300 kHz
Sweep time: 20 msec
Marker frequency: 1.2863E9 Hz
Marker amplitude: 10.50 dBm

1.2.7 Observation

Notice the IM distortion. These distortions are because of M1 and M2 mixers. The levels are low enough not to be critical.

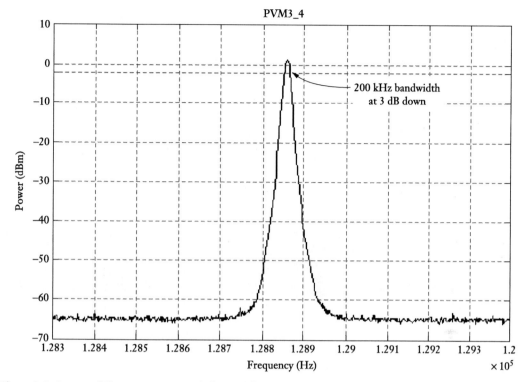

Figure I–8 Output of filter F4 over a 10-min interval (in order to examine frequency drift).

Center frequency: 1288.7 MHz
Frequency span: 12 MHz
Video bandwidth: 300 kHz
Resolution bandwidth: 100 kHz
Sweep time: 20 msec

1.2.8 Observation

It is essential to check the signal quality at this stage prior to giving it to the final carrier "upmixer" M3. This drift, which we have just measured, of 200 kHz has been caused due to drifts in 1 GHz, 2,128 MHz, and 3,640 MHz oscillators.

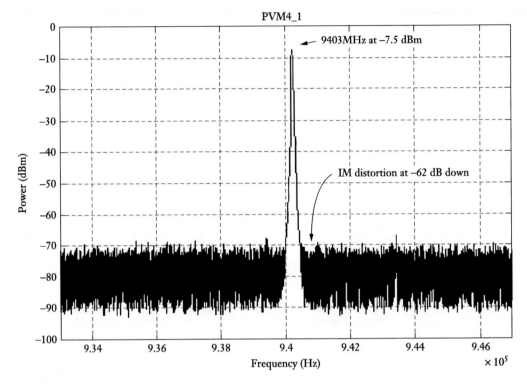

Figure I–9 Output of F5 filter.

Center frequency: 9,402 MHz
Frequency span: 150 MHz
Video bandwidth: 1,000 kHz
Resolution bandwidth: 300 kHz
Sweep time: 20 msec
Marker frequency: 9.4025E9 Hz
Marker amplitude: −7.50 dBm

1.2.9 Observation

This checks out the output of the carrier mixer and the final output before the signal is transmitted. It can be seen that this is a clean signal.

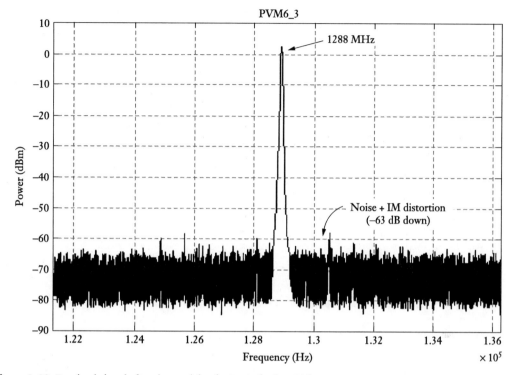

Figure I–10 Received signal after downmixing (output of mixer M4).

Center frequency: 1,288 MHz
Frequency span: 150 MHz
Video bandwidth: 1,000 kHz
Resolution bandwidth: 300 kHz
Sweep time: 20 msec
Marker frequency: 1.2889E9 Hz
Marker amplitude: 2.80 dBm

1.2.10 Observation

This is just after the carrier downmixer M4. The signal will now be processed prior to giving it to the stretch processor M6.

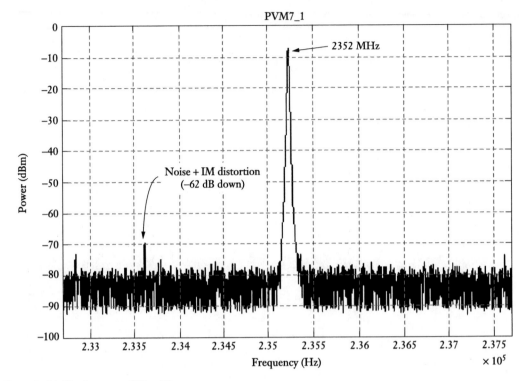

Figure I–11 Final output of filter F8.

Center frequency: 2,352 MHz
Frequency span: 50 MHz
Video bandwidth: 300 kHz
Resolution bandwidth: 100 kHz
Sweep time: 20 msec
Marker frequency: 2.35230E9 Hz
Marker amplitude: −7.70 dBm

1.2.11 Observation

The signal is now ready for the stretch processor M6. It has already been passed through the entire receiver channel, viz., mixers M4 and M5 and filters F6, F7, and F8. The signal can now be given to a correction filter if the side lobe condition so warrants it. We shall now examine this aspect further down in this appendix.

1.3 GROUP DELAY MEASUREMENT ACROSS F3 AND F8

We now measure the group delay across filters F3 to F8. Connect the input of filter F3 in the sweep converter stage to a network analyzer as well as output of filter F8. Measure S21 across the sweep bandwidth, i.e., 1,264–1,312 MHz. The output signal will vary from 2,328 to 2,376 MHz. This measures the actual group delay distortion across the sweep bandwidth. The inverse of this group delay constitutes the correction filter required from K&L, USA. Store the results in files "PVM7_2" and "PVM7_21" respectively.

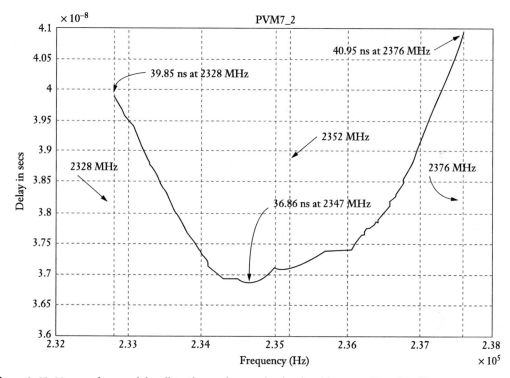

Figure I–12 Measured group delay distortion undergone by the signal between F3 and F8 filters.

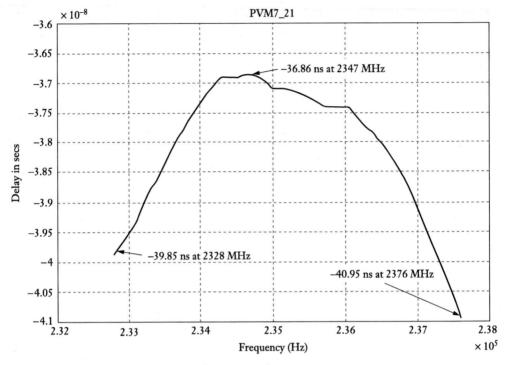

Figure I–13 Correction filter group delay profile (measured).

1.3.1 Observation

The correction filter has been now designed to correct distortions caused due to the signal passage from F3 to F8 filters.

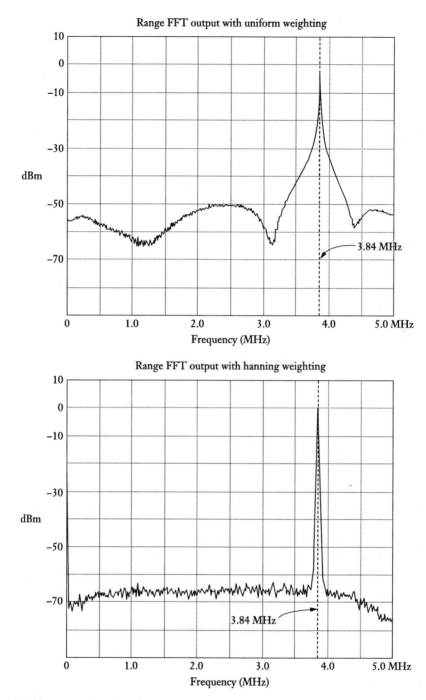

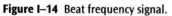

Figure I–14 Beat frequency signal.

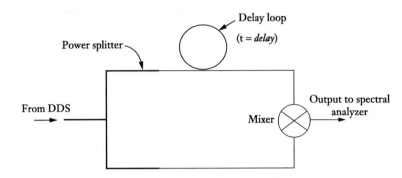

Figure I–15 Toward creating a beat signal.

The testing of beat frequency signals in such radars can be carried out using the concept shown in Figure I–15.

The output of the mixer will give the beat signal. The quality of the beat spectrum will define the upper limit of attainable signal processing, i.e., this is the best one can achieve with the DDS as a source.

In Figure I–14, the sweep bandwidth was 48 MHz from 200 to 248 MHz. This was then fed to the input of the sweep upconvertor and the reference tapped from the output of PD3 as shown in Appendix H, (Figure H–1). This result was obtained using a glass delay line of 80 sec corresponding to a target located at 12 km. We can see the excellent result obtained with Hanning weighting. This means that there is no need for a correction filter for this radar as the side lobes are better than −65 dBc. We can also clearly see the slope of the low pass filter beyond 4 MHz. The RMS value of the correction filter is around 1 nsec or 17°. This means that there can be no side lobe distortion in the beat signal as the quadratic phase distortion is less than the stipulated 22.5° as discussed in Section 8.10 of the main text.

We know from Chapter 8 the following parameters:

Maximum instrumented range = 6,550 m
Sweep time = 1 msec
Sweep bandwidth = 48 MHz
Number of points in the range FFT = 4,096 (2,048 range bins)
Maximum sampling frequency = 4 MHz (max. beat frequency = 2 MHz)

Hence, range bin resolution is

$$\frac{R_{max}}{f_{max}} = \frac{6.55 \times 10^3}{2048} = 3.2 \text{ m and the frequency resolution per range bin is } \frac{2 \times 10^6}{2048} = 976 \text{ Hz.}$$

1.4 NONLINEARITY MEASUREMENT

Pulse spectral width (measured −3 dB down, as is normal for spectral width measurements as shown below) is 33 Hz.

This width was achieved using a sweep bandwidth of 32.736 MHz and a sweep time of 682 μsec. The sweep time was adjusted using an oscilloscope. The cable had a delay of 31 nsec for a target located at 7.5 m.

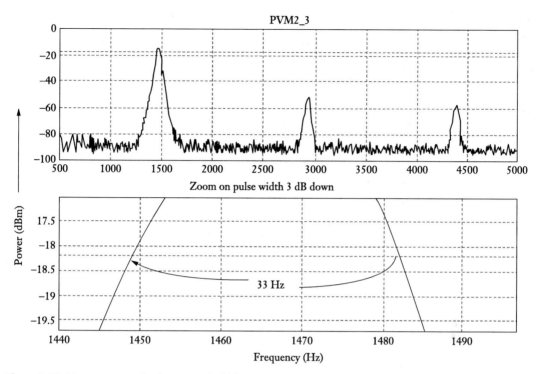

Figure I-16 Measurement of pulse spectral width.

Hence, nonlinearity expressed as a percentage of sweep bandwidth is

$$\frac{\text{Pulse spectral width}}{\text{Sweep bandwidth}} = \frac{33}{32.736 \times 10^6} = 0.0001\% \text{ of } \Delta f$$

We aimed for a nonlinearity of 0.003% in order to satisfy the receiver frequency resolution requirements (see Appendix B). We note that this requirement has been exceeded. Furthermore, a nonlinearity of 0.0001% ensures that with the highest sweep bandwidth of 48 MHz (in our case, as shown above) the spectral width does not exceed 480 Hz. This is less than the frequency resolution per range bin of 976 Hz. This implies that we are assured of 2,048 resolvable range cells (bins) from the FFT processor. This 4K FFT needs to be completed in less than 1 msec. This is within the capability of most signal processors, e.g., SHARC processors, which can easily complete such FFTs in less than 1 msec.

It should be noted that the side lobes shown in Figure I-16 are caused due to truncation during synchronization. The length of the delay line was insufficient. Consequently, the effects of truncation become more pronounced. Hence, other than nonlinearity, we cannot assess the quality of the beat signal in this experiment. The beat signal was measured as in Figure I-15 using a glass delay line of 80 second.

It is interesting to check the output of the M6 mixer when there is no target. This is shown in Figure I-17.

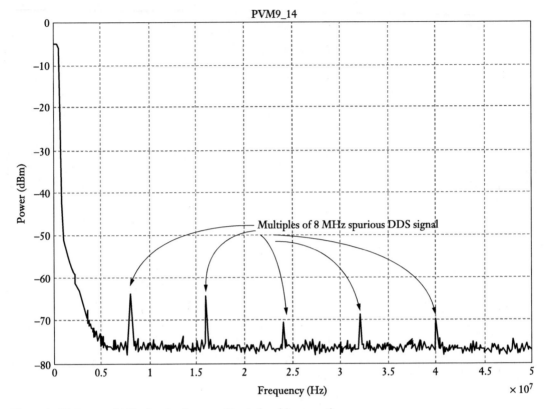

Figure I–17 Output of M6 mixer in absence of beat signal (no target).

Center frequency: 25 MHz
Frequency span: 50 MHz
Video bandwidth: 300 kHz
Resolution bandwidth: 100 kHz
Sweep time: 20 msec
Marker frequency: 8.05E6 Hz
Marker amplitude: −63.80 dBm

1.4.1 Observation

We can see the 8-MHz spurs and their multiples in the above figure. These spurs were generated in the DDS. They will never go away.

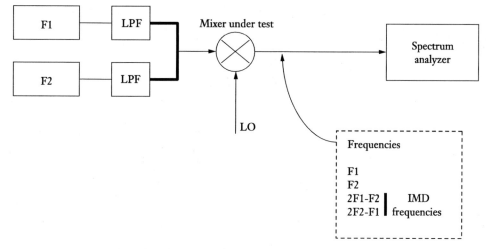

Figure I–18 Measurement of IM distortion.

1.5 MEASUREMENT OF TWO-TONE THIRD-ORDER IM DISTORTION IN M3 MIXER

The Measurement of IM distortion affects mixers M3 and M4. The separation of two-tone signals is 50 MHz as this is the inter-channel separation (guard band, see Section 8.3). The idea here is to ensure that in spite of the guard band, the signal from one channel should not enter the other via the power splitter.

1.5.1 Method

We proceed as follows:

1. The concept of testing is shown below.
2. Depending upon the frequency range of the mixer under test, select two signal sources 50 MHz apart, F1 and F2.
3. Ensure that there are no spurious frequencies after the low pass filters, especially 2nd harmonics.
4. The output of the LPFs is connected to a two-way combiner and then given to the mixer under test.
5. Adjust the spectrum analyzer to clear show the IMD frequencies. Ensure that the IMD levels are as per specifications. These mixers have been selected especially for having very high third-order intercept points. The IMD values should be better than −50 dBm. Store the result in files "PVM10_M3" and "PVM10_M4."

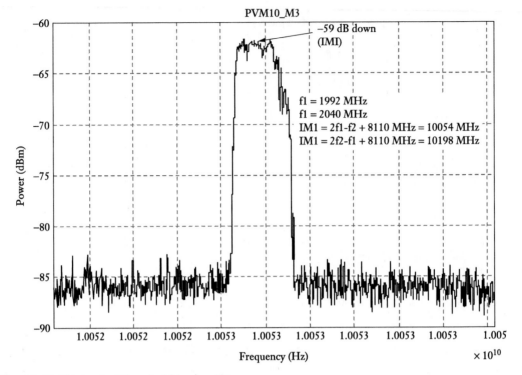

Figure I–19 Third-order IM product for mixer M3.

Center frequency: 10052.718 MHz
Frequency span: 1 MHz
Video bandwidth: 30 kHz
Resolution bandwidth: 10 kHz
Sweep time: 30 msec
Marker frequency: 10.052700E9 Hz
Marker amplitude: −62.50 dBm

1.5.2 Observation

The IM product is −59 dB below maximum. This is adequate. The experiment on mixer M4 was not carried out since it is expected to be less critical as mixer M4 has an even higher intercept point as compared to M3 mixer. Hence, we present results for M3 only.

1.6 MEASUREMENT OF RECEIVER NOISE FIGURE

1.6.1 Aim

The aim of this experiment is to measure the receiver noise figure using the "Y" factor method.

1.6.2 Method

We proceed as follows:

1. Connect the HP noise source to the input of amplifier AMP4.
2. Connect a 28-V DC supply to input of noise source as trigger. When the DC source is 0 V, the noise source is off. When it is 28 V, the noise source comes on.
3. Keep the DC input at 0 V and measure the noise level at the output of attenuator 14. This is the noise level with the source off.
4. Now switch on the 28 V supply and measure the noise level at the output of attenuator 14. This is the noise level with the source on.
5. The "Y" factor is given by $\frac{\text{Source with noise 'ON'}}{\text{Source with noise 'OFF'}}$ in dB.
6. Look up the excess noise ratio (ENR) from the look up table for the noise source. This is given in dB.
7. The noise figure for the receiver excluding the IF amplifier (not fitted in the case of Pandora radar) is given by

$$\text{NF (dB)} = \text{ENR (dB)} - \text{"Y" factor (dB)}$$

1.6.3 Conclusion

We have now measured the noise figure of the receiver tract.

1.6.4 Results

Source with noise "ON" = −129.6 dBm/Hz
Source with noise "OFF" = −143.02 dBm/Hz
"Y" factor = −143.02 − (−129.6) = 13.42 dB
Excess noise = 15.65 dBm
NF = ENR − "Y" = 15.65 − 13.42 = 2.23 dB

1.7 MEASUREMENT OF ADJACENT CHANNEL INTERFERENCE AT THE OUTPUT OF MIXER M6

1.7.1 Aim

The aim of this experiment is to measure the level of adjacent channel interference at the output of mixer M6. This is crucial for the Pandora architecture.

1.7.2 Method

We proceed as follows:

1. Ensure that the input is a CW signal of around −9 dBm at 9,402 MHz, i.e., 1,288 MHz translated through 8,114 MHz from the sweep upconverter. *At this stage, it should be from an oscillator.*
2. Connect this signal to amplifier AMP4 and adjust attenuator 8 till we have a power level of around 9 dBm at the output.
3. Now, connect this signal to filter F6 and adjust attenuator 9 till we achieve a power level of around 7.7 dBm at the output.
4. Check the quality of the spectrum. There should be no spurious signals.
5. Connect the signal to input of mixer M4 and feed 8,114 MHz at 17 dBm *from LO1 source.*
6. Adjust the attenuator 10 till we achieve a power level of around 1.2 dBm at the output.
7. *Now, connect the sweep upconverter stage in lieu of the oscillator.* The output of attenuator 10 constitutes the received signal which is still CW at 1,288 MHz.
8. Connect the signal obtained in the previous experiment to filter F7 followed by amplifier AMP5.
9. Adjust attenuator 11 till we obtain a signal of around −1 dBm at 1,288 MHz.
10. If the spectrum is satisfactory, connect to input of mixer M5.
11. Connect 3,640 MHz at 10 dBm from LO2h-2 to LO input of mixer. Monitor the quality of the signal at mixer output. This is the final downconverted signal prior to stretch processing.
12. Connect the signal to filter F8. Check the quality of the output spectrum. There should be no spurious signals in the passband. Adjust attenuator 12 till we obtain a power level of around −9 dBm.
13. The output of F8 constitutes the desired signal. We now connect an eight-way power combiner at the output of filter F7. Connect any one of the arms to output of filter F7 and the combined output to amplifier AMP5.
14. Feed a frequency of 1,368 MHz to any one of the remaining arms. This constitutes the adjacent channel interference. *The level of this signal should be around 0 dBm, i.e., identical in level to the 1,288 MHz signal.* Terminate the remaining arms with 50 Ohm terminations.
15. Measure the level of this signal, which manifests itself as 2,272 MHz, at the output of filter F8. Subtract this value from the value of 2,352 MHz signal. This gives the level of adjacent channel suppression. It should be better than −50 dB. If not, then there is a need to add an additional F8 filter as stated in the Pandora Phase Analysis report [2]. Store in file "PVM12_1."

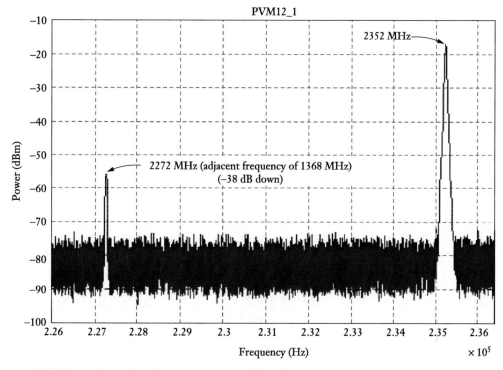

Figure I–20 Adjacent channel interference at output of filter F8.

Center frequency: 2,312 MHz
Frequency span: 104 MHz
Video bandwidth: 1,000 kHz
Resolution bandwidth: 300 kHz
Sweep time: 20 msec
Marker frequency: 2.2727E9 Hz (adjacent channel signal)
Marker amplitude: −55.20 dBm

The original signal from the adjacent channel was 1,368 MHz. This signal manifests itself at the output of filter F8 as 2,272 MHz. We note that it is 38 dB down. To avoid this, that it was decided to add an additional F8 filter. This will ensure further suppression by 25 dB, i.e., this spurious signal will now become 63 dB down. This guarantees channel isolation.

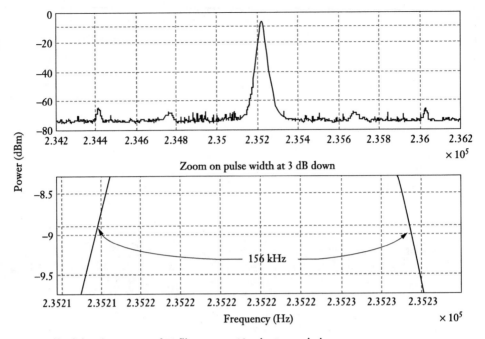

Figure I–21 Drift of signal at output of F8 filter over a 10-minutes period.

Center frequency: 2,352 MHz
Frequency span: 20 MHz
Video bandwidth: 300 kHz
Resolution bandwidth: 100 kHz
Sweep time: 20 msec
Marker frequency: 2.35224E9 Hz
Marker amplitude: −6.00 dBm

We note that the drift is 156 kHz. Compare this drift to the drift of the basic CW signal (see Figure I–2) which is only 4 kHz. This drift of 156 kHz is caused solely due to the 2,128 MHz oscillator, since during the signal processing; this frequency is never cancelled unlike the 8,114 and 3,640 MHz oscillators. This will, however, not affect the stretch processor (mixer M6), because the reference signal will also drift in synchronization with the RF return. Nevertheless, it is recommended that the 2,128 MHz oscillator be PLL-based with high stability.

References

1. Jankiraman, M., Wessels, B. J., and van Genderen, P., "Design of multifrequency FMCW Radar," *Proceedings of the European Microwave Conference EUMC-98*, Amsterdam, 1998.
2. Jankiraman, M., *Pandora Multifrequency Radar—Project Report*, IRCTR-S-014-99, Delft, The Netherlands, April 1999.

Appendix J: 8-Way Combiner Analysis Results

The 8-way combiner intended for the Pandora radar is shown in Figure J–1.

For the purposes of this analysis, the pin numbers are taken from 1 to 8 starting from the left. The output pin is pin number 9. This same combiner is used as a resolver in the receiver channel.

1.1 SPECIFICATIONS

The specifications as supplied by the manufacturer, Pulsar Microwave Corp., USA, are given below.

1.1.1 Model PS8-TBD-454/1S

Freq.: 1,227–2,077 MHz
Ins. loss: 0.8 dB max
Isolation: 20 dB min
VSWR: 1.35:1 max
Ampl. balance: 0.4 dB max
Phase balance: 5.0Y max

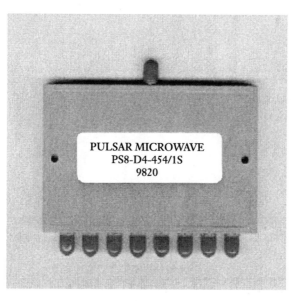

PULSAR MICROWAVE
PS8-D4-454/1S
9820

Figure J–1 Pulsar 8 Way Combiner/Splitter.

Power handling: 5 W CW into 1.2:1 load VSWR
Connectors: SMA (female)

Initially, the reflection loss at pin 1 was measured. This was measured in two ways. Initially, one reading was taken across pins 1 and 9 with the others having a 50 Ω load on them. Next, one reading was taken across pins 1 and 2 with dummy loads on the rest. The reflection losses were plotted for pin 1 (Figure J–2).

The curves are coinciding as expected. The values are better than 20 dB, which is satisfactory. This is the same at all pins.

The insertion loss across pins 1 and 9 were next measured. This is shown in Figure J–3.

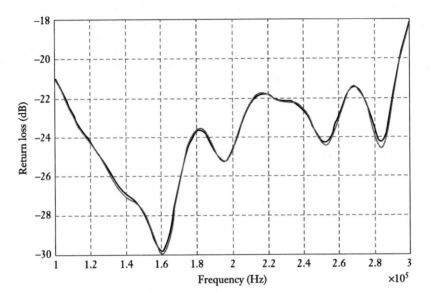

Figure J–2 S22 for pin number 1.

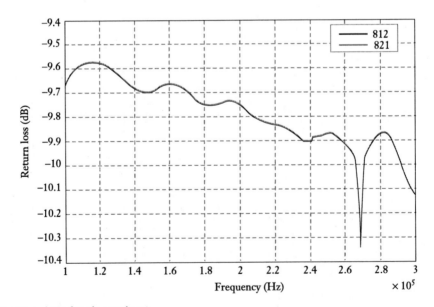

Figure J–3 S12 and S21 for pin number 1.

It can be seen that the variation across the bandwidth of interest 1–2.2 GHz does not exceed 0.3 dB. The graph shows S12 and S21. The basic loss of 9 dB, is due to the 8-way combiner. The amplitude unbalance of 0.3 dB is adequate from the combining requirements of the Pandora and it is less than the specified 0.4 dB. Similar results were observed at the remaining pins.

We now measure the isolation between pins 1 and 2. This is shown in Figure J–4.

It can be seen that the isolation loss is better than 20 dB, which is satisfactory and as per specifications.

We now check the combining capability of the combiner. We mix two CW signals at 1,288 and 2,100 MHz, i.e., across the bandwidth of interest. We can clearly see the output from pin 9 on the spectrum analyzer (Figure J–5).

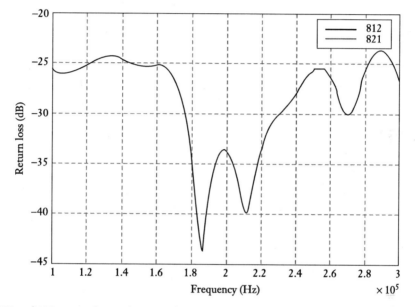

Figure J–4 S12 and S21 across pin numbers 1 and 2.

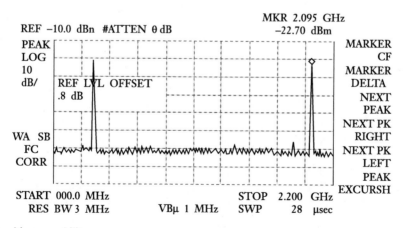

Figure J–5 Combiner capability.

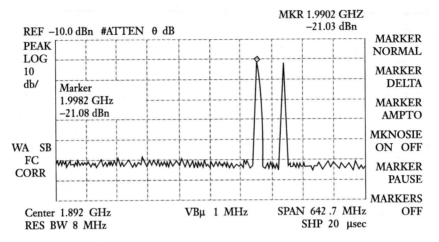

Figure J–6 Combiner capability (cont.).

Both the signals were calibrated for equal input power (−11 dBm). A total of 1,288 MHz was given at pin 1 and 2,100 MHz at pin 8. The lower frequency was found to have a power level of −20.7 dBm, while the upper was at −22.7 dBm. However, the power levels were found to be more identical when the upper frequency was lowered to 2,040 MHz and the lower was placed at 1,992 MHz, as per the sweep requirements of the radar. This is shown in Figure J–6.

The power levels were found to be differing by 0.4 dB, but this was the situation when the signals were given across 8 pins. In reality, we need to ensure amplitude balance across the sweep bandwidth *in one pin*. The results are given in Table J–1.

Table J–1

Frequency in MHz	Power Level in dBm
2,040	−20.7
1,992	−21.1
1,400	−20.6
1,350	−20.6
1,000	−20.6
1,050	−20.6

It can be seen that the variation of power levels measured at pins 1 and 8 *individually* was practically flat. This meets with our requirements that the amplitude unbalance should not exceed 0.4 dB.

Finally, a sweep signal was given at pin 8 and the outputs recorded. This was done at two-center frequencies, 1,288 and 2,016 MHz. The lower end sweep is shown in Figure J–7 and the upper sweep in Figure J–8.

It can be seen from the markers in Figure J–7, that the amplitude unbalance is 0.42 dB, which is acceptable. In Figure J–8, it is even less at 0.2 dB. These two sweeps represent the two extreme ends of the spectrum of interest to the Pandora.

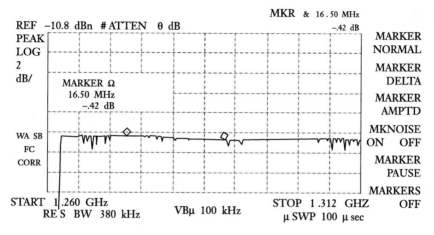

Figure J–7 Sweep range is from 1,264 to 1,312 MHz.

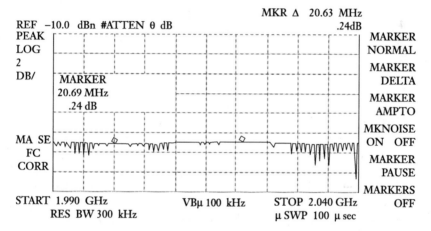

Figure J–8 Sweep range is from 1,990 to 2,040 MHz.

1.2 CONCLUSIONS

The following conclusions are arrived at:

1. The combiner has low amplitude unbalance not exceeding 0.4 dB across each sweep bandwidth. This is sufficient for Pandora.
2. The channel isolation is better than 20 dB.
3. There is a power loss of 9 dB due to the 8-way combiner.
4. The insertion loss across the entire bandwidth varies less than 0.4 dB.
5. The phase distortion as per specifications does not exceed 5Υ. This was, however, not measured as it is not critical since we are in any case correcting for group delays across the filters of the Pandora.

Appendix K: MATLAB Simulation Program

This appendix pertains to a GUI-based program developed by the author and supplied with the accompanying software. It supplies the .mat files and data vectors for running with the program supplied by Prof. Nadav Levanon [1]. The program supplied with this book caters to the following polyphase codes:

1. Frank Codes
2. P1 Code
3. P2 Code
4. P3 Code
5. P4 Code

The reader is advised to go to the website of Prof. Nadav Levanon and to download his GUI-driven program *ambfn7.m* along with the attendant *M files*. The reader is also recommended to read the instructions provided there.

Before using *ambfn7.m* program, we require to generate the *.mat* files for our codes. This is because Prof. Levanon has run only certain codes. The user-driven codes require a *.mat* file. The program *signal.m* is first run by typing *"ui_start"* at the Matlab command line. It is a GUI-driven file and is self-explanatory. Information on the input fields can be obtained by positioning the cursor on the blue colored field headers. The user can choose to name the file in the file field. The file will be stored in the directory holding the program. Levanon's files should also be located here for convenience.

Once the .mat files are in place, type *ambfn7* on the Matlab command line. This brings up the GUI. Select "User Defined" from the drop list. Next, disable the radio button for "frequency." Next, proceed to the slider buttons. These buttons cater to the plot.

The GUI parameters (F*Mtb, T, N, K) are associated only with the plots.

F*Mtb—defines the extent of the plotted Doppler axis in normalized Doppler (Doppler multiplied by the entire duration of the signal (Mtb)). F*Mtb also defines the extent of the frequency axis in the spectrum plot (the lower subplot created when you click on the "ACF. & SPEC Plot."). While the maximum value in the F*Mtb ruler is 60, you can type in a higher value.

T—is the extent of the (positive) delay axis in units of the entire duration of the signal, so it is actually T/Mtb. For example, if you choose the signal "pulse train, 6 pulses" and choose T = 1, the ambiguity plot will include all the five recurrent lobes. If you choose T = 1/6, you will get exactly one repetition period.

N—is the number of grid points on the (positive) delay axis of the plot.

K—is the number of grid points on the Doppler axis of the plot.

The user is advised to initially use the settings provided by Levanon for the P4-coded signal. Once the software becomes familiar, one can experiment with his/her own settings.

To summarize, the following procedure are recommended:

- Type "*ui_start*" on the Matlab® command line. This brings up the GUI called "Ambiguity Function Data Loader."
- The top half of the GUI pertains to data entry and the default data entry values are shown. The type of code to be examined can be selected from the drop-down list box under the heading "Code Generator." The user can be guided on the use of the data loader by the comments appearing on the edit window. On clicking the "Generate" button, **code** *.mat* files are created. Once the code *.mat* files are generated, we then load it on the load edit window, not forgetting to check the relevant load check box. When we press the load button, **another** *.mat* file is then generated. The parameters are stored in this **parameters** *.mat* file called "*af_var.*"
- Close the data loader and type "*ambfn7*" on the command line. This brings up the GUI developed by Levanon.
- Select "User Defined" from the drop list and also deselect the radio button for "frequency." Enter the plotting values as discussed earlier.
- Now click the load button and browse to the location of the code *.mat* file created by the program data loader. Ignore the warnings which appear on the command line. These occur because the parameters *.mat* file (*af_var*) we have generated has not yet been loaded.
- Load the parameters *.mat* file *af_var* on the command line. This will load the following parameters into the RAM, viz., *u_amp*, *u_phase*, and *f_basic*. These are vectors required by *ambfn7.m*.
- We are now ready to go.
- Click "Cal. & Plot sig". button. This loads the plot values.
- Now the user can run any plot desired by clicking the relevant button.

One important point to remember is that while the default value for the number of code phases is 8 for all types of codes, it is 64 for P3 and P4 codes.

References

1. http://www.eng.tau.ac.il/~nadav/amb-func.html.

Appendix L: Level Diagram Calculations—SFCW Radar

This appendix pertains to the level diagram calculations of the Pandora single-channel radar.

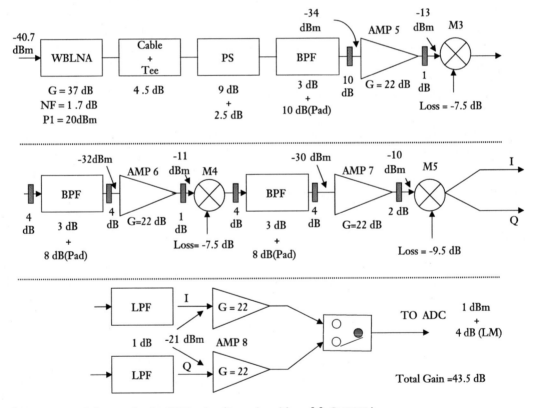

Figure L–1 Level diagram for the SFCW radar. (Reproduced from [1], © IRCTR.)

The following parameters were determined:

1. Noise bandwidth of receiver = 80 kHz
2. Sample rate = 150 kHz
3. Number of bits = 16
4. Maximum input 2.048 V ptp into 50 Ohm = 10.2 dBm
5. Input noise level: 1.5 lsbs = 1.5 × 2.048/65536 = 46.9 μV rms = −73.57 dBm
6. Noise level = 20 log(1.5) = 3.52 dB

Based on the above parameters, we calculate the following:

1. Dynamic range: 16 × 6.02(number of bits) − 6.02 (sign) − 3 (rms sine) − 3.52 (noise level) = 83.78 dB
   Given: Noise Figure of receiver = 3 dB
2. We calculate FKTB = 3 − 174 + 49 = −122 dBm
   ∴ Required gain = −73.57 − (−122) = 48.43 dB
3. We need to provide a linearity margin = 4 dB for the ADC.

Based on the above figures we compute the level diagram for this radar. In making these calculations, we already know the various component losses. This diagram is given in Figure L–1. In the figure, PS is the power splitter of the receiver channel. Even though this was a single-channel radar, this was included in the calculations for a realistic calculation of the multi-channel radar.

The spreadsheet calculations are already given with the accompanying software as *Pandora_spreadsheet.xls*. The spreadsheet is a multiple sheet spreadsheet and it shows how one can calculate the parameters for this radar based on actual values. The readers can see the various losses for the components and as to how they are taken into account. The graphs show the losses stage by stage. The IP3 losses are well above the receiver loss curve. The AGC is set at zero gain (since no AGC was planned in this prototype). The reader is encouraged to change the values and see how it influences the final results.

Reference

1. Jankiraman, M., *Pandora Multifrequency Radar—Project Report*, IRCTR-S-014-99, Delft, The Netherlands, April 1999.

Index